page 94
British supersonic
research
also Soviet union

page 95
Soviet spies
in U.S.

page 98
German research
captured by Soviets

Supersonic FLIGHT

Breaking the Sound Barrier and Beyond

Also from Brassey's

HALLION (ed.)
Air Power Confronts an Unstable World

MASON
Air Power: A Centennial Appraisal
(new edition)

OXLEE
Aerospace Reconnaissance
(Brassey's Air Power Series)

TREADWELL & WOOD
German Knights of the Air, 1914–18
Holders of the Orden Pour le Mérite

Supersonic FLIGHT

Breaking the Sound Barrier and Beyond

The Story of the Bell X-1 and Douglas D-558

RICHARD P HALLION

Brassey's
London • Washington

First English Edition 1972
Revised 1997

UK editorial offices: Brassey's, 33 John Street, London WC1N 2AT
UK orders: Marston Book Services, PO Box 269, Abingdon, OX14 4SD

North American orders: Brassey's Inc., PO Box 960,
Herndon, VA 20172, USA

Library of Congress Cataloging in Publication Data
available

British Library Cataloguing in Publication Data
A catalogue record for this book is available from the British Library

ISBN 1 85753 253 8 Hardcover

Typeset by M Rules in Sabon
Printed in Great Britain by Redwood Books, Trowbridge

Contents

Acknowledgements

The following were the individuals whose assistance was critical to the preparation of the 1972 edition of this work. Time has not dimmed but only increased the appreciation I have for their help.

It would have been impossible to write this book without the generous co-operation of many individuals and organisations. Special thanks are due to the staff of the NASA Historical Office and the Smithsonian Institution for their gracious help through the years. I am particularly grateful to the late Dr Eugene M Emme, the first NASA Historian, his deputy, Dr Frank W Anderson, Lee Saegesser, the NASA archivist, and Carrie Karegeannis, all of the NASA Historical Office; Michael Collins, the Director of the National Air and Space Museum, and his deputy, the late Melvin B Zisfein. I also want to acknowledge the assistance given to me by staff members of the Air Force Historical Office (now the Air Force History Support Office), Air Force Museum, Air Force Systems Command (now Air Force Materiel Command), the NASA Langley Research Center, the National Archives, and Naval Air Systems Command. I wish to thank especially the late Dr Carl Berger, Office of Air Force History; the late Dr Donald R McVeigh, historian of Air Force Systems Command; Robert W Mulac and Neva B Brooks, NASA Langley Research Center; Dr Lee M Pearson, historian of Naval Air Systems Command; and Ralph Jackson and James Love of the NASA Flight Research Center. I especially appreciate the assistance given by Russell Bourne, Consultant on Special Publications, Smithsonian Institution, and Louis S Casey, the former Acting Assistant Director, and Archives Assistant Robert B Wood, National Air and Space Museum.

The faculty of the Department of History, University of Maryland, provided patient counselling and thoughtful advice that I found most helpful. My debt to them is great. I benefited from consultations with Dr Richard E Thomas, former Head of the Department of Aerospace Engineering, University of Maryland, who directed my initial research in this area, and with Dr Jewel B Barlow, Assistant Professor, Department of Aerospace Engineering. Their advice and encouragement were invaluable.

I wish to thank the following participants in the supersonic breakthrough who assisted me in research for this book through interviews, correspondence, or both: Milton B Ames, Neil A Armstrong, John V Becker, Ralph P Bielat, Clinton E Brown, Robert A Champine, A Scott Crossfield, Leo J Devlin, Jr, Captain Walter S Diehl, USN (ret.), Macon C Ellis, Jr, Paul Emmons, Richard H Frost, William G Gisel, Robert R Gilruth, Benson Hamlin, Edward H Heinemann, Robert A Hoover, Dr Abraham Hyatt, Ezra Kotcher, Axel T Mattson, Jean A Roche, Dr L Eugene Root, A M O Smith, Stanley W Smith, Robert M Stanley, John Stack, Dr Floyd L Thompson, Thomas A Toll, Gerald M Truszynski, Harold Turner, Joseph Vensel, Robert A Wolf, and Brigadier General Charles E 'Chuck' Yeager, USAF (ret.)

A Note to Readers of the 1997 Edition

With regard to this edition, I wish to thank those who have encouraged me to – indeed, in some cases, insisted that I – reissue this work, particularly Dr John Anderson, Chairman Emeritus of the Department of Aerospace Engineering at the University of Maryland; Dr Jim Young and the staff of the Edwards AFB history office (particularly Joyce Baker and Cheryl Gumm, and Jim Young's predecessor J Ted Bear); Lieutenant Colonel William 'Flaps' Flanagan, USAF (ret.) of the Northrop-Grumman Corporation; Sharon Wanglin; Harry Kotcher; Tony Landis; Mike Machat; Brian Nicklas; Curtis Peebles; Margaret Woolams; Joe Cannon; the late Milt Thompson; the late Dr Walter C Williams, former director of the NACA High-Speed Flight Station; and – one who played a key role in making this edition possible – Colonel Donald S Lopez, USAF, (ret.), the Deputy Director of the National Air and Space Museum.

Others who were helpful or encouraging, or who furnished critical insights or bits of information include the late Captain Carl T 'Tex' Birdwell, USN (ret.); the late Bob Perry; Roger Bilstein; Colonel Ken Chilstrom, USAF (ret.); Chalmers H 'Slick' Goodlin; Frederick C Durant, III; Alexis 'Dusty' Doster; Frank H Winter; Greg Kennedy; Lou Purnell; Major General Peter 'Peet' Odgers, USAF (ret.); Major General Phil Conley, USAF (ret.); Pete Adolph; Hal Andrews; Harry Gann; Peter Bowers; Diana Cornellise; Dave Menard; Wes Henry; the late Chuck Sewell; Ray Wagner; Bill Immenschuh; Bill Chana; Tom Smith; Colonel William J 'Pete' Knight, USAF (ret.); Colonel Jim McFeeters, USAF (ret.); Tom Oldfield; Air Vice Marshal Paddy Harbison, RAF (ret.); Air Commodore Alan Merriman, RAF; the late Air Commodore Allen H Wheeler, RAF (ret.); Sir Morian Morgan; Professor W A Mair; R P Probert; Roland Beamont; the late Dr John Fozard; Graham Weller; and Darryl Cott. As always, I owe a very special debt to some especially close friends and colleagues: Richard and Jeanne 'Mr and Mrs B' Blalock, and Wilbur 'Bill' Bettis of *Nieuport 17*, and, with particularly fond memories, the late Art Neff of Pasadena, whose never-failing public-mindedness, graciousness and unselfishness made the word 'gentleman' always seem utterly inadequate.

Again, from the perspective of a quarter-century after writing the first edition of this work, I am astonished – and very grateful – for how generous (and patient) the veterans of the supersonic breakthrough were in granting a very junior historian their time and assistance. It is to them, on the 50th anniversary of the first supersonic flight, that this book is dedicated, with affection and always, respect.

Introduction to the 1972 Edition

Michael Collins: Director,
National Air and Space Museum
Melvin B Zisfein: Deputy Director,
National Air and Space Museum

The first supersonic flight and the first trip to the moon had many things in common. Both were produced by the American aerospace industry, both required quantum jumps in many areas of engineering and both required the utmost in pilot training and human and physical support systems. Most significantly, supersonic flight and space travel were both born of the vision of people who dared to assert that 'it could be done', who had to prove repeatedly to sceptics that it might be worth doing and who were harassed from start to finish (but particularly at the start) by many and variously oriented prophets of failure.

In order to appreciate the Bell X-1 achievement, let us examine the problems of flying faster than sound and review the pitfalls as they might have been foreseen on the 'before' side of the achievement. We will find that in the mid-1940s it was not difficult to extrapolate existing knowledge into many fears, few entirely groundless.

We are considering in this study manned, powered flight in a lifting vehicle (an aeroplane) through the air at speeds faster than sound travels through air. Of all the possible limiting speeds, there were good reasons to select the speed of sound as the upper limit. Ahead of a lifting wing there is an upwash, a current of air particles moving upward in anticipation of the approach of the wing. This flow pattern prevails at all speeds below the speed of sound. The particles are 'informed' by pressure waves emanating from the wing. The pressure waves that move the air ahead of the wing travel at the speed of sound.

If the wing in our example then speeds up so that it is moving at or above the velocity of sound, the pressure waves that previously 'warned' the air ahead of the wing's coming are overtaken by the wing itself.

In the early 1940s this chain of reasoning led to dire predictions about supersonic flight. Pilots blamed the sonic 'bunching up' of air for the frightening vibrations aeroplanes encountered when they dived at nearly sonic speed. Engineers knew that a wing moving at supersonic speed created shock waves, violent disturbances in the air which move along with the wing as if attached and in a sense, adjust the airflow for the supersonic wing's lack of advance warning. These shock waves were expected to ruin the air flow, stall the wings, reverse the control forces and otherwise produce serious mischief. Numerous incidents of near catastrophic high-speed buffeting, loss of control and upset encounters as well as some unexplained crashes, seemed to confirm those fears.

Supersonic bullets had been coming out of pistol, rifle, and cannon muzzles for years and (with their shock waves) speeding precisely to their targets. But a bullet is a bullet and a thin skinned, winged, control surfaced aeroplane seemed to be quite another thing.

There were serious predictions that the pilot's voice would get stuck in his throat when he flew faster than sound, ignoring the fact that in the cockpit the pilot is surrounded by air at rest with respect to him. Some even feared that flying supersonically reversed time and that the pilot would turn into a young boy.

If you were a sober analyst you would have noted that Geoffrey de Havilland died in a disintegrating DH 108 while trying to break the sound barrier. The strange buffeting, high structural loads, aerodynamic flow separations, erratically varying control moments, trim changes and stick forces associated with transonic and supersonic flight certainly indicated that achieving supersonic flight was not merely a problem of packing more power into a stronger aeroplane. Supersonic flight was known earlier than the 1940s to be fundamentally different. The aerodynamicist recognised that at supersonic speeds many of the time-tested predictive methods were no longer applicable. At the most fundamental level the governing supersonic flow equation was the Wave Equation, not LaPlace's Equation as at subsonic speeds; all supersonic flow solutions were expected to be different. This difference in air behaviour certainly was fundamental. An unprecedented zone in front of a supersonically travelling body where the approaching body's presence could not be felt was poetically named by Dr. Theodore von Karman, the 'Zone of Silence or Forbidden Signals'. (The Tenth Wright Brothers Lecture delivered by von Karman in 1947 is a monumental paper which summarises the state of knowledge of supersonics at the dawn of the supersonic era.)

A new aerodynamic force, 'wave drag', became important at supersonic speeds, adding substantially to the thrust requirements imposed by the more familiar skin-friction drag, profile drag, lift-induced drag and interference drag. In the mid 1940s it was impossible to be certain that there were no other transonic and supersonic phenomena as yet unanticipated. Even at the most 'practical' level in the 1940s the unnerving aspect of supersonic flight was that there were little 'hard' supersonic test data. It was difficult to assess the accuracy of the various transonic and supersonic flow solution techniques that the more mathematically orientated aerodynamicists had devised. Any predicted sonic or supersonic load or control force might be off several orders of magnitude. Only experience would change this, and the experience might be costly.

During the middle and late 1940s, the authors of this Introduction were at school. The fact that about 25 years later they would become responsible for a collection containing the X-1 was in their own 'zone of forbidden signals'. What they learned collectively during this period included a series of analyses and reasoned hypotheses which indicated that at best, flight faster than sound would be tricky with foreseeable equipment and at worst, impossible in a practical sense. Let us go back to re-create some of these anticipated problems.

First we have the 'sound barrier'. What was it? It was no barrier at all, but what might be considered to be a mathematically-expressed warning of probable physical difficulties. In one sense the 'sound barrier' was a number of subsonic curves of flight parameters that seemed to approach infinity as speed was increased (far beyond the zone of validity of the curves) toward the speed of sound.

Aerodynamicists, particularly those of a mathematical bent, had long been exercising the problem of 'compressibility effects', the term frequently used to describe what happened to aerodynamic properties as a vehicle's speed approached and (if possible) exceeded the speed of sound. The term 'Mach Number' (named after the scientist Ernst Mach) was devised to describe the ratio of aircraft speed to the speed of sound. Hence:

$$M = V/V_s$$

or, Mach Number = Vehicle Velocity ÷Velocity of Sound. (Mach 1 is the velocity of sound, Mach 2 is twice the velocity of sound, Mach 0.95 is 95% of the velocity of sound, etc.).

The theoretical aerodynamicists had for some time examined what happens as a vehicle approaches the speed of sound. Ludwig Prandtl and Herman Glauert derived the approximate factor

$$1 / \sqrt{(1 - M^2)}$$

as a correction factor for pressure or lift. This term shows that for a wing-like body fixed in an airstream (or an airfoil moving through the air at a given attitude), the pressure or overall lift will increase over the low speed value as velocity is increased, and that the rate of increase will accelerate; so that as we approach the velocity of sound (Mach 1), pressure and lift will approach infinity (as the denominator approaches zero). In 1947, the year of the first supersonic flight, one of the first, generally available compressible aerodynamics textbooks, *Introduction to Aerodynamics of a Compressible Fluid* was published, written by Hans Liepmann and Allen Puckett. In this text the excellent correlation of (then) recent NACA experiments with the Prandtl-Glauert factor (with a more refined factor by von Karman and Hsue-shen Tsien) is demonstrated graphically to show the reader this trend of rising pressure leading into a then-unexplored region. (This same trend was also evident in a text published in 1951 by the British aerodynamicist William F Hilton, *High-Speed Aerodynamics*). Such data were generally available in the aeronautical engineering community well before the X-1 flew.

Unfortunately during the 1940s, the period when these comparisons were being made and published in reports, their generality was overestimated. Some people inferred that this increase of 'everything' toward infinity as Mach 1 was approached could be taken literally (forgetting the limitations of the theoretical derivations, clearly stated by the aerodynamicists) and that vehicle drag would increase to infinity. Hence the 'sound barrier'

Let us approach the sound barrier notion from a more physical viewpoint. A lifting wing moving through the air constantly emits pressure disturbances. Each pressure pulse expands outward like ripples in a pool. Each wavefront moves out in all directions at the speed of sound. When the wing is moving at Mach 0.9, it is rapidly overtaking the forward-moving pressure wavefronts. When the wing is moving at Mach 1.0, the pressures from all forward wavefronts pile up in a concentrated front. With a little imagination this front becomes a 'pressure wall' and the sound barrier concept is given a physical interpretation. In his 1947 Wright Brothers Lecture, von Karman called this pressure pile-up phenomenon the 'Rule of Concentrated Action'.

The expression 'breaking the sound barrier' while dramatically edifying, might better have been phrased as 'traverse the transonic tangles and traps'. Uncertainty was the key emotion. As no-one really knew the level at which the steadily increasing subsonic force coefficients peaked out to merge with the steadily decreasing supersonic force coefficients, transonic loads were an open question. Doubts about static and manoeuvring loads were compounded by additional uncertainties regarding the onset of 'shock stall', the possible loss of lift from this phenomenon and the accompanying incremental buffeting loads. Extreme conservatism was called for, but you cannot build an aeroplane like a bridge and expect it to fly.

The flutter implications of not really knowing the magnitude and distribution of these aerodynamic loads were enormous. Flutter is that aerodynamic-structural phenomenon which results in spontaneous vibrations of members of a flying air vehicle sometimes growing rapidly to induce catastrophic structural failure. Transonic wing, tail and control surface flutter predictions were all extremely questionable. A non-linear variant of flutter, aileron buzz, was a particular nuisance all through the early X-aeroplane programmes.

A constant source of danger and a tax on the designer was the expected (and sometimes unexpected) shift of centres of pressure as speed changed from subsonic to supersonic. A centre of pressure might be regarded as the application point of any aerodynamic force, such as wing lift. The change in wing centre of pressure from about a quarter of the wing chord (subsonic) to about a half of the wing chord (supersonic) sounds innocuous. It is not. Particularly with the aeroplane shapes of 1947, huge changes in trim, stability, control forces and manoeuvring loads could result. The designer was faced with accommodating such changes and foreseeing more capricious variants of these that might be associated with the constantly changing shock wave patterns characteristic of transonic flow.

A principal ingredient in all this uncertainty was the 'uncertainty of similitude'.

We simply were not sure how we should test models or interpret the results of tests. Many experts considered a transonic wind tunnel to be forever impossible. (They did not foresee the invention of the porous or slotted wind tunnel throat.) Even if in an expensive free-flight transonic rocket test a shock wave was found to appear on the model, did this mean that one would appear on the full-scale aeroplane? The academic and industrial research and development communities were seriously at work on all of these questions in 1947, but there were no firm answers and little 'hard' data.

Of course there was no impenetrable sound barrier and supersonic flight was achieved through the process that forms the bulk of this book. We hasten to point out however, that the success of the supersonic flight venture leaned heavily on the methodical mathematical aerodynamic studies that constituted the life work of a generation of dedicated scholars, and that readers interested in understanding what level of knowledge existed at the time prior to the first supersonic flights may find the bibliographical references in Leipmann and Puckett's 1947 text particularly useful.

This should in no way detract from the credit due to the first supersonic flight teams or, for that matter, to those who followed the initial flights by working for years to extract more and better data from the X-1 and D-558 flight programmes. These data were used for everything from the confirmation of total aeroplane drag prediction methods to the validation of laminar-to-turbulent boundary layer transition point calculations for the prediction of aerodynamic heating rates. Their effect on subsequent design efforts is great. These tests and their output, in addition to generating considerable primary design method validation data, also helped train the first generation of supersonic aircraft designers.

One of the authors of this introduction, Melvin B Zisfein, had the privilege of working in the midst of the X aeroplane aerodynamics team during his early years at Bell Aircraft Corporation. Even a few years after the initial X-1 flights, the excitement of flight-planning for a new record-setting (and of course, data-gathering) flight was infectious. Some of the flight-planning was desk work, but a good deal of it was centred on two strange-looking analogue computers. At least one of these was known as the BAPA (Bell Aircraft Performance Analyzer); perhaps both were; they will be called BAPA-I and BAPA-II here, although these designations may not be exact. BAPA-I looked like a black girder structure made from some large Erector set. It had dials and knobs on the front and (when operating) a bank of softly glowing light bulbs (for electrical ballast) underneath. BAPA-II was an improved version replacing BAPA-I. It had a shiny grey case, large dials and knobs built into a sloping front and the general air of a submarine control panel in a low-budget motion picture. Each BAPA run simulated, in accelerated time, a possible X aeroplane flight, taking into account launch inputs, fuel consumption, empirical lift and drag variation with Mach Number, altitude, etc. From the results of these runs and related calculations, final flight strategies were selected and flight plans were prepared.

At the beginning of a BAPA run the team assembled in front of the machine with about three seated dial-reading knob-turners and one flight director. On command the aeroplane was 'launched' and all the personnel hand-controlled their assigned knobs to reflect readings of dials showing changes in such variables as weight, Mach Number and altitude. One virtue of the machine was that it was slow, so that input settings could be readily changed by hand to correspond with constantly changing flight variables. Since all input data (such as lift and drag coefficients) were known only as nonlinearly varying empirical functions of Mach Number and other variables, automatic regulation of machine inputs to correspond to these variables was out of the question. Each 'flight' was an odd-looking exercise with a four-man (or so) crew with eyes glued on the dials, following the machine readouts and a string of verbal commands as they turned the appropriate knobs. However, this was reasonably sophisticated at a time of relatively low technological sophistication – and the technique worked.

Today, our level of sophistication is much higher in the hardware and software of both design and flight. Sometimes higher sophistication leads to higher costs, and usually it leads to a better designed, more reliable product. From the pilot's point of view, today's sophisticated aeroplanes operate routinely on either side of the 'sound barrier' and in fact, under some conditions, the pilot must check his instruments to determine whether he is in subsonic or supersonic flight. How disappointing to find that the engineers have learned their lessons so well that they have reduced the pilot's awareness of the 'sound barrier' to a surreptitious glance at the Machmeter! While this may be true of recent aircraft, such as the 'Century Series' fighters, some of the aircraft following closely on the heels of the X-1 were not so well behaved. One of the authors, Michael Collins, recalls his first trip past Mach 1 in a barely supersonic North American F-86 Sabre jet fighter. The trick here was to climb to high altitude and then go into a near-vertical dive. Even with this gravity assistance, it was difficult to obtain the telltale jump of the Machmeter which indicated that the shock wave had passed the static port on the side of the fuselage. In fact, individual F-86s showed pernickety differences. Some were easily handled through Mach 1; others developed an uncontrollable roll and could only make it through into supersonic flight in corkscrew fashion; while some would waver right at the threshold and could not be induced to go beyond Mach 0.96 or 0.97. But this era in aviation was soon over and, with the advent of the North American F-100 Super Sabre, supersonic level and even climbing flight became commonplace for the military fighter pilot. The bomber pilots even had their supersonic fling with the Convair B-58 Hustler and the experimental Mach 3 North American XB-70 Valkyrie which embarrassed the latest fighters by leaving them far behind in a cloud of vapour trails. Then pilots were given a new supersonic option; going straight up on a rocket headed for earth orbit or beyond.

The Mercury, Gemini and Apollo boosters all have similar launch profiles. Having only slightly more thrust than weight as they leave the launch pad, they get off to the excruciatingly slow start we have all seen as the missile inches up our television screen. As the fuel load lightens however, the constant thrust of the engines causes the acceleration to increase rapidly and supersonic speed is usually achieved in a bit over a minute's time at an altitude of around 35,000ft. In other words, we find ourselves in the F-86's domain, but heading straight up instead of straight down, and hopefully without any uncontrollable tendencies. As speed and dynamic pressure build, noise and vibration inside the cockpit (of aircraft or spacecraft) also increase, and reach a peak just as Mach 1 is reached. Beyond that, things get quieter and smoother rapidly, especially in the case of the spacecraft which soon finds itself far above the disturbing aerodynamic properties of the dense atmosphere below.

Although the past 25 years have shown amazing progress, even today only a few can visit the supersonic domain. We have deliberately chosen not to attempt commercial supersonic flight, less because of technological difficulties than concern over economic and ecological factors. It is difficult to predict when these impediments may be overcome, but it does seem safe to look past the Anglo-French Concorde to the day when high Mach number flight becomes commonplace.

Chuck Yeager's flight of 14 October 1947 in the Bell XS-1 is clearly a major milestone on the road from Kitty Hawk to the moon. We join Mr Hallion in recognising this achievement on its 25th anniversary and in saluting those in industry, government, research institutions and universities whose combined efforts gave us the knowledge and machines to fly at supersonic speeds.

Introduction to the 1997 Edition

Richard P Hallion

It is now 50 years since Chuck Yeager and the Bell XS-1 first whiplashed the Mojave desert with a supersonic boom. What was at the time considered impossible or difficult at best became commonplace 25 years later, and now does not seem remarkable at all. For over 40 years, supersonic jet fighters have routinely exceeded Mach 1 in level flight and for over 20 years a supersonic jet transport, the Anglo-French Concorde, has plied the skies. Refined engines, aerodynamics and flight control designs have produced new generations of military aircraft that use the supersonic regime with an assurance and degree of efficiency unknown to the first 'Century series' of supersonic jet fighters and their experimental predecessors. These same developments promise that potential future civil jet transports will avoid some of the environmental and economic problems that worked together to doom American efforts to develop a first-generation supersonic transport.

For example, the low power available from early turbojet engines in the 1940s impelled the first supersonic designers to use rocket engines to exceed Mach 1. Even so, the thrust:weight ratio (the thrust output of the engine in pounds divided by the weight of the aircraft) of these early supersonic aeroplanes was less than one: about 0.49:1 for the X-1 and 0.38:1 for the rocket-powered D-558-2. (In contrast, a contemporary high-performance jet fighter has a thrust:weight ratio in excess of one, and is able to accelerate easily through the speed of sound while climbing.) By the mid-1950s, powerful turbojets could accelerate an aeroplane to just below the speed of sound, and by using a thrust-augmenting afterburner (or 'reheat' in British parlance), a pilot could exceed the sound barrier, reaching speeds of Mach 2. This came at a cost however, namely in soaring fuel consumption. Today, electronically controlled high-performance turbojet engines produce so much power for their weight that they can accelerate an aeroplane to speeds beyond the speed of sound on non-afterburning power alone. Therefore, the afterburner can be limited for combat and take-off assistance only, and supersonic flight can be sustained without recourse to the wasteful and profligate use of limited fuel.

The first supersonic aircraft were largely projectile shapes – one thinks of the XS-1 modelled on a .50 calibre bullet, the Douglas X-3 (aptly nicknamed the 'Flying Stiletto') and the Lockheed F-104 Starfighter, dubbed the 'Manned Missile' – while more recent supersonic aircraft have taken advantage of the power of new engines and refined aerodynamic design techniques to produce evocative and highly efficient aircraft. These aircraft have, for example, far more generous wing areas than their predecessors of the early supersonic era. Sinuousness has replaced crude angularity except where, notably, the quest for a radar-evading, stealthy shape has led to the distinctive faceting of new designs. While the greatest accomplishment of the first generations of supersonic aircraft was merely 'getting to' the supersonic regime, successors could exploit it for military purposes or, in the case of the Concorde, for commercial advantage.

The advent of comprehensive electronic flight-control technology (exemplified by the General Dynamics – now Lockheed-Martin – F-16 Fighting Falcon) when joined with the other advances in aerodynamic and propulsive design, transformed supersonic flight from a frontier of flight traversed rarely to an operational environment exploited often. The author well remembers a very brief supersonic sojourn in a relatively low thrust:weight ratio Lockheed TF-104G Starfighter, capped by a relatively gentle 3g supersonic turn made memorable by a pronounced airframe buffet that severely limited any ability to read the rear-cockpit instrumentation, together with a rapid bleed-off of speed that quickly would have left us on the defensive had we been in air combat. A decade later in 1986, a 9.2 g turn in very high thrust:weight ratio F-16B was by contrast, rock-steady (although the plane was clearly happier at that moment than its crew!) and thanks to a high thrust:weight state there was a tremendous reserve of energy which enabled us to resume an offensive posture virtually immediately in the low supersonic regime. Even so, both of these Mach 2-class designs still cruised subsonically, using supersonic speed only briefly. In contrast, the Lockheed-Martin F-22, the most advanced next-generation fighter now under development, has already demonstrated its ability to cruise at supersonic speeds to an area of operations where it can exploit its extraordinary manoeuvrability, advanced sensor fusion, weapons and – above all – stealth, to guarantee 'first look, first shot and first kill' against an opponent. Such has been the logical outcome of half a century of supersonic evolution.

This book constitutes a 'slice in time': an attempt to look at a period of approximately 15 years during which aviation science passed from the subsonic to the supersonic era. It is about a breakthrough, about the development of a capability to exploit the high-speed frontier. It is not, however, a comprehensive history of supersonic flight since that time. Were it so, it would address other questions, not the least of which is how various nations pursued correctly or incorrectly the grail of supersonic flight and attempted to incorporate it in their own national interest.

The history of supersonic flight has *not* been an unblemished success. For example, the Anglo-French Concorde; an extraordinary technical achievement for its time, nevertheless failed to secure broad international support and though it was a technical success, has at best been only marginally economical, serving more the purposes of prestige than profit. Its Soviet counterpart, the Tupolev Tu-144, could not even claim technical excellence and a tragic accident at the Paris Air Show of 1973 further blackened its reputation. To a large degree a fascination with supersonics (and for that matter, nuclear weaponry) led the United States to pursue a series of over-specialised 'big ticket' acquisition programmes during the 1950s that were increasingly unrelated to the real and valid needs of defence of the time. The following constitute but a partial list of these acquisitions: the proposed XP-92 sled-launched rocket-and-ramjet propelled interceptor, the Convair B-58 Hustler and North American A3J-1 Vigiliante Mach 2 nuclear bombers, the North American SM-64 Navaho Mach 3+ cruise missile, Pluto (a nuclear-powered Mach 3+ ramjet-powered cruise missile), the Republic XF-103 and North American XF-108 interceptors, the Bell BOMI and ROBO proposed rocket-propelled orbital boost-glider bombers and the huge North American XB-70 Valkyrie Mach 3+ bomber. With the sole exception of the multipurpose F-100, five of six of the successful 'Century series' fighters developed in the 1950s – the F-101, F-102, F-104, F-105, and F-106 – were specialised as supersonic-dash interceptors or nuclear strike aeroplanes. The civilian world likewise reflected this history, as the sad saga of the abortive Boeing Mach 2.7 American SST clearly demonstrates.

But when opportunity, perceptive need and appropriate application of transonic and supersonic design theory came together, the results could be at the least most satisfactory and, in some cases truly elegant: the North American F-100 Super Sabre, the Vought F8U-1 Crusader, the McDonnell F4H-1 Phantom II, the Northrop T-38/F-5 family, and the Lockheed A-12/SR-71 Blackbird constitute notable examples. The two great transonic jet transports that revolutionised international air commerce (and put the North Atlantic steamship out of business) in the late 1950s and early 1960s; the Boeing 707 and Douglas DC-8, both benefited from the knowledge-base derived by the national transonic and supersonic ground-and-flight-research programme. The United States was not alone in its mixed history with supersonic design. The Soviet Union and many of the European nations also pursued a mixture of projects, some of which were not at all satisfactory, others of which ranged from satisfactory to truly outstanding; examples of the latter mirroring the same 'late-1940s to late-1950s' perspective include the Saab J-35 Draken, the English Electric Lightning, the MiG-19 and MiG-21 and the Mirage III. Along the way were some that clearly should have been pursued, but were not. The most notable of these was Great Britain's elegant TSR 2, perhaps the saddest of all 'might have been' stories.

About this Book

Supersonic Flight first began in the autumn of 1969 as an undergraduate honours thesis in the Department of History at the University of Maryland. The choice of subject, to say the least, puzzled the majority of the school's mostly traditionally minded (if nevertheless well-meaning) academics. As surprising as it may seem in the post-Tom Wolfe *The Right Stuff* era, in the late 1960s and early 1970s, there was little general awareness about the remarkable events – and even more remarkable people, particularly Chuck Yeager (then an Air Force attaché in Pakistan) – that had brought supersonic flight to the world. But to one raised in the 1950s when it appeared that another headline came out of the Mojave desert every week, it seemed a most natural and suitable topic. After all, aviation had transformed society, high-speed aviation in particular. Thanks to the encouragement of a unique group of faculty members, I was permitted to pursue the subject. These individuals were: in the Department of History, Professors Walter Rundell, Gordon Prange, Wayne Cole, Keith Olson, Adrienne Koch and James Flack; and in the Department of Aerospace Engineering, Professors Richard Thomas (who served as my senior advisor) and Jewel Barlow. Others who played a significant role are listed in the acknowledgements, but one in particular – the late Dr Eugene Emme who was the first NASA historian – deserves particular mention as it was because of his efforts that I was soon confronting the stuff of history: the actual flight reports, pilots' notes, design documents, etc. that made this hitherto neglected story come alive. Interviews and correspondence with key programme participants fleshed the tale out further. To all of these people, many of whom have since died, I was – and always will be – very grateful. Their vital support came at a key point in the development of my interest in aeronautics.

Following graduation, I continued to pursue the subject of supersonic research and in the fall of 1971, I wrote to the energetic new Director of the National Air and Space Museum, Michael Collins, suggesting that the Museum commemorate the 25th anniversary of the first supersonic flight on 14 October 1972. That letter had totally unexpected consequences, both for my career in aerospace history and more importantly, for this book. Collins, the former command module pilot on the Apollo 11 mission to the moon, consulted with his equally dynamic deputy director, Melvin Zisfein who had, early in his career, worked at Bell Aircraft Corporation. Together they invited me to lunch at the then-Museum of History and Technology (now the National Museum of American History) and broached an idea: would I write a book on the subject? To a relatively junior graduate student, the offer was irresistible. I quickly agreed and the result was this work, written over the summer of 1972 and published in September of that year. In 1974, I joined the museum staff as Curator of Science and Technology with responsibility for a new exhibition gallery on flight testing. The next two years were very busy in preparing for the opening of the museum in July 1976, but they had their moments of reward, such as supervising the preparation of the Mach-busting Bell XS-1 and the Mach 2 Douglas D-558-2 for exhibition. (I have to confess that I did take advantage of the opportunity to sit inside them and marvel at how Yeager, nursing broken ribs, could have squeezed into the XS-1 from a small ladder out of the launch B-29's bomb bay, locked the entry hatch and then sat with his back snuggled up against a bitterly cold liquid-oxygen tank or how

Crossfield could have sat in the claustrophobic Skyrocket, his helmet almost wedged in the sharp 'V' of the hooded cockpit canopy). One very great pleasure during this time was making the acquaintance of the late Walter C Williams, the former chief of the NACA's High-Speed Flight Station at Edwards AFB, California. Walt Williams furnished provocative insights into the history of the rocket research aircraft programmes and we remained close friends until his death in 1996.

Since that time my research has taken me in many different directions, but I still follow the supersonic flight testing field with great interest. It is nice to note that a number of works have appeared since this book's first edition that have added to and expanded upon it, particularly relating to foreign flight research activities. I would particularly like to acknowledge the research of Dr Jim Young, John Becker, Don Middleton, Yves Candal, Charles Burnet, Dr Jim Hansen, Louis Rotundo, Henry Matthews, Kenneth Owen, Ben Guenther and Jay Miller. Long-overdue and much-welcomed pilot memoirs from Chuck Yeager, Don Lopez, Milt Thompson, Tex Johnston and Bob Hoover have added further important views and recollections. At various places throughout this edition, I have taken advantage of their work to add to the text or to make some corrections and elaborations. Their works where used, are cited separately in the notes.

In the interest of brevity, I have also deleted a small section of extraneous material from the original introduction by Michael Collins and the late Melvin Zisfein, consisting of a quote from an article in *Reader's Digest*, curves plotting airfoil performance at transonic speeds and a bibliography of purely aerodynamic studies which was reprinted from a 1947 textbook: *Introduction to Aerodynamics of a Compressible Fluid* by Hans W Liepmann and Allen E Puckett. The 1972 edition also reflected the common graduate school affliction of 'footnote-itis'; in this second edition, I have compressed the footnotes and in some cases, gathered them together for ease of the reader. I have also expanded upon the photographs that were used in the first edition, thanks to the willingness of my editors to indulge my illustrative desires. I hope that this reprinted edition will continue to serve as an encouragement and stimulus for further research – worldwide – on the history of high-speed flight.

Prologue

27 September 1946

In the summer of 1946, Great Britain abandoned all efforts to develop manned supersonic research aircraft, citing the postwar financial condition for the decision. The Ministry of Supply announced that henceforth Great Britain would use rocket-propelled models to investigate supersonic flight conditions. But a more dramatic and perhaps pertinent reason appeared in a press interview of Sir Ben Lockspeiser, the Ministry of Supply's Director-General of Scientific Air Research. 'We have not the heart,' he declared sombrely, 'to ask pilots to fly the highspeed models, so we shall make them radio-controlled.'

It is not known what 36-year old Geoffrey Raoul de Havilland, chief test pilot of his father's firm the de Havilland Aircraft Co., Ltd., thought of the Ministry of Supply's decision. He was busy readying the trim de Havilland DH 108 Swallow, a sweptwing tailless aircraft, for an assault on the world's air speed record. Britain actually had two Swallows: the first to check the low-speed behaviour of sweptwings; and the second for high-speed sweptwing research. To complete the two aircraft as quickly as possible, de Havilland had fitted sweptwings and swept vertical tail fins to standard Vampire fuselages containing their Goblin turbojet powerplants. The company had not built the Swallows for government-sponsored transonic or supersonic research, but rather to investigate sweptwing stability and control problems and to provide data for the proposed DH 106 transport. Therefore, the Ministry of Supply's decision did not affect company plans to test the two DH 108 aeroplanes.

In his career as a test pilot, Geoffrey de Havilland had already compiled a distinguished record. He flew the first Mosquito bomber his father's company built, the first Vampire jet fighter, and the first de Havilland Hornet long-range fighter. On the coming speed-record runs, he would have to fly the blue-grey DH 108 over a measured course at an altitude of less than 1,100 feet. On 7 September 1946, Royal Air Force Group Captain E M 'Teddy' Donaldson, flying a modified Gloster Meteor F Mk. IV fighter, had raised the world air-speed record to 615.778 mph, no mean feat in an aeroplane that had shown pronounced tendencies to tuck its nose down and dive into the ground due to airflow effects as it neared the speed of sound while flying close to the ground.

De Havilland would have to exceed Donaldson's record by at least 1% to establish a new record. He planned his record attempt for the end of September 1946. Few doubted that de Havilland would break Donaldson's record and in one small sweepstake, speeds below 640 mph were given little chance of winning. In mid-September he began practising for the record run. Flying the Swallow at increasingly higher speeds at lower altitudes, he unofficially exceeded the Meteor record even on these trials.

At 5.30pm on the late afternoon of 27 September, Geoffrey de Havilland taxied the DH 108 out on the runway at the company's Hatfield Aerodrome, ran up the Goblin turbojet and with the engine shrieking at full power, rolled ever faster down the runway and then into the air. The Swallow disappeared from view to the south, over the Thames estuary into what promised to be a clear, fine evening. The flight plan called for a low-level maximum speed run followed by a maximum speed dive. If all went well, the next Swallow flight would be for the air speed record.

The first reports of disaster reached the company 30 minutes after the swift DH 108 lifted away from Hatfield. While walking along Canvey Island, an RAF Flight Lieutenant and his wife heard the whistle of a jet, a crackling noise and saw three bits of wreckage tumble from the sky into the shallow water of Egypt Bay. Other observers saw the Swallow, flying several thousand feet above the Thames estuary at high speed, disintegrate before their eyes. A group of Mosquito pilots spotted wreckage amid the marshland along Egypt Bay. The Swallow had crashed and Geoffrey de Havilland was missing. Ten days later the pilot's body washed ashore at Whitstable. He was the second of company founder Sir Geoffrey de Havilland's sons to die while test-flying company aircraft.

Accident investigators recovered much of the wreckage, enough to ascertain that the aeroplane had not exploded. Analysis of retrieved recording instrumentation indicated that the plane had fallen victim to compressibility – the hammering and often destabilising airflow changes that occur as a plane transitions from subsonic to supersonic flight. Inadvertently, Geoffrey de Havilland had exceeded the Swallow's critical Mach number –

the point where its drag would rise, its lift would drop and it could go out of control – and, in the dense lower atmosphere, compressibility forces had ripped the DH 108 apart. The accident lent credence to then-current dire predictions about the existence of an impenetrable sonic barrier through which no aircraft could fly and reinforced the British government's determination to avoid manned supersonic research.

For more than 25 years, the effects of compressibility had been a matter of interest and concern to aerodynamicists and engineers, propeller specialists, aeroplane and engine manufacturers, high-speed aeroplane design teams and pilots. Now all of this interest and study was to culminate in the first attempt to exceed the speed of sound in a manned aircraft. For across the Atlantic Ocean, and on the steppes of the Soviet Union, groups of scientists and engineers were readying a special breed of research aircraft for their first trial flights. Despite de Havilland's death they pushed doggedly on, and their resolve to lay to rest the myth of the 'sound barrier' stiffened all the more.

PART ONE
Forging The Tools:

Developing The Bell X-1
and Douglas D-558

1
Making A Revolution:
The Need for a Transonic Research Aeroplane

Today it almost seems quaint: the idea that humanity might never fly faster than sound – as ludicrous as the predictions back in the nineteenth century that riders of steam locomotives would disintegrate into atoms because of the speed. But there *was* a genuine crisis, and it claimed the lives of pilots from many nations. Together with the turbojet revolution, 'breaking the sound barrier' constituted one of the two major revolutions to affect world aviation during the middle of the century. The physical problem was deceptively simple: at low airspeeds, air can be treated as an incompressible fluid (like water) in order to simplify the mathematical treatment of airflow; at high airspeeds however, a compressibility factor has to be taken into consideration. As an aeroplane moves through the air at high speed, it builds up a pile of air ahead of it. At about 450 mph, the accelerated flow above the wing may become locally supersonic, moving as fast as sound, or even faster. At that point, the plane is in the *transonic* region, an area of mixed subsonic and supersonic flow. If the speed is increased further, the complete flow around the aeroplane becomes supersonic, and the plane is in the *supersonic* region, moving faster than sound.

At sea level, the speed of sound is approximately 1,116 feet per second (fps). It varies with altitude, dropping off to approximately 968 fps at altitudes between 36,000 and 66,000 feet. The common unit of measurement for the speed of sound is *Mach number*, in honour of the Austrian physicist Ernst Mach. The Mach number is the ratio of the speed of an object to the speed of sound. Thus, Mach 0.5 is half of the speed of sound, Mach 1 the speed of sound, and Mach 2 twice the speed of sound, etc. During transonic flight, shock waves form on an aircraft, move back and forth and if the plane is not designed specifically for the sonic environment – as was the case with Second World War fighters – violently disrupt the airflow. Critical changes in pressure and loading may occur and the aircraft can undergo severe structural stress. Total airframe drag may rise sharply. The controls sometimes lose their effectiveness, thanks to the shifting position of shockwaves and the distortions in airflow around the aeroplane. It was for these reasons that, just prior to and during the Second World War, 'compressibility' became such a serious problem to designers of high-speed military aircraft that an increasingly public debate raged whether or not in fact a piloted aeroplane could ever safely exceed what was luridly called the 'sound barrier'.

The Beginning of a Concern

In mid-1918, members of the United States Army Air Service Engineering Division could peer up into the sky above McCook Field, Ohio, and recognise a big two-place biplane droning above Dayton. This plane, the USD-9, was an 'Americanised' version of the British De Havilland DH 9 day-bomber, itself an outgrowth of the famed DH 4 then in widespread service on the Western Front. The USD-9 cruised along at 100 mph and had a top speed of 115 mph, propelled by a 12-cylinder 400 hp Liberty engine turning a large four-bladed propeller. The USD-9 represented the peak of the 1918 aeronautical state-of-the-art. A rugged aircraft, weighing well over two tons fully loaded, it was a far cry from the frail open-structure biplanes with which most of the European nations had gone to war in 1914.

To aeronautical engineers, the stimulus of wartime demands had required that they develop aircraft to fly faster, further and higher than those of the enemy. Often they lacked the necessary research equipment to derive useful information that could be incorporated into new or existing designs. Powerplant engineers were concerned with wringing every available horsepower out of aircraft engines and then translating it into thrust from the whirling propeller. One question facing them was whether or not existing information on propeller airfoils – the cross-section of the propeller blade itself – could be applied to propellers whose tips were moving much faster through the air. Though the USD-9, for example, flew at 115 mph, the tips of its propeller scythed through the air at 650 mph, close to the speed of sound. Such matters were important; a propeller is essentially a 'wing with a twist' and its rotation generates a lifting force – along a lift vector – that pulls an aeroplane through the air.

In the fall of 1918, the Army Air Service Engineering Division decided to construct a small wind tunnel, entrusting the design to Frank W Caldwell and Elisha N Fales. Both were graduates of the Massachusetts Institute of Technology; Fales received his degree in 1911, Caldwell his a year later. Upon graduation, both men joined the Curtiss Aeroplane Company in Buffalo, New York. In 1916, Caldwell became foreman of the Curtiss propeller department. A year later, he went to McCook Field. Fales, after a short stint as an assistant professor of mechanical engineering at the University of Illinois, joined Caldwell at the Army airfield in Ohio. By then, Frank Caldwell had become chief of the propeller branch of the Air Service Engineering Division. He and Fales, working in conjunction with C P Grimes, designed the tunnel to operate up to 450 mph so that it could test propeller airfoils at high speeds. The throat section of the tunnel featured a diameter of 14 inches. Both Fales and Caldwell had undertaken preliminary propeller research at MIT from 1909 to 1912 with the assistance of Gaetano Lanza, head of the Department of Mechanical Engineering. Now they built six airfoils with 6-inch spans and 1-inch chords (the distance from the leading edge to the trailing edge), mounted them in the McCook tunnel, and made observations as the airstream passed around the airfoils at various speeds from 30 mph to 450 mph.[1]

Unknowingly, Caldwell and Fales were about to launch an aerodynamic revolution, for during the tunnel tests at 450 mph, they unexpectedly stumbled upon the effects of compressibility on airfoil behaviour, discovering two basic phenomena. At high speeds, the drag of the airfoil increased abruptly while the lift decreased, resulting in a sharp drop in the lift:drag ratio. Further, the airflow separated from the airfoil because of the mixed subsonic and supersonic flow patterns. Such behaviour came as a shock to propeller engineers, who had always assumed that 'we could double our propeller speed just as soon as we were able to double our plane speed and strengthen our engine enough to stand the stresses involved'.[2] Caldwell and Fales' pioneering work was next taken up by Dr Lyman J Briggs and Dr Hugh L Dryden, both of the Bureau of Standards and Lieutenant Colonel G F Hull of the Army Ordnance Department. These men tested the same six airfoil sections used by Caldwell and Fales, but at much higher velocities from 550 to 1,000 fps, using an air jet from a turbine-driven 3-stage centrifugal compressor at the General Electric Company's Lynn, Massachusetts plant. (Dryden, a brilliant young physicist who had received his Ph.D. from Johns Hopkins in 1919 at the age of 20, performed all data reduction and computations.) The findings of Briggs, Dryden and Hull confirmed the earlier work of Caldwell and Fales. During bad weather tests, the men had covered the airfoils with oil to keep them from rusting. Serendipitously, they noted that the oil moved in a pattern indicating separation of airflow at high speeds; likewise, they observed the classic decrease in lift and increase in drag that occurs at transonic speeds. They concluded that propeller airfoils lost efficiency as the tip speed approached the velocity of sound.[3]

Intrigued by the clear loss of efficiency, Briggs and Dryden continued their research at the Army's Edgewood Arsenal using a compressor plant formerly used during the First World War to refrigerate mustard gas. They ran tests on airfoils from Mach 0.5 to Mach 1.08, 563 fps to 1,218 fps. Most importantly, Briggs and Dryden fitted the airfoils with pressure orifices to record pressure distribution over the airfoil surface. In these tests the now-familiar separation of flow, drop in lift and increase in drag occurred at transonic speeds. Most importantly, at Mach 1.08, Dryden and Briggs observed standing shock waves about half an inch in front of the airfoil models. They reported their findings in the annual report of the National Advisory Committee for Aeronautics for 1926. The Briggs-Dryden team next began a comparative study of 24 airfoil sections at transonic speeds, subsequently discovering that thin airfoils (airfoils having a low thickness:chord ratio) retained their lifting properties at higher speeds markedly better than did thicker airfoils. The thick airfoils, in contrast, showed a marked decrease in lift coefficient. In an attempt to improve airfoil efficiency at transonic speeds, Briggs and Dryden undertook analysis of airfoils having the cross-sectional shape of a circular-arc. They found the circular-arc airfoils, while inefficient at low speeds, featured much greater efficiency at speeds above Mach 0.65 than did conventional airfoils and concluded their propeller research programme by recommending that propeller designers incorporate conventional airfoil sections on the inner portion of the blade, but use circular-arc airfoil sections on the outer third of the blade to retain efficiency of the blade at high tip speeds.[4]

Since an airfoil for a propeller behaved similarly to an airfoil for a wing, aeronautical engineers immediately realised that an aeroplane would also encounter a rise in drag, loss of lift and rearward shift in the centre of pressure location if it moved through the air at transonic velocities. However, the question seemed an academic one, since existing aircraft did not fly fast enough to encounter compressibility. This feeling changed dramatically in the late summer of 1931. On 29 September, Royal Air Force Flight Lieutenant George Stainforth went aloft in a sleek Supermarine S.6B floatplane racer to win the Schneider Trophy for Great Britain. When he touched down on the smooth water off Calshot Castle, he had concluded the world's first 400 mph flight. Despite the bulk of its floats and the drag of its bracing struts and wires, the silver-and-blue S.6B had made one circuit of the 3-kilometre course at 415 mph. At these velocities, aeronautical engineers could no longer afford to think of air as an incompressible fluid.

Stainforth's flight was the first obvious signal that aeronautical research would have to look forward to a time when aircraft were flying at 400 mph and above, where the accelerated airflow around the wing might well be even faster, well over 500 or even 600 mph. Could, in fact, an aeroplane be designed to fly so fast?

From Props to Planes: John Stack and the Transonic Aeroplane Idea

One man who believed that aircraft could be built despite the potential problems of compressibility was John Stack, an energetic young junior aeronautical engineer employed at the National Advisory Committee for Aeronautics' Langley Memorial Aeronautical Laboratory at Hampton, Virginia. 'NACA' was the nation's premier aeronautical research establishment, the predecessor of today's space-oriented National Aeronautics and Space Administration(NASA); Langley was its heart, already a centre of towering personalities and expansive facilities devoted to the study of flight. In this heady environment, Stack thrived. Another product of MIT (he had graduated in 1928 with a BS in the still-relatively new field of aeronautical engineering), he made a study of what sort of airframe configuration would be best suited to overcoming compressibility effects. Almost one year to the day after Stainforth's record flight, Dr George W Lewis, the NACA's Director of Aeronautical Research, had stated that 500 mph seemed the practical limit for 'the aeroplane of the present day'. He held out the hope, however, that engineers might well add 200 to 300 mph to the maximum speed of aircraft over the next decade by refined design techniques and careful streamlining.[5]

John Stack, one year after Lewis' disclosure, came up with one such aircraft. In many respects the proposal was as radical in its day as the X-1 was in 1945. Stack envisaged a propeller aircraft just large enough to house a proposed Rolls-Royce R piston engine of 2,300 hp. It was a mid-wing monoplane of cantilever construction, with a fuselage of circular cross section. The wing spanned 29.1 feet. The fuselage diameter was 40 inches. The wing had an area of 141.2 square feet, an aspect ratio of 6, and an average wing chord of 4.85 feet. The wing featured a NACA 0018 symmetrical airfoil section at the root, tapering to a NACA 0009 section at the wing tip. Stack tested a model of the design in the Langley variable-density wind tunnel, with encouraging results. The design reflected Stack's thinking that the aircraft should be as clean aerodynamically as possible, with a fully retractable landing gear and skin-type radiators. The pilot would see out of the aeroplane through transparent fuselage skins or by an indirect mirror reflective system. Stack assumed that the propeller, despite tip losses from compressibility, would retain 75% efficiency. Thus, the aeroplane would have an uncorrected maximum speed of about 566 mph and a landing speed of 103 mph. When Stack took into account compressibility effects, he found that the airspeed of the aeroplane would drop approximately 40 mph to about 525 mph. Stack optimistically concluded that this 'paper aeroplane' (for the aeroplane was never seriously pursued) indicated that aircraft could safely attain much higher airspeeds. He acknowledged that much compressibility research remained to be done in the future. Not all scientists shared Stack's optimism. British aerodynamicist W F Hilton plotted horsepower requirements versus Mach number and found an almost asymptotic relationship; in order to exceed the speed of sound, he concluded, an aeroplane would require over 30,000 hp. The speed of sound, Hilton gloomily predicted, loomed 'like a barrier against higher speed'. The resulting news accounts shortened Hilton's cautious statement to the more memorable 'sound barrier', thus coining a phrase that has lasted to the present day.[6]

In September and October 1935, leading aerodynamicists from around the world gathered in Campidoglio, Italy, for the Fifth Volta Congress on High Speeds in Aviation. Italian dictator Benito Mussolini opened the proceedings on a somewhat sombre note by announcing the Italian invasion of Abyssinia. With the monoplane revolution well underway and speculation about the future of high-speed flight, the Congress understandably devoted its attention to anticipated problems of supersonic flight. The Italian government showed its guests the Guidonia Laboratory, where technicians were building a Mach 2.7 supersonic wind tunnel patterned after an earlier Mach 2 wind tunnel – the first modern supersonic wind tunnel in the world – designed by Jakob Ackeret and completed previously at Zurich, Switzerland. Historically, the Volta Congress had two important influences upon supersonic flight. First and most importantly, it served to stimulate supersonic wind tunnel development. For example, the great Hungarian aerodynamicist Theodore von Karman, director of the Guggenheim Aeronautical Laboratory at the California Institute of Technology (GALCIT), returned to the United States from Campidoglio and immediately began contacting responsible government sources about developing supersonic wind tunnels. (Unfortunately, he met with indifference.) Abroad however, supersonic research moved more quickly. German scientists developed a Mach 4.4 wind tunnel that went into operation at the rocket development centre at Peenemunde in 1941. Second, the Volta Conference marked the first discussion of the sweptwing platform for supersonic flight. (Arguably, the sweptwing, together with the jet engine, were the two technology elements that most visibly, and noisily, in the case of the latter, separated the subsonic from the supersonic era.) The most advanced paper presented at the Volta Congress was that of German engineer Adolf Busemann; its subtle influence was not truly appreciated until after the Second World War. In his paper, Busemann suggested use of 'arrow-wings' (sweptwings) to alleviate the onset of high drag rise at transonic speeds. Early in the pioneer days of flying, John W Dunne employed the sweptwing configuration on a series of British biplane designs as a means of improving stability and control. Busemann's proposal, however, was the first time that the swept configuration appeared in high-speed aerodynamics. The paper went unnoticed by other aerodynamicists, but later stimulated German sweptwing research during the Second World War. The results of this research, together with the wartime research of NACA scientist Robert T Jones – who was unaware of Busemann's work at the time of his own pioneering studies – greatly influenced postwar American sweptwing aircraft development.[7]

Ezra Kotcher and the Call for High-Speed Flight Research

Despite increased awareness of the problems of transonic and supersonic flight, little 'hard' data existed on transonic conditions. Most of the existing data came from the propeller tests that gave some indication of the nature of transonic conditions and from drag-test results of bullets and shells decelerating through the speed of sound. Lieutenant Colonel Heinz Zornig, head of the Army Ballistic Research Laboratories at the Aberdeen Proving Ground, delivered a ballistics lecture in the mid-1930s at Wright Field, Ohio, illustrated by curves that plotted the drag coefficient of blunt, medium and sharp-nose projectiles versus Mach number. The curves did show the abrupt rise in drag close to the speed of sound. But the transonic drag rise, while high, did not exceed by a factor of more than two or three the low-speed subsonic drag value. Sitting in the audience was Ezra Kotcher, a senior instructor at the Air Corps Engineering School at Wright Field, soon to thrust ahead as the Air Corps' – and subsequently the Army Air Forces' – point-man on transonic and supersonic flight. Following graduation from the University of California in 1928 with a BS in mechanical engineering, he had joined the Army Air Corps as a junior aeronautical engineer with a strong interest in high-speed flight. Listening to Zornig's lecture, Kotcher developed an intuitive feeling that the sonic barrier was 'not necessarily a permanent flight barrier, but rather a wind tunnel technique barrier – or a psychological barrier'. Zornig's data impressed Kotcher because, based on full-scale 'real world' dynamic in-flight testing, it gave an inkling of what really happened aerodynamically in the transonic region across the speed of sound. Over the next few years Kotcher frequently thought of the Zornig lecture. In mid-1939, Captain T A Sims requested that he write a short report on future aeronautical research and development problems for transmittal to the Kilner-Lindbergh board that Air Corps Chief General Henry H 'Hap' Arnold had established on 5 May 1939 to investigate future military aircraft development. Kotcher prepared the report and submitted it in August 1939 to the Air Corps Engineering Section which passed it on to General Arnold's staff and the NACA.[8]

The Kotcher report was noteworthy in several respects. Kotcher stated that in order to achieve significant advances in high-speed performance it would be necessary to undertake extensive transonic research. He advocated 'comprehensive flight research programs' to correlate wind tunnel data with full-scale performance. He suggested that gas turbine or rocket propulsion systems be developed eventually to replace propeller propulsion systems because of compressibility limitations upon propeller aircraft at high speeds. Kotcher's strong advocacy of gas turbine or rocket propulsion came at a time when neither was in serious consideration within the United States, save for a small and non-influential group of enthusiasts. Yet overseas, developments in gas turbine and rocket technology were moving ahead at a swift (if unpublicised) pace. In the same month that Kotcher submitted his report, German test pilot Erich Warsitz completed the world's first turbojet flight in the experimental Heinkel He 178, and the British gas turbine developer, Frank Whittle, was well on the way to developing the first British turbojet aeroplane the Gloster E.28/39, which would fly in mid-1941. In the United States, however, scientists did not consider seriously the turbojet engine until the formation of the Durand Special Committee on Jet Propulsion by the NACA in March 1941, a scant two months before the Whittle aeroplane flew. Indeed, in 1938, a Navy board of engineers concluded that gas turbines were unsuited for aircraft. At the time of Kotcher's report, the AAF had just awarded GALCIT a $10,000 contract for research into both solid- and liquid-fuel rocket engines for takeoff assistance. (At this time, the Army Air Corps, later the Army Air Forces, was not receptive to rockets for aircraft propulsion though, thanks largely to Kotcher, this would change.)[9]

The chief importance of the Kotcher report, however, lay in its call for a major high-speed flight research programme. Kotcher stressed the importance of a full-scale flight research programme as a necessary check on available wind tunnel data. The wind tunnels at this time were of the closed-throat type and, as the speed of the airflow approached that of sound, the tunnels 'choked'. Shock waves forming off the test model and its supports would reflect off the tunnel walls, bouncing across the tunnel test section and thus inhibiting accurate measurement of flow characteristics and behaviour around the model. This condition persisted from about Mach 0.7 to about Mach 1.3. At Mach numbers below 0.7 or above 1.3, there was smooth airflow through the tunnel again, and aerodynamicists could make accurate measurements. Therefore, the very area in which scientists were interested, the transonic region between subsonic and supersonic flight, remained a closed area in which the wind tunnel could not operate with precision. In the coming years, wind tunnel advocates would become the strongest supporters of a transonic research aeroplane pending advances in wind-tunnel technology in the late 1940s that would eliminate the choking phenomenon and open up the transonic region to ground-based researchers. (Ironically, the very pressures to build transonic research aircraft would themselves generate greater interest and creativity in solutions to the choking problem.)[10]

The choking phenomenon led to the first NACA interest in transonic research vehicles. At NACA's Langley Laboratory, John Stack brought up the idea of a research aeroplane in conferences with NACA's Director of Aeronautical Research, Dr George W Lewis, when Lewis visited Langley on several occasions in 1941. Stack had foreseen the need for a full-scale flight vehicle for transonic research the previous year when it became apparent that wind tunnels were hampered by choking, and that no advance in wind-tunnel technology appeared likely in the near future. Following his earlier configuration research, Stack had kept in close touch with the compressibility problem through a concerted programme of propeller research. In 1933, he had constructed a 24-inch high-speed tunnel capable of more than 500 mph, parasitically driven from the pressure tank of the Langley variable density tunnel which was a specialised facility that could mimic the atmospheric conditions of high altitude flight. In 1941, he and his associates at Langley began research on propeller configurations for 500 mph operation at 25,000 feet using the Langley 500 mph 8-foot high-speed tunnel that had gone into operation in 1936. In short order they evolved a six-bladed contra-rotating propeller capable of converting fully 90% of its engine power into thrust at 500 mph, an almost unbelievable efficiency by contemporary standards. Tunnel research and development went ahead, though the choking problem remained a serious and disturbing limitation on knowledge.[11]

The Onset of a Crisis

Up until the late 1930s, compressibility had really remained the problem of the aerodynamicist, seeming largely theoretical. Now, dramatically, it became the problem of the pilot. On 17 July 1937, high over the Rechlin flight test centre, Messerschmitt test pilot Dr Kurt Jodlbauer rolled a sleek new Messerschmitt Bf 109 fighter into a vertical dive; it never pulled out, plunging with its hapless pilot into Lake Muritz. From that point on German test pilots described the shaking, bucking ride as they approached the transonic frontier as 'a case of Jodlbauer'. Soon, the problem came home to the United States. Lockheed Aircraft Corporation's young designer Clarence L 'Kelly' Johnson had conceived a radical new twin-engine interceptor, the P-38 Lightning, which began its flight testing in January 1939. Exceptionally streamlined for its time, the P-38 was a case study in compressibility waiting to happen; indeed, the P-38's compressibility problems did not come as a complete surprise to Lockheed engineers, for as early as 1937, Johnson and Hall L Hibbard had stated that compressibility would present a serious problem at speeds up to and above 500 mph. By late 1941, the Army Air Corps already had the first production models in service. Extremely high-powered and 'clean', the P-38 picked up speed rapidly in a dive. On steep, prolonged dives, pilots suddenly found the plane began buffeting and shuddering, with a tendency to tuck its nose under, as if it would bunt, or begin an outside loop. Disconcertingly, the control column froze as if implanted in concrete. After diving to lower, denser attitudes, shaken pilots discovered that the P-38 mysteriously nosed up and recovered. On several occasions, the dive continued until the aircraft broke up, often killing its pilot.[12]

Accordingly, Lockheed began a compressibility investigation using two specially modified P-38E aircraft. One P-38E featured a longer fuselage to accommodate a flight test observer and special research instrumentation, understandably earning the company name *Nosey*. The other had its tail booms aft of the coolant radiators swept upwards so as to place the horizontal stabiliser 30 inches higher. Engineers hoped (incorrectly as it turned out) that this modification would place the horizontal tail surfaces above the disturbed wake of the wing, thus retaining elevator effectiveness and with it, longitudinal control (i.e. control in pitch). In the meantime, developmental testing continued with other conventional Lightnings. On 4 November 1941, test pilot Ralph Virden took off from Lockheed's Burbank plant in the first of Lockheed's pre-production YP-38 Lightnings. The test card for the flight called for a high-speed dive to evaluate some new spring-loaded elevator tabs intended to improve the Lightning's ability to recover from a dive, followed by a low-altitude demonstration before watching AAF officials. Observers on the ground heard the snarling whine of a P-38 at full power; the noise built up to a strident crescendo suddenly broken by the awful thump of an aeroplane striking the earth. The subsequent accident investigation showed that when Virden desperately attempted to pull the P-38 out of its dive – at about 535 mph (Mach 0.7) at 3,000 feet – the structural loads exceeded the strength of the tail, ripping it from the plane.[13]

In December 1941, the AAF requested that the NACA investigate the cause of the P-38's problems. A month later, a P-38 arrived at Langley laboratory where a team of engineers under John Stack mounted and tested it in the Langley 30 by 60 foot full-scale wind tunnel. Langley engineers also tested a 1/6 scale P-38 model in the Langley 8 foot high-speed tunnel at speeds up to 500 mph. The 8 foot tunnel tests revealed that loss of wing lift due to shock waves forming on the upper surface of the wing at about Mach 0.675 – equivalent to 445 mph at 36,000 feet – would cause a loss of downwash on the horizontal stabiliser that, in turn, would essentially unbalance the plane, forcing it into an ever-steepening dive. The violently disturbed airflow streaming from the wing flailed the tail and, in Virden's case, tore it off. Under the pressure of wartime development – for the United States and its allies were fighting with their backs to the wall in the spring of 1942 – Langley engineers found a cure for the P-38 by 4 March. They suggested installation of a dive-flap on the lower surface of the wing which, when extended, would delay the onset of shock wave formation, permitting the wing to retain lift at high Mach numbers, thus enabling the pilot to retain sufficient control of the aircraft to pull out. Swamped with wartime work, Langley turned over the P-38 dive-flap investigation to the NACA Ames Aeronautical Laboratory at Moffett Field, California (Langley's west coast equivalent) where under the direction of Albert L Erickson, engineers checked out the flap in the Ames 16 foot tunnel. The flap idea proved sound, and Lockheed flight-tested them on *Nosey*. Company test-pilots Milo Burcham and A W 'Tony' LeVier (two of the greatest names in the history of American flight testing) found the dive flaps corrected the P-38's behaviour during the course of many high-speed dives, including one by LeVier at over Mach 0.7 followed by a 71/2g pull-out at 1000 feet! Lockheed subsequently installed dive flaps as standard production equipment on Lightnings beginning with the P-38J-25-LO family. Though the flap cured the P-38's difficulties, it did not spell the end of the problem of compressibility. It simply raised the aircraft's critical Mach number – the point where the 'local' airflow over a wing would go supersonic, creating a lift-killing and drag-inducing shock wave that would leave a disturbed wake behind it – so that the aircraft could fly without encountering loss of control due to compressibility effects. The problem of developing new aircraft capable of flying safely at transonic speeds remained.[14]

With the advent of turbojet propulsion, the problem took on new urgency, for the turbojet offered the promise of routine 500-plus mph flight speeds – if an aeroplane could be designed to fly that fast safely – and it did not require a propeller, it eliminated the inefficiencies associated with high-speed propeller design. In March 1941, because of ever more frequent intelligence reports hinting at German research into reaction powerplants, the Army Air Corps had requested that the NACA study jet propulsion. In response, NACA Chairman Vannevar Bush established the NACA Special Committee on Jet Propulsion chaired by Dr William F Durand who, like Bush, was one of the most eminent names in American science and technology. In April, the Air Corps chief, General Arnold, journeyed to Great Britain and learned (to his great surprise, chagrin and indeed, embarrassment after suggesting to his hosts that they investigate gas turbines themselves) that the British were but a month away from flight-testing a turbojet engine, the Whittle W.1, on their experimental Gloster E.28/39. America had lost the turbojet race; indeed, had not even realised until far too late that it was in a race. Fortunately, the victor was a friend and an ally. Arnold returned to the United States, subsequently arranging for the transfer and exploitation of the Whittle engine technology in the United States by the General Electric Company. This ultimately resulted in the AAF placing a contract with the Bell Aircraft Corporation on 30 September 1941 for the United States' first turbojet aeroplane, the XP-59A Airacomet. It also started Bell down the path that would eventually result in the company producing the first American supersonic aeroplane as well: the XS-1.[15]

Within a month of the formation of the Durand Committee, Eastman N Jacobs – at the time the NACA's most influential aerodynamicist and a striking individual possessing a rare charisma – predicted at a committee meeting that with the NACA's new family of high-speed airfoils he believed it would be possible to attain approximately the speed of sound in flight. The P-38 then encountered its compressibility difficulties. To aeronautical engineers, it seemed as if the propulsion limitation on high-speed flight had disappeared, only to be replaced by a design problem of equal magnitude, compounded by the inability of existing wind tunnels to furnish reliable information at Mach numbers between 0.7 and 1.3. Accordingly, in mid-1942, several Langley engineers informally joined together to investigate alternative methods of research. One idea that came up in discussions was the concept of a transonic research aircraft, an idea that John Stack had suggested to George Lewis in the previous year. Stack established a small team of Langley engineers consisting of himself, Milton Davidson, Harold Turner and Walter Williams. This small group began studying possible research aircraft configurations. Prior to this, Jacobs, then head of the Langley Airflow Research Division, had initiated his own study of possible transonic aircraft designs with Milton Davidson and Harold Turner. From the start, Jacobs and Stack were enthusiastic supporters of the research aeroplane concept though, as shall be seen, each had different ideas as to how to proceed and imparted their views strongly to their teams. In the former group, Jacobs assumed overall research direction, Davidson performed the calculations and Turner did the drafting of the proposals. Jacobs, Davidson and Turner investigated possible configurations using the proposed Westinghouse 19A jet engine (a small axial-flow design) but concluded the configurations were so underpowered that they would be incapable of transonic flight. One of the earliest configurations studied incorporated a sweptwing planform and landing skids. The group – as with other NACA proponents – did not initially consider rocket propulsion. Later however, the Stack team investigated both turbojet and rocket-propelled concepts, running wind tunnel tests on some of them.[16]

Both the Stack and Jacobs studies generally concentrated on air-breathing propulsion systems using conventional ground takeoff. At this point however, they abruptly departed. Starting in 1939, Jacobs had experimented with the Campini propulsion system (named after Italian engineer Secundo Campini) in which a conventional gasoline engine drove a compressor. Fuel was then mixed with the airstream and ignited in a burner, emitting exhaust through a nozzle. Jacobs and his two assistants, Macon C Ellis Jr and Clinton E Brown, had high hopes for the Campini system. They built an operating Campini test-bed at Langley that consumed three pounds of gasoline per second, producing a gigantic and spectacular blow torch from the burner that impressed – and often terrified – viewers. Jacobs felt that the Campini system could be incorporated in military aircraft to give them performance superior to that of existing propeller aircraft, as well as long loiter times. His two assistants, Ellis and Brown, conducted a feasibility study that later resulted in design of a Campini-powered aircraft by Ellis, Brown and Fred Daum.

This design proposal featured a Pratt and Whitney R-1535 radial engine, a tricycle landing gear, a high shoulder-mounted wing, and a 'V' tail anticipating the later Beech Bonanza. It had a 15% thickness-chord ratio wing spanning 41.4 feet and possessing an area of 215 square feet. The design team estimated gross weight of the plane at 9,780lbs. They could not estimate the maximum speed of the aircraft owing to uncertain values for drag at speeds over 550 mph due to compressibility effects. What interested them most, aside from the apparent advantages of the Campini system, was that when the aircraft attained Mach 0.75, the engine would be producing three times the amount of thrust necessary to reach that velocity. They did not know what stability and control problems might be encountered above Mach 0.75, but knew that with the available thrust, as one team member wistfully recalled, 'we could have really barrelled into the transonic region'. However, fate lent a cruel hand. In October 1942, with the Langley Campini engine progressing satisfactorily, the Durand Committee expressed interest in it and called Jacobs to arrange for a demonstration run at Langley. The appointed day arrived, the committee members stationed themselves 200 yards behind the burner and the test crew lit the engine. Perversely, the test run was disappointing, marked by unstable combustion that had plagued trials of the Campini engine at Langley the previous year. The timing could not have been worse, for at the beginning of the month, on 1 October, Bell test pilot Bob Stanley had completed the first flight of the XP-59A, its Whittle-inspired American engines shrieking lustily and confidently. Perhaps not surprisingly therefore, a week after the demonstration, word came from NACA Headquarters to suspend all work on the Campini project. The disappointing test run, together perhaps with the distant success of the XP-59A and a growing concern over the huge amounts of precious gasoline being burned in an apparently fruitless attempt to develop a reliable propulsion system, killed the Campini investigation. All hopes of developing a Campini aeroplane that might prove useful for transonic research vanished, and Stack superseded Jacobs as the NACA's most vocal and recognised research aeroplane advocate.[17]

Starting the Search for a Solution

By early 1943, compressibility had assumed critical importance in NACA, AAF, and Navy wartime aeronautical research. Some indication of how prominently it figured in NACA planning can be seen in the formation of a Compressibility Research Division at Langley in 1942 under the direction of John Stack, and in the increasing amount of correspondence between Langley staff members concerning compressibility research at this time. Army Air Forces and Navy combat squadrons suddenly found that their aircraft faced a new and frighteningly mysterious danger aside from the enemy. An epidemic of tail failures during dives appeared in three production combat aircraft, the Republic P-47 Thunderbolt, the Curtiss SB2C Helldiver and the Bell P-39 Airacobra. Extensive tests at Langley and Ames laboratories resulted in recommendations to manufacturers to strengthen the tails to withstand high air loads at high speed. In each case, the turbulent wake behind a wing experiencing compressibility effects induced severe aerodynamic flutter of the horizontal tail, weakening and finally exceeding the tail's structural strength (as one would bend a paperclip until it breaks), destroying it within seconds. The problem of compressibility was all the more vexing because of the complete inability of wind tunnels at the time to serve the needs of the aeronautical engineer in his search for reliable and useful aerodynamic data. Now that the turbojet appeared as a means of overcoming propeller limitations, the only remaining barrier to high-speed flight, including flight faster than the speed of sound appeared to be a knowledge barrier, where aerodynamicists could not predict with any degree of certainty what occurred in the turbulent speeds of transonic flight. So the immediate problem facing aerodynamicists was how to arrive at accurate research tools and methods to derive needed data and strip the mystery from compressibility and transonic aerodynamics.

In mid-December 1943, William S Farren, director of the British Royal Aircraft Establishment, arrived in the United States to present the Seventh Wright Brothers Lecture before the Institute of Aeronautical Sciences in Washington, DC, on 17 December. To people in aeronautics, 17 December assumed the stature of a national holiday because it was the anniversary of the Wright Brothers flight at Kitty Hawk, North Carolina, in 1903. Government and industry aeronautical representatives always came to Washington for the lecture, and the NACA committee – the representatives appointed by the President – always held a special meeting. The topic Farren chose for his talk was 'Research for Aeronautics: Its Planning and Application'. In the lecture, Farren acknowledged the problem of compressibility and stated that if aeronautical engineers planned on attaining high speeds, they would have to devote considerable effort to 'avoiding or reducing the effect of compressibility'. What Farren did not announce to his audience was that the Royal Aircraft Establishment at Farnborough already had, by this time, a comprehensive programme of transonic research well underway. Specially instrumented fighters such as the Mustang and Spitfire were routinely (if hazardously) dived to speeds above Mach 0.8, and, in a precedent-setting event, Great Britain's Ministry of Aircraft Production had *already* awarded a development specification, E.24/43, for an experimental aeroplane to fly at a speed of 1,000 mph at 36,000 feet – approximately Mach 1.5 – to the Miles Aircraft Company over two months previously! (This action generated the exotic if stillborn Miles M.52, abandoned – save for some disappointing air-launched and instrumented rocket-propelled models – in mid-July 1946.)[18]

In keeping with the spirit of the talk Dr Durand, in his capacity as chairman of the NACA Special Committee on Jet Propulsion, called a special conference the next day at NACA Headquarters in Washington on jet propulsion in England. Among approximately 20 people at the meeting was Robert A Wolf, a keen young engineer with the Bell Aircraft Corporation at Buffalo. Since the beginning of 1943, the idea of a high-speed research aeroplane had been germinating in Wolf's mind. Wolf had served as technical adviser and designer of the airframe, cabin pressurisation and propulsion system of the Bell XP-59A Airacomet, and it is indicative of his abilities that, in July 1943, he was sent to England to review the progress being made there in turbojet propulsion. While abroad, the vision of more powerful turbojets with greater thrust stimulated his interest in designing transonic aircraft, provided that the stability and control problems facing high-speed aircraft could be solved. At the 18 December meeting, Durand asked essentially one question: What should the United States do with the turbojet propulsion concept brought to this country from Great Britain? In the discussion that followed, Wolf suggested that the country had the opportunity to construct a special high-speed research aeroplane to provide transonic aerodynamic data for aircraft manufacturers. The data were urgently needed since jet aircraft would soon encounter in level flight the transonic speeds now being approached by propeller aircraft in dives. He then boldly (and in light of subsequent history, presciently) suggested a basic plan of development, in which the AAF, Navy, NACA and aircraft industry would define the type of aircraft needed. The military would procure it, the industry would develop it under contract, and the NACA would conduct the flight research and disseminate information.[19]

Wolf's proposal was but one of many ideas suggested in outline form at the roundtable discussion. Fearing that it might be distorted or lost in the meeting minutes, he wrote to Dr George W Lewis on 29 December, again stressing the need for a transonic research aircraft. His letter stated:

> It appears quite possible to construct a single engine aircraft based on available gas turbine jet power plants which will fly at speeds in level flight exceeding the critical Mach numbers of currently used types of wings. If this aircraft were designed with enough inherent versatility to changes in control surfaces, wings, etc., it should be possible to develop usable control surfaces such as ailerons, dive control flap, tail surfaces, etc., which would work satisfactorily at or above the critical speeds of the wings. Furthermore, this could be done in level flight and would not be subject to the dangers and difficulties associated with the high accelerations encountered in current dive programs.[20]

Lewis responded by writing to Wolf that the NACA was highly interested in procuring a turbojet-research aircraft for investigating compressibility and aircraft stability and control at high speeds, and was giving the matter 'our very serious consideration'.

Simultaneously with Wolf's suggestion, the Army Air Forces were renewing interest in transonic flight research thanks to Kotcher's persistence and the general wartime situation. British intelligence reports indicated that Nazi Germany had rocket and jet-propulsion research projects in an advanced state of development. In fact, had the full state of Nazi Germany's development efforts been known, concern might have become outright alarm. The Messerschmitt Me 262 *Schwalbe* jet fighter was in an advanced state of development and would soon enter combat; an experimental rocket-propelled interceptor, the Messerschmitt Me 163 *Komet* was in flight-testing; even more advanced jet-and-rocket-powered fighter and bomber projects were underway; a variety of supersonic and hypersonic wind tunnels were in use, under active development or marching across drawing boards; precision air-to-surface, air-to-air, and surface-to-air missile programmes proliferated; and V-1 cruise missiles and V-2 ballistic missiles were roaring aloft from Peenemünde's test stands (in fact, the Peenemünde centre itself had already been attacked by RAF Bomber Command in anticipation of the German V-weapons campaign): these were a few of the larger and more impressive products of the Third Reich's engineers and scientists. One Nazi research study had already concluded that:

> Taking into consideration the advantages of rocket propulsion . . . the construction of a supersonic aircraft is possible with the present technology. A proposal has been made to the research leadership to build such a research aircraft and observe in flight the supersonic phenomena which cannot be obtained in wind tunnels.

Had this last aspect of German high-speed research been known, it undoubtedly would have profoundly accelerated Anglo-American research aeroplane efforts. Since 1942, the *Deutsche Forschungsanstalt für Segelflug* – the German Research Establishment for Soaring Flight, an organisation whose interest extended far beyond conventional gliders and sailplanes – had embarked on a multiphase study effort on high-altitude and high-speed flight, including proposing specialised test aircraft to evaluate wing behaviour at high speeds. The DFS programme had a markedly 'applications oriented' flavour; the first product was to be a rocket-boosted long-range reconnaissance aeroplane, the DFS 228. The most ambitious however, was a proposed supersonic rocket-propelled aeroplane, the DFS 346, then in the early design concept stage. Though intended ultimately for military purposes, designers hoped that the initial model of this radical sweptwing aeroplane would nevertheless hopefully furnish much useful transonic and supersonic information. (It would be this aeroplane that would, in fact, give the XS-1 its closest race to be the first supersonic aeroplane in the world, not in the service of the *Luftwaffe*, but rather in post-war Soviet Union.) As with the turbojet engine of less than half a decade before, America and Great Britain were already unknowingly locked into an international race with the Third Reich to be the first to develop a supersonic aeroplane.[21]

Kotcher and the 'Mach 0.999' Aeroplane

Over four years after Ezra Kotcher's original suggestion for transonic flight research in the autumn of 1939, the AAF decided to act upon his recommendations. In mid-January 1944, the Development Engineering Branch of the Materiel Division at AAF Headquarters, Washington, issued Confidential Technical Instruction 1568 that authorised 'the initiation of a study of the possible development of an experimental article for the purpose of investigating aerodynamic phenomena in the range of 600–650 mph'.[22] After 1939, Kotcher had become chief of the Vibration and Flutter Unit of the AAF Aircraft Laboratory at Wright Field, the service's principal aeronautical research and development centre. Pearl Harbor led to his being called to active duty as an AAF captain and in late 1942, he was introduced to the jet-propelled P-59. In 1943, the Engineering Division selected him as project officer on the proposed Northrop XP-79 rocket-propelled flying wing interceptor. The rocket-powered XP-79 programme never reached the hardware stage, which was fortunate for anyone who would have had to fly or to maintain it. A manned subscale prototype of the XP-79, the Northrop MX-324, did, however, make a rocket flight on 5 July 1944, the first flight of an American rocket aircraft designed from the outset for rocket propulsion. Yet the XP-79 programme did subsequently influence American transonic research aircraft developments, thanks to its proposed engine – an engine that, ironically, was never actually built and flown. The proposed powerplant for the XP-79 was a 2,000lb-thrust rocket engine being developed by Aerojet Engineering Corporation. This engine, dubbed the Rotojet, used red-fuming nitric acid and aniline for propellants. When in contact, the two substances were hypergolic: they would combust violently, rendering an ignition system unnecessary. Nor was this the only complication. Aerojet designed the engine with four chambers having canted nozzles to impart rotation to the engine and drive a gear train to power fuel pumps supplying the engine's propellants. Northrop naively designed the basic structure of the XP-79 using magnesium which, when in contact with nitric acid 'fizzed like a piece of zinc in hydrochloric acid,' as Kotcher recalled years later. Most importantly, however, Aerojet planned a 6,000lb-thrust engine, also using acid and aniline as propellants and rotation to drive fuel pumps as a follow-on to the 2,000lb-thrust engine designed for the XP-79. This engine was being developed under Project MX-121, in anticipation of a 'superperformance' aeroplane.

In early 1944, Kotcher decided to make a comparative investigation of the merits of rocket and turbojet propulsion for a transonic research aircraft. This resulted in the influential 'Mach 0.999' study.[23] The term 'Mach 0.999' referred facetiously to the 'impenetrable sonic barrier' of popular imagination. Kotcher requested that the Design Branch of the Aircraft Laboratory at Wright Field investigate two configurations, one using a General Electric TG-180, an axial-flow turbojet of approximately 4,000lb-thrust then under development, the other using the proposed 6,000lb-thrust Aerojet rocket engine. By April 1944, the Design Branch had completed the comparative designs and from every performance standpoint, the rocket-powered aircraft appeared the better way to attempt transonic research. Its chief advantage was its high speed, which rendered dives to high Mach numbers unnecessary. The high thrust of the rocket engine provided good high altitude performance, in contrast to existing turbojets whose thrust was marginal at altitude. The rocket powerplant additionally provided flight researchers with added testing time. Instead of diving earthwards, which put the test plane and pilot at great risk for about 30 seconds, the rocket offered the possibility of high-speed level flight for around two minutes. In light of subsequent developments the Mach 0.999 rocket-research aircraft, the Design Branch configured, makes an interesting comparison with the Bell XS-1. Both aircraft featured bullet-shaped fuselages, mid-wing design, conventional tail surfaces (though the Wright Field design had the horizontal surfaces mounted on the fuselage and not on the vertical fin, as did the XS-1), a smoothly faired cockpit canopy and a liquid-fuel rocket engine of 6,000lb thrust.

Kotcher next showed the rocket design to Theodore von Karman, for the GALCIT director by now was acting scientific consultant to General Arnold. Von Karman concurred in the feasibility of building such an aircraft, approved the design concept and calculated the craft's lift:drag ratio to be three. When von Karman examined the Mach 0.999 proposal, he already knew about AAF's efforts to investigate high-speed flight. Earlier in the year, the chief of the Engineering Division at Wright Field, General Franklin O Carroll, who was a shrewd and highly regarded officer, had already asked von Karman if it were possible to develop an aeroplane to fly at Mach 1.5. The distinguished Hungarian assembled a small study team, and generated a quick report over a weekend which endorsed favourably the feasibility of a ramjet-powered 10,000lb gross weight aeroplane, having a small 125 square foot wing capable of reaching Mach 1.5 at 40,000 feet which could fly at that speed for five minutes. With ramjet technology in its infancy, such a design had little chance of winning development approval. Nevertheless, it was an important psychological encouragement, as was an equivalent ramjet study by NACA's Macon Ellis and Clinton Brown in 1945. As important as von Karman's endorsement of Kotcher's efforts was, it was far more important, however, that General Carroll as Kotcher's commanding officer, both recognised the need for a transonic research aircraft and endorsed Kotcher's approach, for it ensured him the highest support for his efforts within AAF research and acquisition circles.

Stopgap Research Methods

While Wolf's proposal was stimulating interest in the NACA and at Bell, and while Kotcher's design studies were emerging from the drafting boards at Wright Field, NACA scientists were investigating alternative stopgap methods of transonic research in co-operation with their British colleagues. W S Farren, after presenting the Wright Brothers lecture, had discussed the transonic wind-tunnel research problem with NACA researchers. One solution that both the British Royal Aircraft Establishment and the American NACA pursued was dropping weighted bodies from high altitudes. The shapes would then attain velocities equal to or faster than that of sound. Radar and visual tracking from the ground could determine their speed and path. In subsequent discussions between the NACA and British Ministry of Supply, British and American representatives agreed to undertake joint falling-body investigations. The NACA Executive Committee authorised falling-body research in March 1944, and actual tests commenced in May using a B-29 Superfortress carrier aircraft lent by the AAF and a Navy-supplied SCR-584 radar to track the shapes as they fell. At about the same time, fighter pilots putting North American sleek P-51 Mustangs in dives found they sometimes could see shock waves streaming from the Mustangs' wings. To one NACA Langley engineer, the accelerated airflow over the wing of a P-51 seemed to offer a satisfactory substitute for the lack of genuine transonic wind tunnels. Robert R Gilruth, a young engineer in charge of flight research working for Melvin Gough, the crusty chief of Langley's Flight Research Division, realised that when a Mustang dove to about Mach 0.75, the accelerated airflow over the Mustang's wing reached about Mach 1.2, he could use this accelerated flow to test airfoils. He devised a small balance mechanism to fit within the Mustang's gun compartment, and then mounted the test airfoil vertically above the P-51's wing. Though Gilruth encountered scepticism from wind tunnel advocates, his wing-flow method proved useful in deriving basic transonic data on airfoil shapes and the effects of aspect ratio, thickness, sweepback and choice of airfoil section. The falling-body and wing-flow methods – together with the later rocket-propelled model work under Gilruth at the NACA's Pilotless Aircraft Research Division (PARD) on Virginia's Atlantic coast at nearby Wallops Island – provided vital data on drag variation through the speed of sound. These methods, however, also had their limitations as the expense and work involved achieved limited results. There was also still a significant risk to a pilot diving an aeroplane of less-than-ideal qualities at an ever-increasing Mach number. In wind-tunnel testing, the scientist develops a feel for what occurs to the test model. He can take pressure distributions, and look through the tunnel port to observe shockwave formation. All this was lost in the falling-body and rocket-boosted model programme. Pending the development of genuine transonic wind tunnels and wind tunnel test techniques, the research aeroplane appeared as the best possible alternative method of transonic research. As the M.52 and DFS 346 programmes indicated, this had already occurred to British and German researchers.[24]

The Royal Aircraft Establishment at Farnborough, awaiting the imminent development of transonic 'on the level' aeroplanes – as the Langley and Ames laboratories in the United States had done – made do with instrumented propeller-driven fighters, though not with the provision for Gilruth's unique wing-flow research apparatus. Risks were high, even with the most cautious approach. At the Royal Aircraft Establishment, Squadron Leaders J R Tobin and A F 'Tony' Martindale, flying a specially-instrumented Spitfire P R Mk XI reconnaissance fighter, had reached well in excess of Mach 0.80, and Martindale laid claim – his instrumentation package supporting the assertion – to Mach 0.89. The price though, was almost unacceptable. During one dive, Martindale narrowly escaped total disaster when the propeller and gear reduction mechanism separated from the Rolls-Royce Merlin engine with 'a fearful explosion', destroying the engine (which threw a piston rod through its casing for good measure), denting the wing leading edge, buckling an engine bearer and severely straining the horizontal tail. Despite oil and coolant smearing the cockpit canopy and windscreen, he managed a deadstick landing back at Farnborough, after noting that 'the aircraft was gliding very nicely without its propeller'. Clearly, the ability to test in level flight conditions rather than hurling oneself at the earth, promised far greater safety, as well as far greater time to acquire data.[25]

The Path Diverges: Turbojets vs. Rockets

On 16 March 1944, representatives of the AAF Air Technical Service Command, Navy Bureau of Aeronautics, and the NACA gathered at the

Langley laboratory. They held two conferences, one chaired by Captain Walter S Diehl, USN, the other chaired by Colonel Carl F Greene, the AAF liaison officer at Langley, both dealing with development of possible transonic research aircraft. Captain Diehl, a noted aeronautical engineer in his own right who in effect, acted as the Navy's Bureau of Aeronautics' senior representative to the NACA, had earlier discussed the research aeroplane concept with John Stack and Commander Emerson Conlon of the BuAer's Structures Branch. Like Stack and Kotcher, Captain Diehl had felt as early as 1942, that the research aeroplane appeared the only way to convince people that the 'sonic barrier' was just a steep hill, drawing upon transonic shell data – again like Kotcher – to support his contention. These two meetings on 16 March tied together for the first time AAF, Navy and NACA interest in the research aeroplane concept, and offered implicit proof of a growing divergence between the perceptions of what such an aeroplane should be to the AAF, the Navy and the NACA. To the Navy, such a craft should be jet-powered and have possible military utility. The NACA generally preferred a jet-powered design purely for research purposes, but would (very) reluctantly consider a rocket-propelled one. The AAF, likewise, had concerns about a purely research project and was receptive to jet propulsion, but favoured a more radical rocket-propelled approach. At the Air Staff level in the Pentagon, General Oliver P Echols, the assistant Chief of Staff, had questioned the wisdom of procuring a non-military research aeroplane solely as an Army Air Forces project. In response, Ezra Kotcher drafted a letter for transmittal to Dr George W Lewis suggesting joint AAF-NACA participation in developing a research aircraft to support the NACA research mission. For its part, in December 1943 the NACA, had formed – at Lewis' suggestion – a special five-man panel to co-ordinate all NACA high-speed research. This panel consisted of John Stack, Russell G Robinson, H Julian Allen, R E Littell and Eastman N Jacobs, with Stack *de facto* in the role of 'first among equals'. At the initial panel meetings, Stack suggested that NACA sponsor rocket- or turbojet- propelled transonic research aircraft as the immediate focus of new initiatives. At the March meeting, echoing Kotcher's earlier letter to Lewis, NACA representatives suggested a joint NACA–military development programme to construct a turbojet-powered research aeroplane.[26]

NACA, despite being located at the AAF's Langley Field (today's Langley Air Force Base), always had a closer working relationship with the Navy (perhaps due to the proximity of nearby Norfolk) than the AAF. (Indeed, most of the laboratory's test pilots from Mel Gough down were ex- or reserve-naval aviators.) Not surprisingly then, the NACA, in consultation with the Navy's Bureau of Aeronautics (BuAer), decided to detail

Milton Davidson, a NACA engineer familiar with the research aeroplane concept, to BuAer in Washington. Davidson would collaborate with Ivan Driggs, a noted pre-war aircraft designer and now the head of BuAer's Aviation Design Research Branch, on configuring and preparing specifications for a Navy-sponsored transonic research aircraft. NACA's actions and search for a joint partner were hardly unexpected, since the agency had little other choice: unlike NASA today, it did not then possess the facilities nor the administration to handle such a programme on its own. Stack, perhaps concerned with placing the onus of development on a single service, used this climate of joint service interest to urge Colonel Carl F Greene and Jean Roche of the AAF liaison office at Langley to convince the service to develop a transonic research aircraft, citing the prominent role the AAF had played in procuring the XP-59A.[27]

Despite the low-thrust values for existing turbojet engines, NACA engineers generally favoured turbojet propulsion over rocket propulsion. Their reasoning did not stem from an unwillingness to innovate, but rather from the cautious approach that was a hallmark of NACA research and which, indeed, would be used by critics in 1957–58 as a rationale for reorganising and restructuring the agency into NASA. They recognised that rocket technology was in its infancy and that rocket engines, circa 1944, were often tricky, unreliable and dangerous. Above all, they did not like the idea of a manned research aircraft, heavily loaded with explosive fuels and test instrumentation, lifting from a runway with the pilot relying on a rocket to keep him aloft. They realised that if the rocket cut out, even fractionally, the transonic research aeroplane programme could end in a tower of smoke above a crater off the end of the runway. Some of the NACA research pilots at Langley did not support the idea of a transonic research aircraft in the first place. They felt they were being asked to risk their lives because wind-tunnel researchers were unable to do the necessary work on the ground. More than once, Mel Gough stated emphatically that no NACA pilot would ever fly a rocket aeroplane. Ezra Kotcher thought otherwise. In early May, the Army Air Forces' Materiel Command Headquarters at Wright Field ordered him to Langley Laboratory to discuss high-speed research with the NACA. Kotcher attended a meeting on 15 May at Langley, and presented the results of the Wright Field Mach 0.999 study, including a drawing of the rocket aircraft. At this meeting, NACA representatives agreed to submit a transonic aircraft design to the AAF for approval and construction. Not surprisingly, when NACA did submit a design proposal to Wright Field two months later on July l0, it stipulated turbojet-power. It was, in effect, a flying laboratory, with provisions for 400 lbs of research instrumentation similar to an installation on a NACA P-51 Mustang flying at Langley in a compressibility research programme involving high-speed dives. Bell Aircraft Corporation had sent Eastman N Jacobs and Dr George W Lewis a three-view drawing of a tentative turbojet-powered research aircraft in April 1944. It is unclear whether or not this proposal in any way influenced the subsequent NACA proposal submitted to the AAF in July, but it is probable that it did so. In any case, the AAF rejected the NACA turbojet design out of hand as being too conservative, particularly since the turbojet engine did not offer the plane a sufficiently high speed in level flight.[28] By mid-1944, Kotcher had succeeded in persuading the AAF to support a rocket-propelled design. If the AAF were to participate, NACA would have to accept a rocket engine as the price of that co-operation. At this point so close to attaining his goal, he had to turn to other tasks far removed from supersonics.

With a war on, Kotcher did not have the luxury of seeking the research aeroplane grail as his sole quest. He was quite busy, for in addition to his role in developing a transonic aircraft, he directed many other combat aircraft development programmes, including the production Bell P-59 Airacomet, the experimental Northrop XP-79 and the rapidly progressing Lockheed P-80 Shooting Star, and even a feasibility study on flight-refuelling Boeing B-17 Flying Fortresses by Consolidated-Vultee B-24 Liberator tankers as a possible means of reaching long-distance Japanese targets from the Aleutians and China. In late June 1944, General Carroll appointed Kotcher project officer on the Republic JB-2 missile project, an attempt to copy the German pulsejet-powered V-1 'buzz-bomb' robot missile. This occupied Kotcher over the next several months – including a trip to Europe where he prowled the Channel coast with British commandos to pick up bits and pieces of crashed missiles from German-held territory—until a climactic launching at Eglin Field, Florida on 12 October 1944 set the stage for his return to supersonic thought. After working all night preparing a JB-2 for a scheduled demonstration before assembled officials including Theodore von Karman and Hugh L Dryden, Kotcher gave the firing order and the JB-2 took off in a cloud of smoke with a loud hiss, flying several miles before splashing inauspiciously into the Gulf of Mexico. The launch was sufficiently successful to release him from his task and so, four months after Carroll's call, Kotcher returned to Wright Field and resumed his deliberations on supersonic aircraft development. He began looking for a likely contractor to develop the aircraft, a task that would occupy the next six weeks.[29]

Meanwhile, a new player on the scene – a Bureau of Aeronautics military engineer – had taken steps to initiate a Navy-sponsored high-speed research aeroplane. First Lieutenant Abraham Hyatt, a Marine Corps officer and graduate of the Georgia Institute of Technology with a BA in aeronautical engineering had worked before the war with the Curtiss, Martin and McDonnell aircraft companies. In 1944, he entered the Marines on assignment to Ivan Driggs' design branch. As Kotcher, Wolf, Stack, Diehl and others had before him, Hyatt saw that the lack of high-speed aerodynamic knowledge posed a serious handicap to the aircraft designer. Likewise, he saw the research aeroplane conception as the most feasible and convenient method of acquiring high-speed data. He concluded that the Navy should sponsor such an aircraft in order to ensure that American naval aviation would retain its high standard of world leadership. Accordingly, on 22 September 1944, Hyatt issued a memorandum for circulation through the Bureau on a proposed turbojet-propelled high-speed research aeroplane. Such an aircraft, the memo stated, could furnish knowledge on transonic drag, flight loads and stability and control. Further, the turbojet configuration would permit acquisition of data on engine thrust and duct inlet design. Evaluation flights by Navy and Marine combat pilots could yield useful tactical information for future high-speed Navy combat aircraft. In part the memo read,

> All branches of aeronautical engineering are confronted with a serious lack of knowledge of engineering laws which govern the design of airplanes at speeds above 500 miles per hour.
>
> The present trend of military aeroplanes is toward super speeds. Even greater competition in super speed airplane design will exist among nations as the war progresses and after the termination of the present war.
>
> It is proposed that the Bureau of Aeronautics take the necessary steps to obtain an airplane with the following specifications:
>
> (a) Minimum high speed of 650 miles per hour at sea level.
> (b) As clean as special care in manufacturing can accomplish.
> 1. Extra heavy covering for wings, tail and fuselage.
> 2. Special care in scaling.
> (c) All around vision sacrificed for cleanliness by making pilot's enclosure contained in fuselage.
> (d) Low thickness ratio of wings – possibly 10% at root section.
> (e) All armament and equipment other than required by pilot to fly in vicinity of airfield be eliminated.
> (f) Fuel capacity determined by maximum time aloft required for a given test.

(g) Landing and take-off limited by ability to do so from a class 'AA' airport only.
(h) Provision for complete instrumentation for flight characteristics and flight loads.
(i) Design yield load factors of 10 to 12 and high strength tail.
(j) Jet propelled.[30]

On the Verge of Finding a Builder

Over 20 years had passed since the first tentative high-speed aerodynamic research on propeller airfoils by Caldwell, Fales, Briggs and Dryden. The speeds of the new turbojet aircraft were over four times those of the fastest First World War fighters and bombers. As a sampling, the intervening years between 1918 and 1945 witnessed the introduction of the controllable-pitch propeller, the supercharger, the retractable landing gear, cantilever construction, the NACA cowling and the turbojet – all of which increased high-speed aircraft performance. As the autumn of 1944 deepened, new turbojet-propelled fighter aircraft – for example Great Britain's Gloster Meteor, Nazi Germany's Messerschmitt Me 262 and America's experimental Lockheed P-80 Shooting Star – routinely attained level flight speeds only reached by conventional propeller-driven fighters during maximum power dives. Pilots, engineers, designers, the military services, governmental research agencies and the aircraft industry – in short, the whole international aeronautical community – suddenly realised (by the force of events if not from shrewd insight) that aviation science demanded reliable information to reduce or overcome the effects of compressibility and thereby permit designers to craft more efficient and safer high-speed aircraft. This demand required immediate solutions. Pending the development of satisfactory transonic wind tunnels, the best immediate solution seemed the development of a transonic research aircraft, a recognition evident in Britain's M.52 and the Nazi DFS 346 already underway. If in the United States, however, the AAF, Navy and NACA recognised the desirability of a research aeroplane to probe the sonic frontier, they nevertheless had sharply differing views of what it should do. Since the services would finance and procure the aircraft, the ultimate establishment of performance goals, selection of configuration and detail design clearly lay with the services and industry even though both services and industry itself relied to a great degree on the technical acumen of the NACA for advice and suggestions.

No issue more markedly delineated the differences between the two services than their position on propulsion and, as a result, top-end performance. Hyatt's proposed Navy aeroplane obviously represented the more conservative approach towards the problem of high-speed research; an approach more in keeping with the views of NACA-Navy engineers. Ezra Kotcher and the AAF on the other hand, favoured a more radical rocket-propelled aircraft. Both men agreed that the research aircraft should be overstrength to withstand high load factors and unencumbered by military requirements. They differed considerably on the desired propulsion and performance characteristics of the aircraft, however. Hyatt wanted turbojet propulsion and 650 mph-Mach 0.85 performance. Kotcher wanted rocket propulsion and 800 mph, so that the aircraft could exceed the speed of sound. At 35,000 feet, the altitude Kotcher felt desirable for flight testing, the rocket aircraft would attain just under Mach 1.2. Both men's visions laid the groundwork – literally – for the Navy's Douglas D-558-1 and the AAF's Bell XS-1, the first of America's transonic-supersonic aeroplanes borne of the 'compressibility crisis'.

2
Winged Bullet:

The Design Development of the Bell XS-1

Few American aircraft companies could lay stronger claims to creative ingenuity and expertise than the Bell Aircraft Corporation, one of two major aircraft manufacturers in the Buffalo area (the other being Curtiss-Wright). Not quite 10 years old by the end of 1944, Bell traced its origins back to the Consolidated Aircraft Corporation. When Consolidated's president, Reuben Fleet, decided to move the firm west to San Diego, one of his vice presidents, Lawrence Dale 'Larry' Bell, elected to remain at Buffalo, New York and head his own firm. So it was that on 10 July 1935 Bell, together with former Consolidated officers Robert J Woods and Ray P Whitman, formed the Bell Aircraft Corporation. Within two years the firm ran up sales totalling nearly $2 million, mostly from Consolidated subcontracts. The firm soon displayed a flair for the radical, the unconventional, and the untried. This tendency which they impressed upon company engineers, stemmed from the character of Bell and Woods themselves, for at heart both men were genuine revolutionaries, involved with aviation almost at ground floor level. They even looked and acted somewhat alike: portly but energetic, dapper dressers and gracious, though possessed of a pointed and at times disconcerting directness. Sadly, both also suffered from heart problems that would claim them prematurely. Lawrence Bell entered the aeroplane business in 1912 as a 17-year old mechanic for two exhibition pilots. One was famed stunt pilot Lincoln Beachey. The other was his brother Grover Bell, who tragically perished in an accident a year later. Bell considered leaving aviation after his brother's death, but decided to remain with it. He developed a makeshift bomber for Pancho Villa, the Mexican revolutionary general, then joined the Glenn L

Martin company. He became, in succession, factory superintendent, general manager and later Martin vice-president. In 1925, he went to work for Consolidated, staying with that firm for the next decade.[1]

When Lawrence D Bell formed the Bell Aircraft Corporation in July 1935, Robert J Woods was a 31-year old project engineer with Consolidated, brimming with ideas and ambitions. Rather than go to San Diego, he immediately left and joined Bell. Woods had graduated from the University of Michigan with a BS in aeronautical engineering in 1928. He had studied under the legendary Professor Felix 'Pavvy' Pawlowski, (arguably the father of aeronautical engineering in the United States and a man who had taught himself to fly in a Bleriot monoplane while studying in France before the First World War), and also under Lawrence Kerber, the school's first Guggenheim Professor of Aeronautics, formerly director of flight research at Wright Field. In August 1928, Woods joined the NACA at Langley Laboratory, working in the variable density tunnel, where he shared the same desk with John Stack who was a new arrival from another Guggenheim school, MIT. The two men struck up a friendship that lasted until Woods' death in 1956. Ever restless, Woods soon left the NACA as he was eager to develop some of his own designs and see them fly. He went with the Towle Aircraft Company in Detroit and developed a twin-engine amphibian. After a brief stint with the financially ailing Detroit-Lockheed company, he left to work for Consolidated. There, he designed a two-seat monoplane fighter prototype, the Y1P-25, based on an earlier Detroit-Lockheed design, powered by a 600 hp Curtiss Conqueror liquid-cooled engine equipped with a supercharger and with advanced design features such as a retractable landing gear and a controllable-pitch propeller. Consolidated developed this aircraft into a very successful and influential fighter, the P-30. After joining Bell in 1935, Woods' engineering talent flourished, always with an unconventional flair. In June 1936, together with Harlan M Poyer, Woods began development of what became the P-39 Airacobra, a single-seat fighter with the engine behind the cockpit driving the propeller via an extension shaft. Obsolescent when the Japanese attacked Pearl Harbor, the P-39 nonetheless served well as a ground strafer and fighter-bomber, both with the air forces of the United States and the Soviet Union. In 1937, the Woods-designed Bell FM-1 Airacuda appeared, an unorthodox five-place twin-engine fighter powered by two Allison V-1710 engines driving pusher propellers. One 37mm cannon projected from the front of each pusher engine nacelle. The Airacuda did not enter active Air Corps service, but it anticipated the development of the long-range escort fighter and the more recent 'gunship'. During the war, Bell not only produced and developed its own aircraft, but those of other manufacturers as well. A Bell plant at Marietta, Georgia assembled Boeing-designed B-29 Superfortresses, and the Bell Buffalo plant turned out Boeing B-17 parts and assemblies. Bell produced advanced versions of its P-39 as well as a new fighter, the P-63 Kingcobra, most of which went to Russia under Lend-Lease. The company also retained its interest in the radical and the unorthodox. It furnished the AAF with its first turbojet aeroplane, the P-59 Airacomet and branched into helicopters (which is what the firm is primarily associated with today). Woods designed a prototype experimental lightweight fighter, the XP-77, though it did not secure a production contract. In 1943, the company began development of the XP-83 which was a disappointing attempt to produce a long-range turbojet-powered escort fighter based on the company's experience with the P-59. By late 1944, Woods was casting about for another project.[2]

Woods and Kotcher Launch Bell on the Path to the Mach 1

On 30 November 1944, while on a trip to the Air Technical Service Command at Wright Field, Woods stopped at an old friend's office after hours. The old friend was Major Ezra Kotcher. Woods did not know that 11 months earlier, chief research engineer for Bell, Robert Wolf had put a research-aircraft proposal to Dr George W Lewis, NACA Director of Aeronautical Research, for Wolf was now occupied in the problems of helicopter development. Kotcher, since coming back to Wright Field from the JB-2 tests at Eglin Field, Florida had resumed his search for contractors to develop an AAF-sponsored research aeroplane. As engineers passed through Wright Field to discuss new fighter proposals, Kotcher asked them whether or not their firms would be interested in developing a research aeroplane, preparatory to developing new advanced combat aircraft. All of them agreed that such a programme was desirable, but they emphasised to Kotcher that their firms were production-orientated, and would not undertake such a 'one-off' programme. What began as a friendly chat between two engineers unexpectedly developed into a serious discussion on the problems of high-speed aircraft. Cautiously, Kotcher advanced the research-aircraft concept, asking Woods if he might be interested. When Woods demonstrated a desire to involve the company, Kotcher responded in detail, telling Woods that the AAF wanted a special nonmilitary-type high-speed aeroplane, unencumbered by military requirements and built well overstrength for safety. Such a plane, Kotcher said, should be rocket-powered for maximum performance: attaining 800 mph at 35,000 feet with an endurance of two minutes. If Bell accepted the offer to develop the aircraft, the company would only have to guarantee a safe and controllable aeroplane up to Mach 0.8. Beyond that, the AAF would not hold the Bell company responsible for aircraft behaviour. Woods agreed to commit Bell on the spot: the AAF had found its contractor. Woods' action caught the company president by surprise. After leaving Kotcher's office, Woods telephoned Lawrence Bell, who was in Louisville, Kentucky and said, 'You'd better sit down and relax, I've got some news. I've just committed you to the production of an 800 mph plane'. There was no answer. Finally, Woods said, 'Operator, I've been cut off'. Bell replied, 'No, I just said "what have you done?"' As there were strong bonds of mutual respect between the two men for the other's ability as well as personal friendship, Bell decided to let the commitment stand. The next day, Woods informed Kotcher that the Bell Aircraft Corporation would begin design studies of a transonic research aircraft for the AAF.[3]

Though Ezra Kotcher recommended rocket propulsion for the aeroplane Robert Woods, like the NACA, favoured turbojet propulsion. The first sketches Woods drew of the proposed aeroplane incorporated a turbojet General Electric 1-16 (J 31) powerplant. Considering long-term production possibilities from the research aeroplane, Woods also included provisions for guns, should the AAF be interested in a combat derivative. To win him over, Kotcher showed Woods intelligence reports on the German rocket-propelled Me 163 *Komet* that by the winter of 1944 was occasionally engaging American bombers over Germany. Here was a combat aircraft that demonstrated successful rocket-engine operation. Woods reluctantly agreed to evaluate the Aerojet 6,000lb-thrust rocket engine then under development as the possible powerplant for the Bell research plane. Woods returned from Dayton to Buffalo by train, drawing sketches of various configurations on the way. On his return, Robert M Stanley, another Bell engineer, called together a design team for the aeroplane. The team consisted of Stanley, Benson Hamlin, Paul Emmons, Stanley Smith and Roy Sandstrom. Each of the team had broad experience in aircraft design and operation, and Stanley, Hamlin and Smith also had strong flying and flight testing backgrounds. Roy Sandstrom was, like Smith and Woods, a graduate of the University of Michigan with a BS in engineering. He had started his career as a stress analyst with the Curtiss-Wright corporation before moving on to Bell in 1938 where he presently served as the chief of preliminary design (he would eventually become vice president for engineering). With a tweedy, thoughtful expression and ever-present pens in his suit pocket, he looked every bit the part. Robert 'Bob' Stanley, a former Navy pilot and holder of a BS in aeronautical engineering from the California Institute of Technology, had been the first American to fly a turbojet aircraft. Before joining Bell, Stanley – a nattily dressed man of rugged good looks whose pencil-thin moustache gave him an uncanny resemblance to Zack Moseley's comic strip aviator 'Smilin' Jack' – had worked as a sales demonstrator and test pilot with the United Aircraft Corporation. Benson Hamlin, the youngster of the bunch, held a BAeE from the Rensselaer Polytechnic Institute, and had been one of six outstanding engineering students selected from across the nation in 1937 to work with Chance Vought Aircraft. Hamlin's youthful appearance and pleasant, perpetually quizzical expression disguised his reputation as an old hand. He was an expert on flight testing who had cut his teeth as a flight test engineer with Boeing and Lockheed-Vega on the B-17 programme. He had joined Bell in 1942 and had managed the flight test activities on the turbojet P-59 at Muroc Dry Lake, California during 1943. Stanley W Smith received a BS in aeronautical engineering from the University of Michigan, and had also been the National Soaring Champion for 1933. He worked as a designer for the Bowlus-du Pont sailplane firm, then as a teacher and supervisor of the New York state aviation schools from 1936 to 1939. He joined Bell as a stress analyst and preliminary design engineer. In short, it was a strong team, and it could draw on the strengths of other Bell staff members as required.[4]

Defining Bell's Speedster

In December 1944 the AAF, Bell and the NACA prepared final general specifications for the transonic research aircraft. In a joint AAF – NACA conference held at Langley Laboratory on 13 and 14 December, the AAF re-emphasised its support of developing a rocket-propelled rather than turbojet-propelled aircraft in the face of strenuous arguments by NACA representatives over safety and complexity. At this two-day conference, delegates summarily rejected a hastily prepared proposed design entry, the MCD-520, by the McDonnell Aircraft Company of St Louis, for reasons of safety, research utility and operational complexity. As far as the AAF and NACA were concerned, the proposed research aeroplane would be a Bell product. At the same time, Langley instrumentation engineers prepared specifications for the instrumentation that the aircraft should carry. They concluded the plane should carry 500lbs of instrumentation equipment, comprising of 370lbs of instruments and approximately 130lbs of wiring and tubing. The instrumentation should fit within a space allotment of 9 cubic feet and would include 60-cell manometers, control position and control force recorders, an accelerometer and a Miller 12-element recording oscillograph that would replace the two 60-cell manometers when it was used. This specification formed the basis for the future instrumentation package installed in both the future XS-1 and

D-558 aircraft. On 20 and 21 December, Kotcher met with Bell representatives Woods, Emmons, Hamlin and Dr. Vladimir Morkovin at Air Technical Service Command headquarters at Wright Field. They discussed the instrumentation requirements of the plane and agreed that the aeroplane should be as uncomplex and straightforward as possible. They decided that the pilot should be seated and not lie prone, as in the proposed XP-79, and that the aeroplane should be capable of taking off at 150 mph within 5,000 feet of a 7,000-foot runway, and land at 90 mph. Both the NACA and the Air Force insisted that the aircraft be designed for an 18g ultimate load factor; this figure represented a 50% higher load factor above existing fighter aircraft, and largely reflected ignorance of what aerodynamic loads the aircraft might experience in the transonic region. The NACA and AAF Aircraft Laboratory decided to build for a higher load factor to ensure the plane would hold together during its turbulent passage from subsonic to supersonic flight. This requirement complicated the already difficult task facing Bell engineers of designing a transonic flying laboratory.[5]

Bell began development of the research aeroplane as the Bell Model 44 under AAF project MX-524 (later MX-653). The first task for the team was to assemble adequate data from which to draw up a suitable airframe. Hamlin the design engineer and Emmons, project aerodynamicist, left Buffalo and toured the research institutions in the United States that might have developed suitable data that would be able to give the Bell team advice. Among the installations Emmons and Hamlin visited were the California Institute of Technology – where they discussed the project with Dr Clark B Millikan – and the Army Aberdeen Proving Grounds. The trip was a disappointment. At each facility, Emmons and Hamlin asked what wing section they should use, what type of control surface, what wing and fuselage shape. Each time they received a new set of answers. 'Various people told us the various thoughts that they might have had,' Hamlin recalled later, 'but every single one of them also added in the same breath, "we really can't tell you because we really don't know"'. At Aberdeen, Army scientists suggested that the two men visit the Ballistics Laboratory at Wright Field, so Hamlin and Emmons left for Ohio. The two Bell engineers had earlier concluded that the only objects they knew that moved at supersonic velocities all the time were bullets. They wondered why a .50 calibre bullet had the shape it did. At Wright Field they asked ballistics experts the question, and what a .50 calibre bullet's transonic drag values were. The ballisticians did not know the drag, but they did explain the shape. What mattered was the dispersion pattern. In tests where researchers fired bursts of bullets from a machine-gun at a target, bullets having an ogival shape produced the best pattern. The .50 calibre bullet had such a shape, selected because of the test results. Hamlin and Emmons obtained some Schlieren photographs* and sketches of decelerating .50 calibre bullets for use in designing the Bell research aeroplane. The two engineers then returned to Buffalo. The trip convinced both men that Bell was strictly on its own in developing the aeroplane: no one really had any concrete knowledge of how to design a transonic research vehicle.[6]

The Bell design team made this feeling clear in a team conference held on 15 January 1945. The subsequent report stated:

> Generally, the reaction from most people approached is that very little is known about design or conditions for supersonic flight, especially stability and control problems. All felt that a project similar to MX-524 is highly desirable and the need for it is keenly felt throughout the industry. Such a flying laboratory would be in great demand by nearly all research and development groups as an aid in studying many different individual problems and phases of research.
>
> Insofar as concrete information and suggestions have not been forthcoming we should continue our design studies based on our own ideas and at the same time keep abreast of any information which may be of value.[7]

By this time, Nazi rocketeers at Peenemünde were, in fact, on the verge of flying the world's first winged supersonic missile, the so-called A-4b. Like the XS-1, it featured a projectile shape – in this case, the V-2 terror weapon's basic 'body of revolution' design. Designed by Ludwig Roth, the A-4b was a winged variant of the V-2, intended to use aerodynamic lift upon re-entry to extend the range of the missile now that Allied armies had overrun its launching sites in Holland. On 8 January 1945, the first A-4b roared off its launch pad at Peenemunde, almost immediately wandering out of control and exploding violently. The second, on 24 January, took off successfully, accelerated past the speed of sound while in its climb – the first winged vehicle ever to exceed Mach 1 – and then arched high into the upper atmosphere. As it plunged into the increasingly denser lower atmosphere during its ballistic free-fall back to earth, it orientated itself nose

---

* Schlieren photographs are taken by a process in which the differing refractive properties of varying air densities are used to show shockwave formation and flow patterns. The term Schlieren is from the German *Schliere*, meaning a flaw or streak in glass.

down and began a steady supersonic deceleration, moving at Mach 4. The high re-entry air loads caused the A-4 to shed one wing, breaking up and scattering debris over a wide area. It was not an auspicious introduction to supersonic flight, and observers might understandably have wondered if the same would occur to a piloted aeroplane as well; no additional A-4b's ever flew.

Of Engines and Wings

Benson Hamlin now began tentative designs of various configurations. The first configuration to be investigated was a research aeroplane powered by an I-16 jet engine for the climb to altitude, but using the Aerojet 6,000lb-thrust rocket engine for transonic and low supersonic research. After the pilot completed the research portion of the flight he could return to base on the power of the turbojet alone. As the study progressed, Hamlin and the rest of the Bell team concluded that the plane would have a poor rate of climb due to a decrease in turbojet performance at altitude. This required additional turbojet fuel, which added to the estimated size and weight. The plane would reach its test altitude of 35,000 feet at low airspeed, requiring large amounts of rocket fuel to accelerate to transonic velocities. This also added to the size of the plane, and the combination turbojet and rocket installation resulted in an extremely complex design. Hamlin decided to abandon this configuration and investigate other possible layouts. He next studied the Kotcher concept – configurations using all-rocket propulsion – determining the shape, the weight and loadings of the aircraft and making a step-by-step analysis of their performance capabilities, dubbing the results 'super aeroplanes'. The first two designs did not achieve the desired performance requirements. 'Super aeroplane No. 3' did. The No. 3 design had a gross weight of 21,000lbs, and utilised a 6,000lb-thrust rocket engine with 6.5 minutes worth of fuel. On takeoff, the thrust of the engine would be augmented by four assist rockets that the pilot would jettison after take-off. The assist rockets boosted the weight to 24,640lbs. Hamlin's performance calculations indicated that the aircraft could take off in about 3,060 feet, climb to altitude at a climb rate of between 16,000 and 20,000 feet per minute – a figure five times higher than that of existing aircraft – and attain 35,000 feet with 171 seconds of rocket time remaining. When Robert Woods reviewed Hamlin's performance estimates, he still expressed reluctance to develop an all-rocket aircraft, requesting that the young engineer investigate an all-turbojet alternative. Hamlin and the Bell team members examined a turbojet aircraft, and ultimately derived a configuration capable of attaining 725 mph at sea level. This figure was still below the 800 mph that the AAF wanted. The performance of the aircraft also fell off at altitude, so that any transonic flight tests of this turbojet configuration would have to be made at low altitudes where the air loads on the aircraft were highest. Not surprisingly, Hamlin regarded this as 'very dangerous', and Bell engineers searched for ways to improve engine performance so that tests could be run at higher and safer altitudes. In particular, team members held meetings with General Electric representatives and asked whether GE could boost the thrust of its projected TG-180 turbojet. GE declared that such a boosting was not possible. At last, this led Woods to drop all thought of using turbojet propulsion in the research aircraft; company engineers did not give consideration to ramjet propulsion since the AAF desired that the plane should be capable of taking off under its own power, and a ramjet would require some form of booster propulsion until the aircraft attained sufficient velocity for the ramjet to operate. Ramjets were even more of an unknown quantity than rockets, and engineers knew little about their characteristics and design.[8]

The NACA and the Air Technical Service Command answered two other questions that the firm's design team had. The first was what wing section to use on the aeroplane. Engineers within the NACA were split as to whether or not a thick or a thin wing was more desirable. Both John Stack at Langley and aerodynamicists at Ames Laboratory in California favoured a wing of about 12% thickness:chord ratio in order to attain the critical Mach number more quickly and create transonic flow. Robert Gilruth of NACA's Langley Laboratory strongly disagreed, however. Using his wing-flow method of research, Gilruth found that thick wings lost their lift in the transonic region, while thin wings retained their lift in the transonic region. Gilruth realised that the success of the transonic research aircraft depended on a thin wing section. The decision on which section to use, either a thick section or a thin one, rested in the judgement of Floyd L Thompson, Assistant Chief of Research at Langley Laboratory, a tall, spare researcher whose shrewd and perceptive nature was masked by a country-boy demeanour. As Dr Lewis, NACA's Director of Aeronautical Research, had clamped a lid of secrecy on early wing-flow-test results, discussions on the data were awkward; indeed, within NACA, only Gilruth and Thompson knew of the results at that time. Thompson examined the data collected by Gilruth and concurred that a thin wing would be more desirable, rejecting the thick wing option. Thin wing advocates suggested using wings of as low as 5% thickness:chord ratio, but Langley engineers decided on a compromise whereby the research aeroplane would have two sets of wings. One set would have a thickness:chord ratio of 8%; the other would have a thickness:chord ratio of 10%; research studies would be made on both. In conferences with Bell, NACA representatives reported their decision on the thickness question. (Some observers, while accepting the aerodynamic reasons for selecting the configuration, had little confidence that Bell designers could fabricate an 8% wing with the desired 18g load factor.) This decision – to use a thin as opposed to a thick wing – was one of the most critical in ensuring that the subsequent XS-1 would, in fact, be a success.[9]

The second question decided for Bell was that centring around the Aerojet 6,000lb-thrust 'Rotojet'. The Rotojet would utilise red fuming nitric acid and aniline as propellants. These two were hypergolic, meaning that whenever they mixed, they reacted violently, thus obviating the necessity of providing an ignition system for the engine. Wondering just how violently the two liquids reacted, members of the Bell design team purchased a bottle of each from a drug store. They taped the bottles together, went to the gun butts at the Buffalo plant where armourers sighted-in the machine-gun installations on Bell aeroplanes and threw the two bottles against a rock. The glass shattered and the contents erupted in flame. Shaken by what they had seen, the Bell engineers returned to their drawing boards. All they could envisage was the plane having a mishap, or having to land with propellants aboard and the fuel tanks bursting on landing, instantaneously obliterating the pilot, plane and anything else close by. It meant that Bell would have to design the plane to extreme requirements to prevent such a possibility from occurring: an almost insurmountable task. At the same time, the AAF Air Technical Service Command was likewise having second thoughts about the Aerojet engine. Its development was not moving smoothly due to technical complications and, like Bell, the AAF recognised the hazardous characteristics of using a hypergolic fuel combination on a manned aircraft. The Aerojet 6,000lb-thrust engine dropped from consideration. Fortunately, a substitute for the Aerojet engine was already at hand: another 6,000lb-thrust engine being developed for the Navy by Reaction Motors, Inc. of Pompton Plains, New Jersey. The Reaction Motors company was truly a shoestring operation. Four pioneers of early American rocketry had formed the organisation in December 1941, shortly after Pearl Harbor. Lovell Lawrence Jr, H Franklin Pierce, James H Wyld and John Shesta were members of the American Rocket Society, a Jules Verne-like group composed of serious researchers, fanatics and visionaries, amateur and model rocket enthusiasts and space buffs. (By coincidence, in 1939, James Wyld had worked together with John Stack at Langley Laboratory on high-speed airfoil research.) The company first set up shop in a garage in North Arlington, New Jersey but moved to a rented store – 'Pat's Tailor Shop' – in Pompton Lakes, moving again in 1943 to a former night-club located in Pompton Plains.[10]

In the same way that the AAF had contracted with GALCIT for development of take-off assist rockets, the Navy contracted with the tiny firm of Reaction Motors for the same purpose. Drawing on their previous American Rocket Society experience, RMI members developed a 1,000lb-thrust rocket engine and tested it in November 1942. By May 1943, they had a 3,400lb-thrust engine undergoing tests. Emboldened, they then moved on to a 6,000lb-thrust engine, the LR-8, for the Navy. Navy Commander C Fink Fischer, an equally active supporter of naval rocket research as Kotcher was for the AAF, had prepared the specifications for the latter engine. (Not content with the laboratory perspective, Fischer, a naval aviator, had successfully piloted a potentially risky rocket-assisted takeoff trial from Anacostia Naval Air Station in May 1942 in a Brewster F2A-3 Buffalo using five British solid-fuel antiaircraft rockets jury-rigged under the plane. One imagines what the nation's lawmakers in the Capitol might have thought – not to mention the President in the White House – had they realised what was occurring within less than a mile of the two leadership centres.)

The RMI engine consisted of four separate rocket cylinders, each of which produced 1,500lb-thrust. The company designation of the engine was 6000C4: 6,000lb-thrust, four cylinders. The propellants were liquid oxygen and alcohol diluted with water. Each cylinder consisted of an igniter and combustion chamber, and the engine, with the exception of the controls, piping and wiring, was of welded stainless steel construction. The operator could not throttle the engine, but could ignite or shut down each cylinder separate from the others. Thus the engine could run at 25, 50, 75 or 100% power. So effective was the engine's regenerative cooling system – whereby propellants were circulated through the cylinder and nozzle walls to absorb the heat of combustion – that the external cylinder temperature rarely exceeded 140°F, though the internal temperature attained 4,500–5,000°F. Lawrence met with Kotcher and arranged a demonstration of the prototype 6,000lb-thrust engine which RMI engineers had disassembled and placed in storage. Engineers hurriedly reassembled it, adding a new ignition system, propellant lines, valves and painting it black, giving it the nickname *Black Betsy*. The demonstration came off without difficulty: RMI was now the manufacturer of the Bell research aeroplane's rocket engine, and Bell detailed engineer William M Smith to the RMI firm to work on making the engine flightworthy.

The decision to use the RMI engine removed the anxiety that had surrounded the Aerojet engine and its hypergolic fuel combination. Bell had already rejected such propellants as hydrogen peroxide which requires special handling and nitromethane, whose detonation characteristics were not fully understood. The company also dismissed a gasoline and liquid oxygen fuel system: gasoline was not suitable for regeneratively cooled engines, and the plane would have required a tank of water to aid in cooling. The RMI engine propellants, liquid oxygen and alcohol, were reasonably safe to handle, non-hypergolic and readily available. Like the decision on wing-thickness, selection of the RMI engine constituted a milestone in the development history of the XS-1 aeroplane.[11]

On 10 March 1945, the AAF Air Technical Service Command notified the NACA that the AAF was awarding Bell a contract to develop a rocket-propelled straight-wing research aeroplane using a NACA 66-series airfoil section. The NACA and the AAF would work out joint guidelines for performance, stability and control requirements, wind-tunnel testing, flight testing, radar tracking and instrumentation. Six days later, the AAF issued contract W33-038-ac-9183 authorising $4,278,537 covering design and construction of three transonic research aircraft under project designation MX-653. The AAF assigned the aircraft the designation XS-1: Experimental Sonic-1. The United States had formally joined Great Britain and Nazi Germany in the race to develop a piloted supersonic aeroplane.[12]

Why Not a Sweptback Wing?

It is interesting to note that the XS-1 contract stipulated a straight-wing configuration, for, by this time, the AAF already knew of the potential advantages of wing sweep as a means of alleviating compressibility shock. The knowledge did not come from rediscovering Busemann's pioneering Volta paper in 1935, but rather from NACA Langley researcher Robert T Jones who had independently discovered the benefits of sweep while investigating slender-wing applications to some guided weapons. While working with Kotcher on the JB-2 programme, Jones mentioned his theory to the AAF engineer who in turn, brought it to the attention of Theodore von Karman and his chief assistant, Dr Hsue-shen Tsien. Von Karman was then heading a new Scientific Advisory Group that General Hap Arnold had formed on 7 November 1944. Both Karman and Tsien thus learned of wing sweep roughly six months before they toured the shattered remnants of Germany's wartime aircraft industry and discovered the German fruits of Busemann's sweptwing research. Indeed, before he went overseas, von Karman arranged for comparative tests in April 1945 of an experimental sweptwing and a conventional straightwing in a supersonic tunnel at the Army's Aberdeen laboratory, at speeds up to Mach 1.72; the sweptwing easily proved superior, clearly (and independently) validating Jones' work. The reason why the AAF did not employ sweep on the XS-1, or at least let the contractor choose between a swept and nonswept configuration, was that the Air Technical Service Command still considered the straight wing configuration as a viable high-speed aircraft planform in the future and desired the XS-1 to provide data applicable to conventional aircraft design, as typified by the Lockheed P-80 straight-wing fighter. The sweptwing, while promising performance improvement, remained an unknown quantity, not verified by experimental application. The discovery of wholesale sweptwing research in Germany triggered a belief – incorrect as it turned out – that the straight wing configuration was obsolete, a mindset that pervaded both the AAF and the NACA immediately after the war. In this new climate, some AAF officials regretted that the service had not stipulated a swept planform for the XS-1. One of the officials, General Alden R Crawford, then Chief of the AAF's Production Division, wrote to Dr Jerome Hunsaker, the NACA Chairman, in October 1945 stating that the NACA's withholding of sweptback information that could have resulted in a sweptwing XS-1 design had delayed the North American P-86 and Boeing B-47 programmes, then being redesigned from straight to sweptwing configurations. Replying for Hunsaker, Langley Assistant Chief of Research Floyd Thompson wrote somewhat testily (perhaps reflecting Hap Arnold's post-Whittle criticism of the agency for not picking up on the significance of the turbojet engine in the late 1930s), that to recommend a sweptwing configuration on the XS-1 at a time when the sweptwing was essentially an unknown quantity might have resulted in a 'blunder of the greatest magnitude'. (The NACA subsequently tested XS-1 configurations with both swept-aft and swept-forward wings, anticipating not only the future X-2 but the Grumman X-29A forward sweptwing technology demonstrator flown almost four decades later. At the time, for the record, Jones thought the XS-1 should be a sweptwing design.) In any case, the discovery of German sweptwing work, following in Jones's wake, caused Bell engineers to study a possible sweptwing modification of the basic XS-1 design. One day in the autumn of 1945, Robert Stanley called Benson Hamlin from Wright Field asking him to start work immediately on a sweptwing version of the XS-1. Try as he might, the gifted Hamlin could not develop a workable swept version of the already-designed-and-being-manufactured plane, for the sweptwings demanded an entirely new centre wing section and rearrangement of the complex propellant system. He told Stanley that Bell would have to develop an altogether new aircraft from the start. Bell decided to approach the AAF on this entirely new aircraft, featuring rocket propulsion, air-launch capability and a sweptwing planform. Thus the Bell X-2 was born, the world's first Mach 3 aeroplane.[13]*

Designing the Tail Surfaces

Early in the design of the aircraft, Bell considered developing the first XS-1 with a conventional tail, and the second XS-1 with a 'V' tail, for NACA studies indicated that the 'V' tail – looking somewhat like a butterfly's wings – might alleviate the problems associated with wing wake interference on the horizontal tail surfaces. Bell had already experimented with the 'V' configuration on several modified P-63 Kingcobra fighters used for experimental purposes, but after detailed consideration the XS-1 team dropped the 'V' tail from consideration and decided to develop the aeroplanes with conventional tail surfaces. In conformance with AAF wishes, Bell wanted to keep the aircraft as simple as possible, so that it could return information useful to conventional-type aircraft. For this reason Bell (at AAF and NACA behest alike) rejected other possible layouts-including the 'tail first' canard configuration and wing sweep. Important questions remained, however, about how the conventional tail surfaces should be designed, and here the NACA played an important role. As has been seen, the NACA had developed basic design criteria for a transonic research aircraft, such as the decision to employ a thin wing section and approximately 500lbs of research instrumentation. One of their most important decisions concerned design of the horizontal tail surfaces. John Stack and other Langley engineers concluded that the XS-1's horizontal stabiliser should have a lower thickness:chord ratio than the wing, for if the wing encountered severe compressibility effects, the horizontal tail would not experience the same problems simultaneously. By using a thinner airfoil section on the tail surfaces, the critical Mach-number of the tail – the velocity where high drag rise occurred – would be higher than the critical Mach number of the wing. Thus, if the pilot encountered severe stability

---

* Sleek, glamorous and so deadly that it killed three men, the Bell X-2 did not make a powered flight until 1955, having been delayed seven years due to developmental difficulties with its temperamental Curtiss-Wright XLR-25-CW-1 15,000 lb-thrust engine. In May 1953, the second X-2 (46–675) blew up under its Boeing B-50 mothership over Lake Ontario during a propulsion systems test, killing Bell chief test pilot Jean Ziegler and engineer Frank Wolko. On 27 September 1956, the first X-2 (46–674) became the first aeroplane to exceed Mach 3, but went out of control and crashed on the same flight, killing its pilot, Captain Milburn 'Mel' Apt, USAF.

and control problems, he could retain control over the aeroplane until the plane decelerated to a safer lower Mach number. If both the wing and the tail had the same airfoil section, both would experience the problems of compressibility simultaneously and with equal severity, and the pilot might find he had no control over the aircraft, possibly resulting in loss of the pilot and aeroplane. To ensure adequate longitudinal control during transonic flight, Stack, Gilruth and others further suggested that the horizontal tail be all-moving. In other words, rather than have a fixed horizontal stabiliser surface with a movable elevator, NACA recommended that the stabiliser itself be adjustable by the pilot while in flight. For ordinary subsonic flight, therefore, the pilot could control the aircraft through the elevator. If the need arose, he could move the stabiliser as well as the elevator, in effect making the horizontal tail all-moving. NACA likewise furnished Bell with a 'very definite' location for the location of the horizontal stabiliser, for Langley engineers believed that one of the principal difficulties facing the aeroplane in transonic flight would be impingement of the wing wake upon the horizontal tail surfaces. Accordingly, they asked Bell to place the horizontal tail as high as possible to position it above the wing wake, which resulted in Bell placing the horizontal stabiliser high on the vertical fin. (The Germans, incidentally, had reached a similar conclusion for their own DFS 346 programme, giving it a 'T-tail' configuration.) Though tail location was important, the decision to use an adjustable horizontal stabiliser proved critical. It gave the XS-1 excellent transonic and supersonic control, and pointed the way towards the all-moving tail, a hallmark of subsequent supersonic aeroplanes world-wide.[14]

Detail Design Questions

The Bell Aircraft Corporation design team had to develop a configuration that had enough excess power to pass quickly through the speed of sound, was strong enough to withstand transonic buffeting, and with more control power than conventional aircraft. Aside from general guidelines for the design of the aeroplane – the type of propulsion, type of wing configuration and desired performance goals and load factor, for example – the AAF wanted to give Bell complete freedom in designing the aeroplane. When the AAF wrote up the XS-1 specification, it stipulated that the standard specifications for aircraft design would not apply to the design of the XS-1. This included the AAF designers' 'bible', the *Handbook of Instruction for Aeroplane Designers,* 8th Edition, 1 July 1936, commonly referred to as the 'Handbook'. Using the data they had accumulated on .50 calibre bullets, Hamlin and Emmons began design of the XS-1 aircraft. They decided to shape the fuselage just like the bullet. With this decision made, they next devoted attention to the design of the cockpit. From a drag standpoint, a protruding cockpit seemed undesirable. It would create a shock wave, and from a design standpoint it seemed a good idea at least to have a similar protrusion below the fuselage for symmetry. After serious consideration, Hamlin and Emmons decided to eliminate these protrusions from the aeroplane and incorporate a cockpit canopy mounted flush with the ogival surface of the nose. This led to yet another problem, this one from a pilot's standpoint. The flush canopy design furnished very poor forward visibility for the pilot on takeoff and landing. Bell built a wooden cockpit mockup, and Hamlin and Emmons asked Bell test pilot Jack Woolams, already selected as the XS-1 project pilot, to sit inside and decide whether he had sufficient visibility from the aeroplane on takeoff and landing. Woolams approved the canopy design, but Bell did modify the window shape to improve the visibility from the pilot's standpoint. The modifications did not change the ogival shape of the nose. Hamlin considered the design of the XS-1's wings possibly the greatest difficulty that the Bell design team faced; he wanted to design the wings to a more typical 7g load limit, but the requirements set forth by the NACA and the AAF stipulated an 18g loading. Since the XS-1 was to have two wing sections, one plane with 10%-thick wings and the other two with 8%-thick wings, Hamlin doubted that the required strength could be built into such thin wings. (Bell eventually managed to incorporate the 18g load factor, a considerable accomplishment itself.)[15]

Air-Launching or Ground Takeoff?

One question that raised some disagreement among Bell engineers was whether to design the XS-1 for air launching. From the start, Benson Hamlin wanted to design the aeroplane with skids, so that a launch aeroplane could carry it aloft, launch it and the pilot could fly the mission and then come in for a landing on the skids. Hamlin believed that air launching was the best way of getting high performance, since the rocket aeroplane would not need to consume rocket propellants except during the brief climb from launch height to the 35,000 foot test altitude. He felt it unnecessary to design the aeroplane with a retractable landing gear, since the gear would occupy valuable space and contribute undesirable weight, both deficits which would eliminate possible fuel reserves. Stanley and others agreed that the skid landing gear seemed the best technical approach, but the formidable Robert J Woods was strongly opposed. Woods was interested in the X-1 as a possible first step towards development of a manned rocket-propelled interceptor, and wanted to include a retractable landing gear so that the aeroplane could take off from the ground. At meetings with the NACA at Langley Field, Hamlin expressed his desire to incorporate a skid landing gear. Woods disagreed. At the same time, a new factor entered the air launch controversy: the question of safe aircraft operation. As a pilot, Bob Stanley recognised that the first few seconds after takeoff would be extremely hazardous, inasmuch as the rocket plane would be heavily laden with a full fuel load, be at a low altitude and at a low airspeed. He recommended in discussions with Ezra Kotcher at Wright Field that the aeroplane be air launched from a mothership, such as a four-engine Douglas C-54 Skymaster transport. Woods did not favour the launch aircraft idea, stating that it would not produce an operational-type aeroplane. The controversy remained unresolved until arbitrated by the company president, Lawrence D Bell who decided both in favour of a retractable landing gear and Stanley's arguments in favour of air launching the aeroplane. Bell transferred Woods to the company's corporate staff and his subsequent technical intelligence trip to Germany after the surrender to examine the Nazi aircraft industry removed him from the crucial detail design of the XS-1, which took place during the summer of 1945. By that time, delays with the proposed turbopump fuel-feed system for the XS-1 had confirmed the earlier wisdom of Stanley's decision in favour of air launching, as will be seen.[16]

Feedback into Wind Tunnel Development

Before it even flew, the Bell XS-1 significantly advanced aeronautical technology by stimulating development of new wind-tunnel techniques. The principal reason behind the development of the XS-1 was, of course the inability of existing wind tunnels to furnish satisfactory and reliable transonic aerodynamic data. In addition to using Bob Gilruth's wing-flow test method of wing and body shapes to furnish information useful for the X-1 (and later D-558), the urgency of research aeroplane development stimulated researchers to explore new methods of transonic wind tunnel testing. Not surprisingly, the quest for better wind tunnels had predated the research aircraft programme by several years. In 1944, during one of John Stack's frequent visits to the Langley 8-foot wind tunnel, John V Becker, holder of a BS in mechanical engineering from New York University and a brilliant young engineer who headed the tunnel, suggested using a 'splitter plate' support system in the tunnel to permit accurate small-scale wing model testing to above Mach 0.8. This method subsequently furnished useful data for high-speed turbojet bomber development. Later, spurred on by the XS-1 and D-558 programmes, Langley engineers hastened their efforts to develop satisfactory methods of transonic wind-tunnel research. Adopting Gilruth's wing-flow methods, wind tunnel researchers constructed specially shaped plaster throats to generate accelerated flow, a method that became known as the transonic bump technique. They also developed an entirely new method of testing model high-speed aeroplane configurations using special sting supports with internal balances. The so-called sting support did away with braces and struts attached to a model that could themselves generate or reflect disruptive shock waves across a test section. By using small models and the non-choking sting support system, even before the XS-1 flew, Langley engineers could test the XS-1 (and the rival D-558) design to above Mach 0.9, almost up to the speed of sound, before getting the now-classic jump to Mach 1.2. Stimulated by the XS-1 programme, NACA engineers now had new methods of acquiring data that went far beyond anything possible in the 1930s. It did not eliminate the transonic knowledge gap in wind tunnel design, but it did narrow it considerably. Eventually, with the development of the slotted-throat wind tunnel Langley engineers solved the transonic experimental problem on the ground, but it can be argued, only from having the stimulus of the XS-1. Its design required sound development, which in turn added urgency to the efforts of ground researchers to develop adequate wind-tunnel techniques. (This serendipitous aspect of research aircraft development, highlighted by the interplay of wind tunnel deficiencies and research aeroplane information pressures, has been seen more recently in the interplay of simulation for flight, and feedback of flight information into the simulation process.)[17]

Crisis: Turbopump Delay Forces Propulsion Redesign

By May 1945, Bell had the design of the XS-1 well in hand. Bell engineers had established the general configuration and layout of the aircraft, though detail design of the plane was not yet complete. The design team had already predicted the performance of the aircraft based on a launch weight of 13,034lbs and a fuel load of 8,160lbs. These estimates were far beyond the performance capabilities of any existing aircraft. Bell estimated that if ground-launched, the aeroplane could attain 1,100 mph at approximately 65,000 feet. If air-launched with the same 8,160lb fuel load, it could streak to over 1,600 mph at the same altitude. As originally planned, the XS-1 design featured two large cylindrical fuel tanks, one forward of the wing for liquid oxygen, and one aft of the wing for the alcohol. Bell planned to incorporate a turbine-driven fuel pump to suck the propellants from the tanks and supply them to the engine, a low-pressure fuel feed system. Technical difficulties now arose to threaten the whole XS-1 programme. The Bell design team had long recognised that the turbopump was, as Stanley and Sandstrom subsequently recollected, ‘one of the items most likely to interfere with the early flight of the airplane'. By late April, the turbopump was falling rapidly behind schedule, and Stanley realised that continued reliance upon the turbopump would necessitate delays in delivery of the aircraft and initiation of flight testing. Boldly, rather than risk delay, Stanley decided to drop, at least temporarily, the proposed turbopump installation; instead, Bell would install a pressurised 'blow down' fuel system, in which high pressure gas would expel the liquid oxygen and alcohol from their tanks, forcing them into the engine under pressure. The AAF received notice of the Bell decision on 3 May 1945.[18]

The decision to switch to a pressurised fuel system automatically necessitated air-launching the aircraft. Whereas previously air launching had emerged primarily as a safety consideration, it now became an absolute requirement because of a marked decrease in the XS-1's estimated performance. Using a pressurised fuel system necessitated storing a pressurised gas under high pressure to force the fuel into the rocket engine. Bell selected nitrogen for the pressurised system, requiring numerous small spherical high-pressure storage bottles. Since the fuel tanks holding the liquid oxygen and alcohol now required pressurisation to expel their contents into the engine, they had to be constructed to withstand the pressure. This necessitated that they be of spherical shape (rather than cylindrical shape), with extremely thick strong tank skins. The effect of the pressurised fuel system decision upon the XS-1's design was two-fold. The landing weight jumped by one ton, and the volumetric reduction in available propellants – caused by the nitrogen bottles and spherical propellant tanks, both of which took up or wasted space, like tennis balls in a canister – reduced the fuel load from a maximum of 8,160lbs to 4,680lbs. This cut engine time from 4.1 to 2.5 minutes. If the XS-1 now flew from the ground, the smaller fuel load gave the plane only subsonic performance, thus not meeting the contract requirements. Air-launching now became an absolute necessity. With air launching, Bell estimated that the aeroplane could attain just over 800 mph at 50,000 feet. Absolute ceiling dropped from around 140,000 feet to 87,750 feet. Both airspeed and altitude capabilities were still well within those desired by the NACA and the Air Force, and still remained about double those of existing aircraft. However, to attain this performance, the XS-1 had to be air launched; any thought of going back to ground takeoff would mean that the aircraft could not fulfil its design mission of transonic flight research. Subsequent events proved Stanley's decision to utilise a pressurised fuel system as sound. Bell completed the first two XS-1 aircraft with the pressurised fuel system, but delayed completing the third XS-1 until mid-1951 when the low-pressure turbopump installation was ready. It featured much higher performance capabilities than its two sister ships, but development of a reliable turbopump had consumed five years of research. By the time it arrived at Edwards, the original two MX-653 aircraft had retired, their flight test programmes complete, with the first one already on exhibit at the Smithsonian Institution.[19]

Where We Stand: von Karman's Forecast for Postwar Aviation

By 1 August 1945, Bell's XS-1 team had completed detail design for the XS-1. Their accomplishment coincided with the publication of a seminal report prepared by Theodore von Karman, entitled *Where We Stand*.[20] In November 1944, Hap Arnold had instructed the Hungarian scientist to prepare a detailed report on the current status and future of aviation technology, boldly stating that, in the postwar world, 'supersonic speed' was a 'requirement'.[21] Though busily assessing the state of Nazi aeronautics, von Karman nevertheless wasted no time in preparing two key reports. The first of these was *Where We Stand*, a assessment of the current state of aeronautical development. The second was the visionary and multi-part (and subsequently better-known) *Toward New Horizons*, which forecast the future of aviation and suggested bold courses of action. Both appeared in 1945, the former as Japan was atom-bombed and the latter in December, as Bell was rolling the first XS-1 out of its Buffalo, New York plant.

Where We Stand noted that supersonic flight had appeared 'a remote possibility' before 1940, but that as the result of 'bolder and more accurate thinking . . . this stone wall . . . will disappear in actual practice if efforts are continued'.[22] Von Karman went on to note that 'we were slow in recognizing the necessity of supersonic wind-tunnel research'. By 1945, few American supersonic test tunnels existed, while Germany had no less than eight in service in four research complexes, six of which could exceed Mach 3 and one of which could exceed Mach 4. He stressed that the AAF would have to acquire similar facilities and also noted the potential value of the sweptwing, an acknowledgement both of Robert Jones' work at the NACA, and the discovery of the fruits of Adolf Busemann's in the rubble of Germany's aircraft industry. He recommended development of large supersonic wind tunnels, using transonic and supersonic research aeroplanes to substitute for the lack of tunnel measurement capabilities in the transonic region, advocating vertical take-off rocket-boosted fighters and using forward-firing rocket thrusters for deceleration of high-performance aircraft prior to landing. If von Karman's vision was not completely perfect – as witnessed by his advocacy of vertically launched rocket fighters, rocket decelerators, and even nuclear-powered supersonic aeroplanes and guided missiles – his final conclusion was, without question, unquestionably accurate: 'We cannot', he wrote, 'hope to secure air superiority in any future conflict without entering the supersonic speed range'.[23]

Taking the XS-1 from Drawing Board to Flightline

It was, of course, to that end that Bell was pressing on with the XS-1 programme. During early discussions with AAF representatives, a question had arisen on the availability of a suitable launch aircraft while the war raged. The AAF needed every available Boeing B-29A Superfortress bomber – the first choice for launch aircraft by Stanley and Hamlin – for the strategic air offensive against Japan, and the second choice, the Douglas C-54 Skymaster transport, was in demand by the AAF Air Transport Command. When the two atomic bombs and the subsequent Japanese surrender ended the war in the Pacific, however, the problem of securing a suitable launch aircraft disappeared amid the hundreds of B-29s returning to the United States for scrap or storage. With an ample bomb bay that could house the rocket plane, good load-carrying characteristics and good altitude performance, the B-29 was a natural – the most logical choice. (Ironically – unbeknown to the Bell-AAF-NACA XS-1 team – in the Soviet Union, at exactly the same time, a team of Russian and captured German engineers were reaching precisely the same conclusion, planning to use one of three B-29s that had force-landed in Soviet Asia following raids on Japan to haul the captured DFS 346 to altitude on its first high-speed trials!)

The smooth bullet contours of the XS-1 design hid an extremely crowded fuselage.[24] Of semi-monocoque 24ST aluminium construction, the fuselage contained two propellant tanks, 12 nitrogen spheres, the pilot's pressurised cockpit, three pressure regulators, a retractable landing gear, the wing carry-through structure, the rocket engine and the flight research instrumentation. Ahead of the XS-1's wing was the spherical liquid oxygen tank. It could hold 311 gallons – 2,920 lbs – of the supercold liquid. Bell designed it of annealed stainless steel 3⁄16 inch thick to avoid the danger of embrittlement at –300° F and to withstand an internal pressure of approximately 350 psi. The alcohol tank, designed to hold 293 gallons – 2,100 lbs – of fuel did not have to withstand the same low temperatures as the liquid oxygen tank, so Bell designed it from normalised SAE 4130 steel with a wall thickness of 1⁄8 inch. To prevent fuel-sloshing from affecting the plane's centre of gravity location, Bell engineers designed the tanks with internal transverse baffles.

The nitrogen pressurisation system gave the Bell design team the greatest difficulty in design. Engineers discovered that liquid oxygen readily absorbed gaseous nitrogen. This forced Bell to install a nitrogen system for very high internal pressure in order to ensure that the XS-1 had sufficient nitrogen on board to force the liquid oxygen and alcohol into the rocket engine. They designed sufficient tankage to contain 17.5 cubic feet of nitrogen at 4,500 psi, a total of 301lbs. Engineers stored the nitrogen in 12 pressure spheres scattered throughout the fuselage, their location depending on available internal space. Bell planned to fabricate the spheres from steel ranging in thickness from ¼ to ⅝ inches depending on the size of the tank. Engineers placed one nitrogen sphere directly ahead of the cockpit, clustered seven in a ring behind the cockpit and in front of the liquid oxygen tank, installed two more below the wing behind the liquid oxygen tank and placed two behind the alcohol tank. They incorporated a common manifold system connecting the 12 nitrogen storage vessels together, using welded high-pressure joints rather than detachable fittings in order to prevent leakage. Three pressure regulators reduced the 4,500 psi source pressure to usable levels. Regulator development posed additional design difficulties due to the extreme pressure and low temperature requirements that the system had to satisfy. One regulator reduced the 4,500 psi source pressure to 1,500 psi for operation of the landing gear and wing flaps, simplifying the operation of the aeroplane and removing the need to incorporate batteries or hydraulic pumps for these functions. Two other regulators each reduced the 4,500 psi source pressure to 340 psi before it entered the propellant tanks to pressurise the propellants. The XS-1 pilot could control the three regulators from the rocket aeroplane's cockpit.

Adding to the difficulties of designing the nitrogen pressurisation system was the fact that nitrogen is sold commercially in cylinders at no more than 2,200 psi pressure. Bell had to have nitrogen at twice this pressure with the gas uncontaminated by oil droplets, or else Bell ran the risk of having an explosion when the contaminated nitrogen mixed with the liquid oxygen. The requirement for pure nitrogen ruled out existing methods of deriving pressurised nitrogen, so Bell engineer Lloyd Bevin developed a specialised nitrogen evaporator: Liquid nitrogen passed out of a 36 inch sphere, then boiled off as a gas that could be piped into the plane's nitrogen tanks at the requisite 4,500 psi pressure. The 36 inch sphere, the heart of the evaporator unit, consisted of two hemispheres fabricated from 3 inch thick stainless steel and hydrostatically tested to withstand 9,000 psi. (During tests, engineers noted liquid dripping from the frosted lines through which the liquid nitrogen flowed. They puzzled over the phenomenon before discovering that the cold nitrogen lines were condensing liquid oxygen out of the air.) Interestingly, the Bell XS-1 design team also used nitrogen to pressurise the cockpit, thus necessitating that the pilot to wear an oxygen mask at all times while flying the aeroplane.* Bell designed the pressure cabin to maintain an internal pressure of 3 psi above the atmospheric pressure in order to ensure the safety of the pilot at high altitude; tests showed the pressure cabin of the aeroplane leaked no greater than 1 psi per hour, a completely satisfactory figure given the planned short duration of the XS-1 flights. Before the launch from the B-29, the rocket pilot would enter the XS-1 through a small hatch located on the right side of the fuselage nose. An assistant would lock the hatch panel itself in place, sealing the XS-1 cabin from the outside.

The XS-1 design contained a number of other interesting features besides the pressurisation system. One was the thin wing planform, whether of 8% thickness-chord ratio or of 10% thickness:chord ratio. The 8% wing was only 5.94in thick at the wing root. The 10% wing was a little larger at 7.42in. It was difficult enough for Bell to design and fabricate such a thin wing in 1945, particularly for an 18g load factor, but the NACA further complicated the task by having Bell cut 240 pressure orifices and install 12 strain gauges in the left wing to acquire pressure distribution and air loads information. Milling the wing skins proved difficult, for they had to be exceptionally thick (tapering from ½ inch at the wing root to conventional thickness at the tip) to provide the necessarily stiff structure, as well as retaining accurate surface contours. As might be expected, the thicker 10% wings were ready before the thinner 8% wings. Bell did not boost the control surfaces, for design-team members felt that the small size of the aeroplane made high stick forces unlikely. The adjustable horizontal stabiliser Bell designed for the aeroplane could move through an angle of 15° at a rate of 1° per second, if needed, for transonic control to overcome rapid trim changes. An electric motor operated a screw jack to change the angle of attack of the stabiliser. Instead of incorporating a traditional fighter-type control stick, the team designed the XS-1 to use a control wheel, a method more commonly employed on bombers or transports. Bell incorporated the wheel control so that, if the need arose, the pilot could use both hands on the wheel for greater effectiveness. Accordingly, engineers installed all the major controls – thrust selector,

---

* This use of nitrogen to pressurise the cockpit nearly led to the loss of the first X-1 in 1949, when the pilot plugged his oxygen mask into the nitrogen outlet rather than oxygen outlet. He was able to rectify the error before he passed out from lack of oxygen.

instrumentation switches, stabiliser control and power shutoff – on the control wheel so that the pilot could operate them without taking his hands from the wheel.

From Dream to Reality

After completing design of the XS-1, Bell engineers constructed a wooden mockup of the proposed aeroplane, showing the internal structural layout and serving as a check on the engineering drawings. On 10 October 1945, Bell held the mockup inspection on the design, with representatives of the AAF's Air Technical Service Command and NACA attending. No major changes resulted from this inspection, though Bell engineers believe some minor technical changes in the design may have occurred as a result of mockup inspection comments. The mockup inspectors approved the design and the Bell company turned to fabricating the aeroplanes, virtually handcrafting each one. By mid-December 1945, the airframe of the first XS-1 was nearing completion. The AAF had assigned a block of serials for the three aircraft; the first was 46-062, the second 46-063, and the third 46-064. Reaction Motors, Incorporated, did not have the first of the 6000C4 rocket engines – which the AAF had designated as the XLR-11 – ready for installation in the aeroplanes, so Bell completed the first XS-1 without its engine. Technicians then painted the bare aluminium aircraft a gleaming saffron orange, stencilling its serial in black on the vertical fin above the horizontal stabiliser. On 27 December 1945, the first XS-1 rolled out of the Bell plant, looking like a sleek saffron bullet with thin knife-like wings. From the nose jutted a long pitot tube for the pilot's instruments. Another airspeed head for the research instrumentation emerged from the left wing tip, while a boom from the right wing tip supported a sideslip angle transmitter.[25]

Sitting on the ramp of the Bell company at Buffalo, the orange XS-1 displayed a marked and somewhat disconcerting contrast to the propeller-driven Bell P-39 and P-63 fighters, and even the turbojet P-59 and P-83, as if one of Flash Gordon's rocketships had somehow come to exist. All of its smoothly flowing lines bespoke of hidden power and speed, an effect heightened by the low-slung thrusting posture the plane assumed on its short tricycle landing gear. The coming weeks would witness whether or not the plane fulfilled its promise. Indeed, just the conception and construction of the XS-1 had provided challenge enough, as one Bell engineer later recalled,

> It required an unhesitating boldness to undertake a venture so few thought could succeed, an almost exuberant enthusiasm to carry

> across the many obstacles and unknowns, but most of all a completely unprejudiced imagination in departing so drastically from the known way.[26]

Meanwhile, 2,000 miles from Buffalo, a group of equally dedicated and imaginative engineers in California were busy developing another transonic research aeroplane (but a very different concept), this one for the Navy.

3
Flying Test Tube:

The Design Development of the Douglas D-558

Abraham Hyatt's September 1944 memorandum recommending procurement of a Navy-sponsored transonic research aeroplane had crystallised the sentiments expressed in the Navy's March 1944 meeting with the NACA at Langley Laboratory; namely, that the research necessary for development of high-speed aircraft and missiles required procurement of a special transonic research aeroplane. In mid-December 1944, Commander Emerson Conlon of the Bureau of Aeronautics' Structures Branch drafted a letter to the NACA stressing the urgency of acquiring more high-speed aerodynamic data. Captain Walter S Diehl, supporter of the research aeroplane concept since 1942, added two paragraphs asking that the NACA give high priority to a long list of projects including a new test station at Wallops Island, Virginia, a new 6-foot supersonic wind tunnel and a research aeroplane. The letter, which the Bureau of Aeronautics sent on 19 December 1944, informed the NACA that the Bureau of Aeronautics believed that procurement of a research aeroplane should be expedited, and that the Navy would take steps to procure the aircraft once the Bureau of Aeronautics had received drawings and specifications from the NACA. The BuAer request came three days after Hitler sent three armies into Belgium and Luxembourg, opening the last great German offensive of the war, the Battle of the Bulge. The suddenly worsening war situation 'brought forth orders to BuAer to leave no stone unturned in the development of new and improved weapons'. As previously recounted, in March 1944 the Navy and NACA had agreed to detach NACA engineer Milton Davidson from Langley Laboratory to Washington to work with Ivan Driggs in the Bureau of Aeronautics on the preparation of specifications for a proposed transonic research aircraft. Fighting a global war, the Navy did not think it could justify a pure research aircraft, while the NACA was more research-minded. Davidson disclosed to other NACA engineers that the Navy favoured an aircraft capable of meeting military requirements. NACA favoured turbojet propulsion and, like the later XS-1, a 500lb instrument package. Additionally, NACA engineers felt that the plane should be as efficient aerodynamically as possible, with a simple nose intake for the turbojet engine. The Navy, on the other hand, favoured side inlets so as to free the nose for a possible armament installation.[1]

In the 1940s, there were four major Navy contractors: Grumman for fighters and torpedo bombers; Martin and Consolidated-Vultee for patrol bombers and seaplanes and Douglas for dive bombers. It was likely that any research aeroplane would be produced by one of these four firms, especially Grumman or Douglas, due to their history of high-speed aeroplane design. Like Bell, the Douglas firm was one of distinguished lineage, created by a true giant of the American aircraft industry, Donald W Douglas. Tall, poised and self-assured – the very picture of a successful corporate executive – Douglas had attended the US Naval Academy, but transferred to the Massachusetts Institute of Technology, where he studied under Jerome Hunsaker, receiving a BS in aeronautical engineering in 1914 – one of the very first awarded in the United States. In 1916, he joined the Martin company, hired by none other than Larry Bell. After leaving to serve as chief civilian engineer with the US Signal Corps during the First World War – the Signal Corps ran military aeronautics at that time – he formed the Douglas Aircraft Corporation in 1920. Since then the company had designed a number of notable aeroplanes, including the World Cruisers that had flown around the world in 1924, the DC-2 and DC-3 transports that had revolutionised air transport worldwide and the speedy and aptly named A-20 Havoc attack bomber. During the Second World War, the company turned out over 10,000 C-47 Skytrain transports, military versions of the DC-3, as well as over 6,000 Havocs, 5,500 SBD Navy dive bombers, as well as producing B-17s and B-24s designed by Boeing and Consolidated-Vultee. Some of the company's aircraft were virtually legendary even at the time: one such was the DC-3/C-47 series of transports which, by 1945, were serving in every war theatre as the logistics backbone of the Air Transport Command. Another was the SBD dive bomber. It was the SBD that had presented the United States with its victorious turning point in the Pacific war, the Battle of Midway. Single-handedly, SBDs from the carriers *Enterprise*, *Hornet*, and *Yorktown* sank the core of Japan's naval strength, the four carriers *Akagi, Kaga, Soryu* and *Hiryu,* as well as the cruiser *Mikuma*. The Japanese never recovered from the shock of the loss, particularly that of well-trained and combat-tested naval airmen. Since that epochal battle in June 1942, the SBDs had continued to serve as the key naval air weapon. This aircraft was so effective that naval dive bomber squadrons displayed a marked preference for it over its chosen successor, the Curtiss SB2C Helldiver. The secret of Douglas' success as an aircraft manufacturer lay in the high-quality engineers and designers that the firm employed. One such individual was Leonard Eugene Root, holder of MS degrees in mechanical and aeronautical engineering from the California Institute of Technology. After graduation from Caltech, L Eugene Root joined Douglas as assistant chief of the aerodynamics section at Douglas' Santa Monica plant, working under Dr W Bailey Oswald. In 1939, he became chief of the aerodynamics section at the Douglas El Segundo plant, working under the plant's chief engineer, Edward H Heinemann.

L Eugene Root Launches the D-558

For many years, Douglas had been a contractor of naval dive bombers, and by 1944 the company realised that the high speeds attained by new combat designs in dives, particularly turbojet aircraft, made it possible for aircraft to encounter severe compressibility effects, particularly general instability and control reversal. Like the AAF, Navy, NACA and Bell, the Douglas engineers felt it desirable to have an aircraft capable of exploring in level flight the high-speed range expected to be encountered in the future by new aircraft designs. Since 1941, when the Douglas company first studied supersonic flight, company engineers had evaluated high-speed aerodynamics, principally under the direction of Frank N Fleming of the Douglas Santa Monica Division.* In 1944, the company held several

---

* Eventually, the Douglas-AAF studies led to the issue of contract W33-038-ac-10413 on 30 June 1945 to cover development of a supersonic aeroplane under AAF Project MX-656. The plane was to be capable of Mach 2.0 at 30,000 feet, and able to take off and land under its own power from a conventional airfield. This resulted in the development of the turbojet X-3, by a design team under the direction of project engineer Frank N Fleming. An excellent design, the X-3 (aptly named the 'Flying Stiletto') unfortunately failed to achieve its desired goals because of inability of the engine manufacturer – Westinghouse – to deliver the planned J46 powerplants within required size and performance specifications. Lower-rated Westinghouse J34's had to be substituted in place of the J46's, resulting in subsonic-level flight performance. (Interview of Frank N Fleming by the author, 25 May 1972; also 'Development History of the Douglas Aircraft Corporation X3 Airplane', n.d., from Air Force Museum Al X-3/his file.)

conferences with representatives of the AAF Air Technical Service Command on the general aspects of supersonic flight. As part of his duties at Douglas El Segundo, L Eugene Root consulted with the NACA, AAF and Navy to obtain ideas on what areas of aeronautics to develop next. On one trip to Washington late in 1944, Root dropped in on the Bureau of Aeronautics to visit Captain Walter S Diehl, Ivan Driggs, Abraham Hyatt and Commander Bill Sweeney. Sweeney, a Caltech alumnus and old friend of Root's, headed the Bureau of Aeronautics' VF (Fighter) Desk. During the discussion, Sweeney reached into a drawer and pulled out a very preliminary research aeroplane specification that Driggs, Hyatt and Sweeney were studying. He asked Root if Douglas would be interested in working on it. 'I said, "You bet,"' Root later reminisced, 'grabbed it, and ran with it'. Back at El Segundo, Root broached the Navy study to chief engineer Edward H Heinemann, Leo J Devlin, A M O Smith and Robert C Donovan. Heinemann, a tall, lanky Swiss-American, already possessing an international reputation for an uncanny ability to design superb aircraft; given to common sense judgements (one of his favourite aphorisms was 'simplicate and add lightness'), was more than interested. After serving as a draftsman with Douglas and the International Aircraft Corporation in the 1920s, Heinemann went with Northrop Aircraft as a design engineer, then with Douglas at El Segundo in 1932. In 1936, he became chief engineer of the El Segundo plant and supervised the design of the A-20 and A-26 attack bombers and the Navy SBD Dauntless. Devlin held an ME in aero engineering from Stanford University and was a member of the NACA aerodynamics committee. Smith, while a graduate student at Caltech, belonged to the original Caltech rocket research group of Frank Malina, Weld Arnold, John Parsons, Edward Forman and Hsue-shen Tsien. Then he left for Douglas, but in 1942 joined Aerojet as chief engineer, staying until 1944 when he rejoined Douglas as assistant aerodynamicist at El Segundo under Root. Robert C Donovan, a former associate with Ben O Howard in the design of Howard's excellent little racing planes (the *Pete, Ike,* and *Mr Mulligan*), had served as project engineer on the Douglas A-26 Invader attack bomber development programme. In short, as with the Bell team, these were men of broad and varied experience, well-suited to the task the Navy was asking of them.[2]

Navy and NACA specifications for the aeroplane did not constitute design requirements as much as general guidelines for Douglas to follow in designing their aircraft, for, like Bell and the AAF, the Navy wished to allow Douglas a free hand in the design. They stipulated that the aeroplane should not be designed as a combat or service type but that it should be capable of taking off and landing under its own power and have such good low-speed handling characteristics that data gained from its flight programme could be directly applied to the design of combat aircraft. They specified that the aircraft should be manned and capable of carrying 500lbs of research instrumentation and that it should be able to attain the maximum velocity possible using available powerplants. Heinemann decided to develop a tentative design for the aircraft (which, he hoped, might serve as the basis for a follow-on production fighter), and submit it to the Navy and NACA for approval with Leo Devlin as his assistant. Root and Smith would cover aerodynamics with Donovan as project engineer. As Smith, in charge of establishing the general planform and appearance of the plane recalled, the team favoured a small aeroplane 'wrapped . . . around the largest engine we could find'. Smith selected three wing configurations: the basic wing section was a NACA 65-110 airfoil with a 10% thickness:chord ratio, but the two other sections chosen for the design were a NACA 16-008 section at the root tapering from 8% thickness:chord ratio to a 6% thickness:chord ratio 16-006 section at the wing tip and a 17% thickness:chord ratio NACA 2417 airfoil at the root tapering to a 13% 2413 airfoil at the wing tip. The design team decided to incorporate a General Electric TG-180 turbojet (later known as the J35) then under development for the AAF as the basic engine for the planes. Team members felt that the Navy should procure six aircraft, equipped with various wings and engine duct configurations. These six aircraft could acquire aerodynamic data in level flight to approximately Mach 0.89. Later, Douglas would install Westinghouse 24C turbojets and supplementary rocket propulsion units in two of the aircraft and fly them at level flight speeds approaching Mach 1. Due to the NACA's interest in flush and semiflush air intakes, the team configured the basic aeroplane with a semiflush air intake with an alternate arrangement where the aeroplane had a simple nose intake and side inlets. It also, of course, met Navy desires that the nose be free for possible armament installation in any subsequent derivative.[3]

Gathering Steam: From Proposal to Programme Approval

Douglas designated the proposed aeroplane the Model 558 High Speed Test Aeroplane and early in January 1945, Robert G Smith, a talented aeronautical artist as well as designer, laid out and drew up the preliminary design of the trim aircraft, which spanned 25 feet with a length of 35 feet. The fuselage was just large enough to house the TG-180 turbojet, which obtained its air supply from two semiflush intakes located on either side of the fuselage just aft of the cockpit. The pilot sat ahead of the engine with a small semibubble canopy protruding slightly above the fuselage for vision. The instrument payload was ahead of the cockpit section, and the nose tapered down to a bullet point. (It was easy to imagine a time when the instruments might come out and armament might go in). In contrast to the straight, mid-wing XS-1, the D-558's wing had pronounced dihedral and emerged low from the fuselage sides. The design featured a conventional tail assembly, with the horizontal stabiliser located roughly in line with the top of the fuselage. Douglas transmitted preliminary engineering proposals to the Bureau of Aeronautics and the NACA a month later, followed by Smith and Devlin journeying to Washington to attend several conferences chaired by Commander Conlon, where they presented the proposal. At the second conference, they discussed the availability of the AAF-developed TG-180 to the Navy and its development status. Four days later, at the third conference on 28 February 1945, Smith and Devlin met with Navy representatives and John Stack and Milton Davidson of the NACA. Still exploring what the nature of the programme should be, the two Douglas engineers stated that Douglas could submit the first aeroplanes with TG-180 engines, nose and side intakes, airbrakes and 10% thickness:chord ratio wings, followed by a new aeroplane for Mach 1 performance; Douglas company pilots would demonstrate both the Mach 0.89 and Mach 1 aircraft. The Navy and NACA responded that Douglas should demonstrate the aircraft to its maximum level flight speed and to a structural loading of 8g at a lower, unspecified Mach number. More importantly, concerned that the Douglas design was too much of a compromise between what a research aeroplane should be and what a production aircraft might require, during these February meetings both the NACA and Navy forced Douglas to redirect its design more towards a pure research type. In this instance, the strong partnership of Captain Diehl, Milton Davidson, and John Stack worked to the NACA's benefit.[4]

Despite its misgivings, NACA was firmly on the side of Douglas for getting a development contract. Early in March 1945, with Douglas in the midst of redesign, the NACA belatedly replied to the Navy's 19 December 1944 letter in which the Bureau of Aeronautics requested that NACA submit drawings and specifications for a transonic research aeroplane so that the Navy could expedite procurement of such an aircraft, recommending the Navy procure the Douglas D-558 proposal in both its turbojet and turbojet-rocket forms. The committee letter further stated:

> The Committee is certain that the procurement of these two models of high-speed research airplanes will permit making a large advance in aerodynamic knowledge in the transonic region of flight and every attempt should be made to make these airplanes available to the NACA for flight research as soon as possible.[5]

Slightly over a month later, on 13 April 1945, Douglas submitted a contract proposal to the Bureau of Aeronautics outlining a three-phase development programme for the now purely research aeroplane. In the first phase, Douglas would construct six TG-180-powered aircraft to obtain aerodynamic data to Mach 0.89. In the second phase, Douglas would modify two for Westinghouse 24C turbojet engines and rocket propulsion so that these two aeroplanes could investigate aerodynamic conditions around Mach 1. In the third phase, Douglas would prepare engineering proposals and construct a mockup of a combat-type aircraft using results gathered during the testing of the six original D-558 aircraft. As with Bell and the XS-1, Douglas would follow traditional design specifications and criteria only where practicable. The company estimated total cost of the programme at $6,888,444.80. The company did not set firm delivery dates for the aeroplanes because of the highly experimental nature of the project. Nevertheless, Douglas representatives believed that the first aircraft could be ready for flight one year after the date of contract with the other five aircraft available six months thereafter. In its conclusion to the proposal, Douglas urged the Bureau of Aeronautics to take prompt action on the proposal, stating, 'The urgent need for reliable data covering airplane speeds at high Mach numbers and the importance of this data in connection with current as well as future military aircraft cannot be overemphasized'. The company's preliminary design met with quick BuAer approval. On 26 April 1945, Captain Theodore C Lonnquest, USN, director of the Bureau's Engineering Division, recommended the procurement of the six research aircraft, stating that the Model 558 proposal represented a 'well conceived and practically executed design'. He did caution that the proposed rocket-propelled phase be delayed, pending further information on the structural details of the rocket installation. Likewise, he advised that the third phase, design and mockup of a proposed fighter developed from the D-558 programme, be set aside until the Navy and Douglas acquired flight test information from the phase one research aircraft. Lonnquest formulated a two-phase flight research programme in which both Douglas and the NACA would fly the aeroplanes. The Douglas company pilots would seek data directly applicable to the design of fighter and dive bomber aircraft. The NACA research pilots would concentrate on acquiring basic and fundamental aerodynamic data at high Mach numbers, such as information on air loads, aircraft stability and control, flutter, and power plant behaviour. Lonnquest advocated procurement of all six D-558 aeroplanes to perform the various research investigations with a minimum of delay. But he also stated gravely, 'Furthermore, when flying experimental airplanes into the high Mach number region, there is an almost certain probability that one or more airplanes will be lost as evidenced by the loss of three [Ryan] XFR-1 and several P-80 airplanes'. On 9 May 1945, the Bureau of Aeronautics and the Assistant Secretary of the Navy (Air) formally approved the Douglas proposal. The D-558 programme had passed its first critical hurdle.[6]

Detail Design: From Drawing Board to Mockup and Programme Branching

To ensure that the D-558 did not merely duplicate the AAF XS-1 programme, the Bureau maintained close contact with AAF personnel at Wright Field. For its part, the NACA had little doubt it could justify development of both the XS-1 and D-558. The Navy aeroplane represented a more conventional design than did the XS-1, and had significant differences in design detail, propulsion and planform. On 22 June 1945, the Navy Bureau of Aeronautics issued a letter of intent to Douglas for the development of the D-558 aircraft under Contract NOa(s) 6850, which provided for construction of six model D-558 Phase One aircraft of alternate configurations, including two alternate wing sections and two alternate nose configurations. Phase Two provided for rocket-boost modification, and Phase Three concerned the proposed design and mockup of a combat-type aircraft. Later that month, Douglas began tests of a ¼-scale model of the D-558 in a wind tunnel at the California Institute of Technology for low-speed stability and control data. In May 1945, Root and Smith had temporarily left El Segundo to go to Germany as part of the Naval Technical Mission, Europe, and evaluate German wartime aeronautical developments. Van Every, holder of an ME in aero engineering from Stanford and assistant chief of the aerodynamics section at El Segundo, took the place of both men on the D-558 design team. Strangely, Root and Smith's trip to Germany had an extremely important impact upon the D-558 programme and led directly to the D-558-2. As early as March 1945, the Douglas design team had fixed the basic design of the aeroplane. The plane would have an ultimate load factor of 18g, a design gross weight of 7,500lbs, an overall length of 35 feet and 10 inches, a wingspan of 25 feet, an aspect ratio of 4.2 and a wing area of 150 square feet. The pure-jet models would have a maximum airspeed of 625 mph at sea level and Mach 0.84 at 25,000 feet. The plane's fuel supply would be carried in an integral wing tank, the pilot's compartment would be pressurised to an equivalent altitude of 15,000 feet when the aeroplane was flying at 30,000 feet and the whole nose could be jettisoned from the remainder of the aircraft in case the pilot had to abandon the aeroplane in flight. The fuselage was designed so that alternate wing and tail surfaces could be installed and tested and the aircraft could be flown with a nose air intake or side air inlets. Finally, the aeroplane would have an adjustable horizontal stabiliser for increased control at high speeds. Douglas planned to construct the aircraft as follows:[7]

Aeroplane	*Engine*	*Nose Scoop*	*Side Scoop*	*Wing*	*Model*
1	TG-180	No	Yes	65-110	558JS1
2	TG-180	Yes	No	65-110	558JN1
3	TG-180	No	Yes	65-110	558JS1
4	TG-180	No	Yes	65-110	558JS1
5	TG-180	No	Yes	2417 (to 2413)	558JS3
6	TG-180	No	Yes	16-008 (to -006)	558JS2

Note: N = nose scoop.
S = side inlets.
J = turbojet (two would have been converted to rockets – R)
1 = NACA 65-110 airfoil section
2 = NACA 16-008 at root, tapering to a 16-006 section at tip
3 = NACA 2417 at root, tapering to a 2413 section at tip

With the basic design established, the Douglas team, minus Root and Smith, devoted their attention to the detail design of the aeroplane. Team members gave considerable thought to methods of pilot escape at high speed. They doubted that the pilot could survive by the traditional method of jettisoning the cockpit canopy and then falling clear of the aeroplane. At first, the team considered incorporating an ejection seat, but found that the force needed to throw the seat and pilot clear of the high vertical fin would be in excess of human physiological limits. (This was, of course, in the era before the rocket-powered 'zero altitude-zero airspeed' seat, when seats were rudimentary and still used cannon cartridges to expel the pilot from the plane.) Heinemann then thought about using a jettisonable nose capsule, the only possible alternative at the time. After separating from the aeroplane, the capsule would decelerate to a lower speed. He estimated that this would be no faster than 350 mph – enabling the pilot to make a normal bail-out from the nose section. At first, Douglas investigated slowing the capsule with speed brakes or a suitable drag parachute, but dropped both suggestions because the brakes added complexity, and no suitable high-speed parachute existed. Finally, the team chose a simple jettisonable nose that would slow due to aerodynamic drag. The pilot pulled a release handle located above the instrument panel, which released four bomb-rack-type hooks that disengaged, separating the nose section from the rest of the aeroplane. Incorporation of the jettisonable nose seriously complicated design of the aeroplane's control system, but design team members displayed considerable ingenuity in coping with this problem. The difficulty involved designing control lines and electrical lines that would separate cleanly when the nose left the aeroplane. Douglas engineers equipped the electrical lines with quick disconnect pullout plugs. The push-pull controls were more complex, since engineers had to retain the movement of the cable, yet equip it so it could separate in an emergency. The design team devised grooved drums on a common shaft that passed through the fuselage bulkhead into the nose section of the plane. One drum was in the nose section, and the other was in the rest of the plane behind the fuselage bulkhead. The drums translated the linear motion of the push-pull controls into rotary motion and on the other side of the fuselage bulkhead, the other drum translated the rotary motion back into linear motion. Finally, the drum shaft had a tongue and groove so that if the pilot jettisoned the nose, the shaft would separate as easily as a sword slipping from a scabbard.[8]

The design team constructed a mockup of the D-558 aeroplane, and held the first mockup inspection from 2–4 July 1945. Among the personnel who attended this mockup meeting was NACA's chief pilot, Melvin Gough, who questioned the strength of the canopy, suspecting it to be less than the 18g load limit for the aeroplane itself, and criticised the view from the cockpit as well. The only really important change as a result of this first mockup conference, however, was the dropping of side inlets from the programme. Although the agency had strongly favoured side inlets, its engineers now suggested the design incorporate only a nose inlet, since the aerodynamics of this configuration were better known and understood. A month later, from 14–17 August 1945, Heinemann held a second mockup conference in his offices at El Segundo. In contrast to the transonic capabilities of the aeroplane under consideration, the trip to Douglas by Navy and NACA representatives took many hours. Milton B Ames Jr who was on the technical staff of NACA Headquarters, left Washington with a group of Navy representatives aboard American Airlines' DC-3 'Mercury' flight. The plodding DC-3 droned along for several hundred miles, then would land and refuel before taking off again. 'We flew all night, and we were bushed the next day,' Ames recollected, 'but we went right out to the plant and checked over the mockup. The job I had there was to see that the cockpit and canopy changes were made in accordance with the NACA's requirements, to provide high visibility yet be safe enough to withstand the high wind loads. The visibility was okay. I remember sitting in it, and I had to be propped up to seat level to check visibility.' The most important outcome of the second mockup conference was the branching of the D-558 programme into both a straight-wing turbojet aircraft, the D-558-1 and a sweptwing turbojet- and rocket-propelled design, the D-558-2. The two aircraft retained some structural similarities, but differed markedly in appearance, capabilities, and design mission.[9]

The D-558-1: Designing a Flying Test-tube

Once the D-558 design team had decided to develop a minimum size aeroplane utilising the most powerful turbojet available, the next question became one of ensuring that the aeroplane and its components had high critical Mach numbers in order to delay high-speed drag divergence – i.e., compressibility effects. This required carefully designing the wing, the fuselage and especially the wing-fuselage fillet. With the assistance of the NACA and the Navy Bureau of Aeronautics, Douglas had selected the 65-110 section for the wing because of its proven high Mach number for drag divergence and also because the company's aerodynamicists already had experience using it on the A-26 attack bomber. With the airfoil shape selected, the design team turned to the planform of the wing itself. They chose a 'relatively low' aspect ratio of 4.17 to save weight and ensure satisfactory low-speed flying qualities, and avoided incorporating any wing twist – i.e., subtly varying the wing's angle of incidence over the span of the wing – to minimise potential aerodynamic and structural problems at high speeds. The chief problem in designing the fuselage was keeping the frontal area of the body at a minimum in order to retain a high fineness ratio, and choosing a body shape that would not produce accelerated airflow, hence an abrupt drag divergence with shockwave formation, in the vicinity of the wing. The design team selected a cylindrical body shape just large enough to enclose a TG-180 turbojet, thus keeping frontal area at a minimum. (They were helped by the TG-180's design itself: an axial-flow engine. The TG-180 had minimal frontal area compared to many early turbojets, which used the much 'fatter' Whittle-type centrifugal-flow compressor design.) At the same time, the straight cylindrical section above the wing minimized velocity increases. In designing the wing-fuselage fillet, Douglas engineers had to choose a configuration that would have a critical Mach number as high as the wing, but that would not degrade the plane's stalling characteristics at low speed. After exhaustive tests in NACA and Caltech wind tunnels, they finally evolved such a configuration, though at some penalty in maximum lift coefficient. Despite their differences, the D-558-1 shared much with its XS-1 rival. As with the XS-1, on NACA's recommendation, Douglas placed the horizontal stabiliser higher on the D-558-1's vertical fin than engineers had originally planned. They put it high on the vertical fin to remove it from the wing wake. Likewise, to retain control effectiveness at high speeds, the stabiliser airfoil section on the aeroplane was thinner than the wing section, with the wing a 10% thickness:chord ratio, and the horizontal tail an 8% section. Finally, the D-558-1's stabiliser was adjustable in flight, just like the XS-1's, for increased control effectiveness in the transonic region.[10]

As part of the development programme, Douglas and the NACA at Langley Field initiated extensive wind-tunnel tests of the proposed aircraft shape. Douglas constructed several ¼-scale models and tested them in the Southern California Cooperative Wind Tunnel at Caltech for low-speed stability and control characteristics, engine intake airflow behaviour, tip tank aerodynamic characteristics and characteristics of various proposed flap configurations. Douglas technicians also built a special 1⁄50-scale model of the D-558's jettisonable nose. Smith, recently returned from his trip to Germany, tested the configuration in the company's Airflow Generator to obtain separation force data and film records of flight paths. Technicians also constructed a ⅙-scale high-speed model of the D-558 nose for tests in the Cooperative Wind Tunnel. All in all, during the six months from May to October 1945, Douglas ran more than 223 hours of tests on the D-558-1 and its components in the Caltech Cooperative Tunnel. NACA, who were deeply committed to the D-558 programme, tested six 1⁄16 solid wood models of the D-558 configuration in the Langley 8-foot high speed tunnel using the new sting support system. The Langley Flight Research Division made P-51 wing-flow dive tests using D-558 models. Finally, in a curious example of 'jointness', NACA engineers used a spin model of the Bell XS-1 configuration to simulate the D-558 in tests in the Langley Spin Tunnel. All this activity made the D-558 one of the most tested aeroplane configurations at the time of its first flight. Satisfied with the progress of the D-558 design, the Navy Bureau of Aeronautics followed up their June 1945 letter of intent with a formal contract issued on 28 March 1946, calling for construction of six D-558s . With the detail design of the aeroplanes complete, Douglas now began construction of the first three aircraft, the three D-558-1s. Following the company postwar policy to incorporate the word 'Sky' in the names of new Douglas planes, the firm gave the D-558-1s the appellation 'Skystreak'.[11]

The Douglas design team had to grapple with a number of technical problems during the construction of the Skystreak.[12] Despite the team's avowal to adopt a conventional approach to avoid any difficulties that might arise from an unconventional or radical design, they often had to find unorthodox solutions. One such problem concerned landing-gear storage. With the fuselage crowded by the engine, flight-research instrumentation, controls and cockpit, the team found only sufficient remaining space to stow the retractable nose landing gear. The main landing gear would have to go into the wings, even though the wings were, for their day, very thin and not readily suited to gear storage. The design team had the Goodrich Corporation develop special 20 × 4.4-size wheels with eightply nylon tyres. Though much smaller and thinner than Douglas would have desired for a plane of the Skystreak's size and weight, these special wheels could fit within the 10% thick wing, avoiding the undesirable alternative of fuselage or wing redesign. To solve the problem of fuel storage, the design team decided to make the forward 50% of the wing fuel-tight so that the wing could act as an integral fuel tank. Technicians coated the inside of the wing with a synthetic rubber sealer that required five weeks for curing and drying. Thus sealed, the Skystreak's wing could carry 230 gallons of kerosene fuel. The plane also had provisions for additional fuel in jettisonable wing-tip tanks, though these were not used extensively during the subsequent flight test programme.

Of semi-monocoque construction, the Skystreak's fuselage consisted of aluminium alloy frames without stiffeners or stringers covered by a heavy magnesium alloy slab approximately $\frac{1}{10}$ inch thick. This method of fuselage construction increased usable internal space provided a smoother external surface due to the heavy gauge slab and countersunk rivets, and resulted in a weight saving of 60lbs. The wings and tail surfaces used conventional rib and stringer construction fabricated from high-strength 75S aluminium alloy. As the wing had insufficient space for conventional lead aileron balance weights, Douglas engineers selected heavier tungsten alloy counterweights that could fit in the small space available. To ensure that the external surface of the aircraft would be as smooth as possible, the company constructed exceptionally rigid assembly jigs. Technicians fitted the wing and fuselage skins to contour frames then assembled the internal structure directly to the skin, thereby reducing surface irregularities. All the Skystreak's flight controls operated without power boost, aerodynamic balance or use of control tabs. The setting of the horizontal stabiliser could be adjusted by electric actuators controlled by a switch on the pilot's control wheel. Like the Bell XS-1 design team, Douglas chose a control wheel rather than a control stick for the pilot, presumably – as with the XS-1 – to allow the pilot to use both hands effectively should the need arise. To minimise the danger of control surface flutter, the Skystreak incorporated dampers for all control surfaces.

Though the NACA had primary responsibility for Skystreak instrumentation, Douglas prepared their own instrument package for the first D-558-1 to be used in a contractor evaluation programme. Douglas cut 400 pressure orifices into the wing and tail surfaces to record pressure distribution patterns on a special pressure-measuring system of automatically recording manometers developed by the NACA. Strain gauges would measure air loads on the wing and tail, and Douglas had the William Miller Corporation of Pasadena develop a specially built 30-channel oscillograph to record control forces, flight operating conditions and data from the power plant installation. To ensure satisfactory operation of the instrumentation under flight conditions, Douglas subjected all the instrumentation to vibration tests and 12g accelerations in a centrifuge at the University of Southern California. The design team also developed a complex but highly efficient refrigeration system to provide cool air to the cockpit and instrumentation compartment. The efficiency of this equipment was such that it took air from the engine compressor at 450° Fahrenheit and cooled it to 60° F. before discharging it into the cockpit. Likewise, technicians painted the interior of the Skystreak's fuselage with heat-reflecting paint to prevent absorption of engine heat.

The Bureau of Aeronautics assigned the three D-558-1 aircraft serials 37970, 37971 and 37972 and by October 1946 all three aircraft were well underway, being 90%, 70% and 60% respectively complete. Douglas had, by this time, completed design development of the aeroplane. The design team made some final design changes to the plane, including installation of speed brakes. Early in the design, the team had made provisions for speed brakes to permit the pilot to decelerate to slower, safer airspeeds if he encountered severe compressibility effects. As a result of wind tunnel tests, Douglas subsequently dropped the brake idea, since the tests indicated that the efficiency of the brakes disappeared at high speeds. In further discussions between the El Segundo engineering department and company test pilots, however, the pilots expressed a strong desire for the brakes as an added safety feature and requested that the engineers design them to slow the aeroplane by 50 mph in 15 seconds at maximum speed. Due to the advanced development of the plane, the engineers had to add the brakes externally on either side of the fuselage just aft of the wing. The brakes could open 60° in three seconds at Mach 0.9 at 25,000 feet and had a 'blow-back' feature so that they would open partially if the air loads were too high for full opening. NACA verified brake operation during tests in the Langley 8-foot high speed tunnel. As a result of static load tests on the Skystreak, Douglas added local spanwise stiffeners to the stabiliser, a very difficult modification since technicians had completed installation of pressure tubing to the pressure orifices in the stabiliser. Finally, everything was complete. On 10 April 1947, the first D-558-1 Skystreak, BuNo. 37970, arrived at Muroc Dry Lake from El Segundo for its initial flight tests.

Company technicians had painted the plane a rich glossy red for high visibility. The Skystreak's coloration made it the most visually exciting of the research aeroplanes, and earned it the nickname (somewhat to Ed Heinemann's annoyance) the 'Crimson Test-tube'. Lacquered and polished to a high gloss, the sleek research plane possessed the illusion of flying while on the ground. Its long cylindrical fuselage joined a well-faired tail assembly. A long streamlined clear cockpit canopy surmounted the red fuselage. Aft of the wing, the crisp blue and white of the national insignia's star and bar overlapped the speed brakes and the nose bore the legend 'Douglas Skystreak'. The scarlet fuselage ended abruptly in a gleaming silver tail cone for the jet exhaust. Only the three spindly landing struts with their extremely small wheels broke the clean lines of the aeroplane. Tests in coming months would reveal whether the attractive aeroplane could withstand the rigours of transonic flight or whether it, like Geoffrey de Havilland's ill-fated Swallow, would end up as a mass of twisted wreckage.[13]

The D-558-2: Crafting a White Rocket

Like the far less successful Bell X-2, the story of the sweptwing D-558-2 Skyrocket has its origins in both American and German sweptwing research. The catalyst for the Douglas-Navy-NACA decision to construct a sweptwing research aeroplane came from the trip that Root and Smith took to Germany to investigate wartime German aeronautical research. In May 1945, Root and Smith left El Segundo for Europe; they were under contract from Douglas to the Bureau of Aeronautics to join the Naval Technical Mission in Europe. Root and Smith first joined the mission in Paris, then left for Germany under orders to investigate aerodynamic reports from the Göttingen laboratory, the *Aerodynamische Versuchsanstalt* (*AVA*). Using their command of scientific German, the two aerodynamicists painstakingly read through the countless documents in the laboratory files that fortunately were largely intact. The data they analysed concerned high-speed wing and airfoil theory, including sweptwing research on wings and complete models having various sweepback angles. Root and Smith immediately realised the importance of the data, and, recognising the necessity of getting it to high-speed aircraft designers as soon as possible, requested that all the Göttingen material be microfilmed. (The microfilming of the reports took three weeks by five shifts working night and day). Root wanted to get the material back to the Bureau of Aeronautics, the NACA and Douglas as soon as possible and after making a preliminary survey, 'NavTechMisEu' transmitted the microfilmed reports to the Bureau of Aeronautics. Dr Clark Millikan, who was, no less, acting director of the Guggenheim Aeronautical Laboratory at Caltech, served as courier, escorted by Abraham Hyatt. Root and Smith then tried to get the sweptwing data back to Douglas before the second D-558 mockup conference, for both believed it highly desirable to obtain flight-test data on the swept configuration. Smith subsequently arrived back at El Segundo in the early part of August while Root remained in Europe for the next several months. As noted previously, early in 1945 Robert Jones had informed his superiors of the sweptwing research he had made at Langley Laboratory. During the early summer, John Stack requested the Douglas company to incorporate 35° wing sweep on the D-558 configuration, but the Navy decided to proceed with the original straight-wing planform and approach wing sweep cautiously. Nevertheless, after several discussions between the Bureau of Aeronautics and the NACA, and after having had a chance to evaluate the microfilmed data transmitted to the company from captured German reports, Douglas agreed to explore a sweptwing research aeroplane as part of the D-558 programme.[14]

At the second mockup conference on the D-558 programme from 14–17 August, the Navy and NACA formally requested initial design studies of a sweptwing D-558. From the start, Douglas, the Navy and the NACA planned to incorporate rocket propulsion in the plane, since they felt there was little point in testing the sweptwings on a purely turbojet aeroplane. Existing turbojets lacked the thrust capability to propel the plane to sufficiently high Mach numbers where the sweptwings could be fully evaluated, but a rocket engine could thrust it beyond Mach 1. NACA engineers, whose reluctance towards rockets had been somewhat overcome now that the XS-1 was far along, stipulated that the plane should have sufficient rocket fuel for two minutes of operation at 4,000lb-thrust to allow sufficient time for the plane's internal recording instrumentation to stabilise and take accurate measurements. The NACA later modified this requirement to 100 seconds of engine operation at 6,000lb-thrust or slightly over two minutes of operation at 4,500lb-thrust.[15]

Douglas and NACA divided up responsibility for aircraft design between them. The NACA would have complete responsibility for high-speed wind-tunnel testing, while Douglas would have complete responsibility for low-speed tests. The Navy had directed Douglas to develop the aeroplane with the equivalent stalling and low-speed characteristics of a conventional straight-wing aeroplane which was a particularly demanding requirement. In response, Douglas technicians mounted a sweptwing on a D-558 wind-tunnel model and tested it for low-speed stability and control data, concluding that the company could build the sweptwing plane with

TABLE 1.

Fuselage diameter	40 in.
Wing span	29.1 ft.
Wing area	141.2 sq.ft.
Wing chord (average)	4.85 ft.
Aspect ratio	6

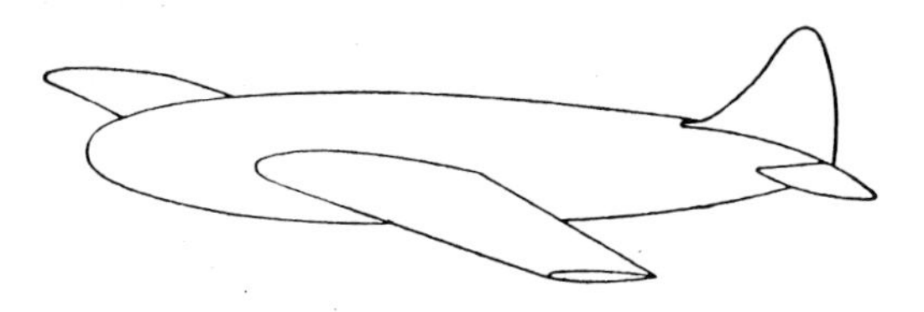

Plate 1: In 1934, inspired by Reginald Mitchell's Supermarine S.6B, John Stack of the National Advisory Committee for Aeronautics (NACA) Langley Memorial Aeronautical Laboratory designed a hypothetical propeller-driven high-speed aeroplane capable of speeds exceeding 520 mph. Though not pursued, this was the world's first proposal for a special research aeroplane to study the problems associated with compressible (i.e.: transonic) airflow. Drawing from 'Effects of Compressibility on High Speed Flight', *Journal of the Aeronautical Sciences*, January 1934.

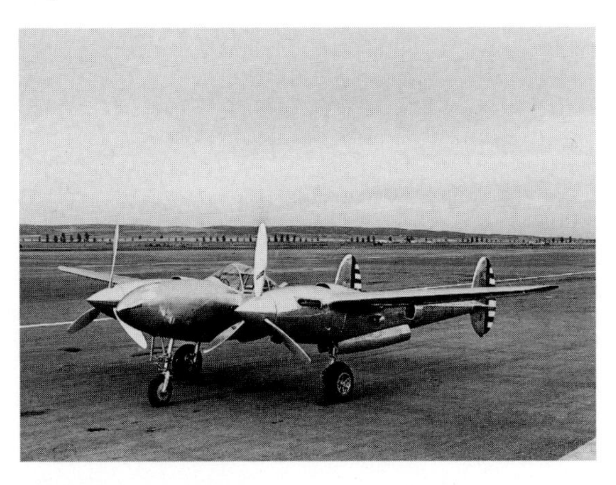

Plate 2: In the late 1930s, aircraft designers began manufacturing aeroplanes that could approach transonic velocities during high-speed dives, such as the 400+ mph experimental Lockheed XP-38 Lightning of 1939. Less than two years later, accidents to pre-production YP-38s dramatically revealed that flight at such speeds was anything but benign. *USAF*.

Plate 3: At the same time, the advent of the turbojet aeroplane – such as Great Britain's Gloster-Whittle E.28/39, which first flew in May 1941 – promised the ability to fly at speeds faster than 500 mph. *British Information Service*.

Plate 4: Inspired by two decades of aviation developments that had gone from the era of the 100 mph biplane to the 500 mph jet, newspaper and magazine writers seized on the advent of the jet engine and advances in rocketry to enthusiastically forecast a near-term future of high-speed flight, such as this fanciful 1942 German conception of a high-speed aeroplane. *Hallion collection.*

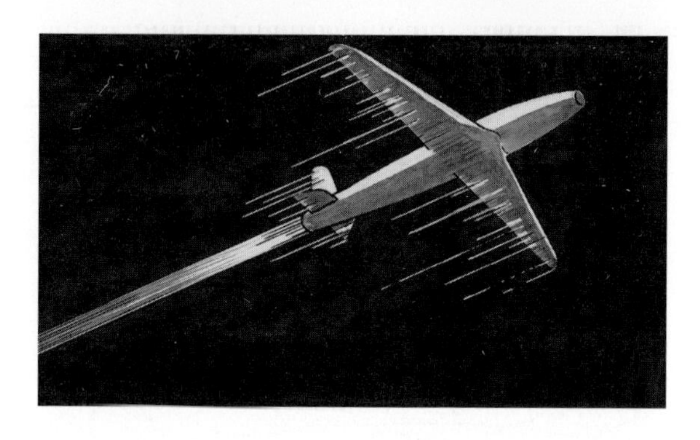

Plate 5: In reality, compressibility problems seriously limited the usefulness and endangered the safety of the new jet aircraft flying beyond 500 mph, particularly the world's first two operational jet fighters, the Gloster Meteor of the Royal Air Force (with a limiting Mach number of 0.74, in large measure due to poor engine nacelle design), . . . *British Aerospace.*

Plate 6: . . . and the *Luftwaffe*'s Messerschmitt Me 262 *Schwalb* [Swallow]. The Messerschmitt's cleaner configuration and slightly swept-back wing gave it a significantly higher limiting Mach number of 0.86 but, even so, deteriorating flying qualities above Mach 0.8 led one German test pilot to call Me 262 dives 'a game for nerves of steel'. *USAF.*

Plate 7: Compressibility researchers soon recognised that high-speed design information was not easy to obtain. Tests of new aircraft designs in conventional wind tunnels – such as this model of the XP-51 Mustang in North American Aviation's company wind tunnel – proved unable to furnish reliable information on transonic flow conditions. This situation, when coupled with mysterious losses – worldwide – of new fighter planes in high-speed dives, precipitated a genuine aeronautical crisis and a quest to 'break' the 'sound barrier'. *Hallion collection.*

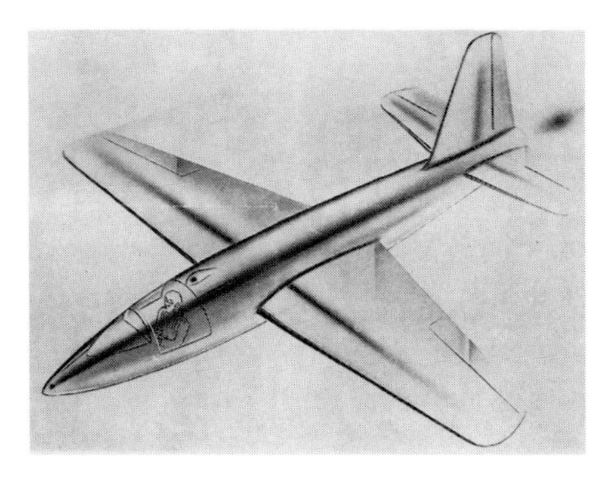

Plate 8: The quest began with theoretical studies for possible supersonic aeroplanes. Here is Ezra Kotcher's 1944 'Mach 0.999' Wright Field study that launched the Bell XS-1. *Hallion collection.*

Plate 9: Other studies favoured jet propulsion, such as this proposed 1945 design by NACA Langley engineers Macon Ellis and Clinton Brown for an alluring (though fearsome!) ramjet-powered supersonic research airplane promising Mach 1.4 flight speeds. Since a ramjet lacks a means of compression unless ram air is actually moving through the inlet duct, this design would have been towed aloft, and then accelerated by small rockets to speeds capable of sustaining ramjet ignition speeds. Note the razor-like wings and tail surfaces (with distinctive raked tips), and the simple body-tube layout. *NASA.*

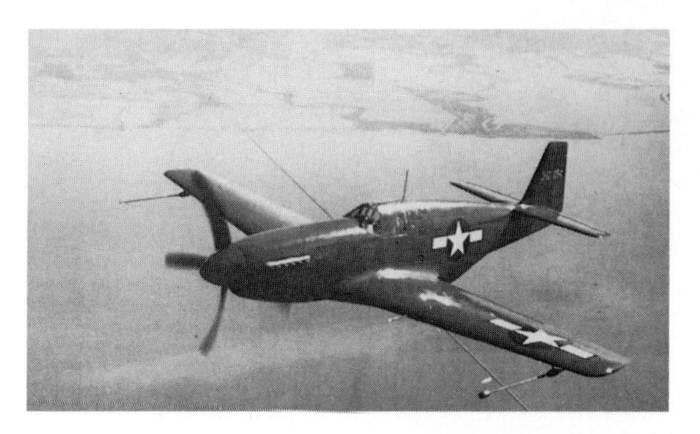

Plate 10: In the interim, both in America and abroad, test pilots flew highly instrumented fighters such as this North American P-51B Mustang to velocities above Mach 0.8 in risky tests to develop rudimentary transonic lift-and-drag data. *NASA*.

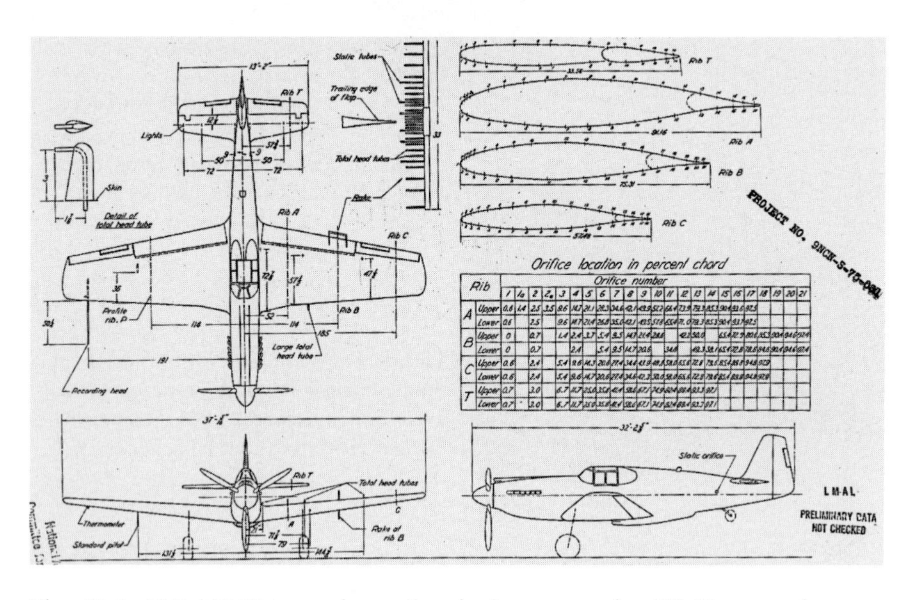

Plate 11: In 1943, NACA researchers at Langley instrumented an XP-51 to record pressure distribution at 136 separate locations on the wing and tail in an effort to understand the mechanics of transonic phenomena. *NASA*.

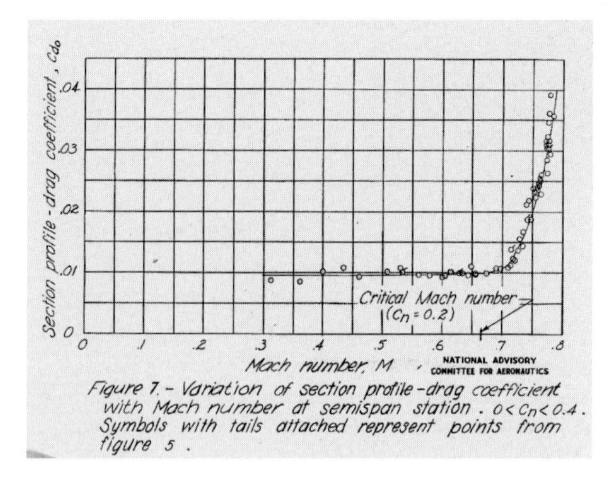

Plate 12: Subsequent tests generated disturbing data that raised far more questions than were answered. Here, for example, is a drag-plot of the diving XP-51 taken from NACA Technical Note 1190. Notice the abrupt overall drag rise as the plane goes from Mach 0.66 to approximately Mach 0.79, with its ominous implication that drag would increase to infinity the closer one approached the speed of sound. *NASA*.

Plate 13: Other stop-gap measures tried by governmental and industry researchers included dropping instrumented transonic and supersonic test shapes and models from high-flying airplanes. The amount of information that could be acquired and passed was, of course, limited. *Lockheed*

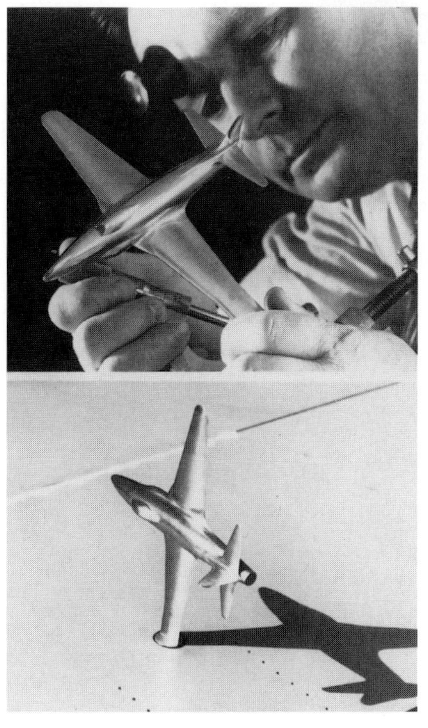

Plate 14: *Left*: As a later (and quite ingenious) measure, National Advisory Committee for Aeronautics craftsmen at the Langley Memorial Aeronautical Laboratory made small research models (top) and installed them on complex lift-and-drag-measuring balances buried in the converted right wing machine-gun bay of a P-51B Mustang (bottom). When the diving Mustang reached Mach 0.75, the accelerated airflow passing over its wing (and around the test model) would actually be already supersonic – in excess of Mach 1.0 – permitting useful (though limited) high-transonic testing. *NASA*.

Plate 15: *Right*: One of the most spectacular expedients were small, instrumented rocket-propelled models, such as this model of the Douglas D-558-2 being readied for launch at NACA's Pilotless Aircraft Research Division at Wallops Island, Virginia, in 1948. Here the benefits of free-flight data outweighed the disadvantages of brief flights (with durations measured in seconds), limited data and models lost to the deep sea. *NASA*.

Plate 16: Though wind tunnels were unable to furnish reliable transonic information, the need for some basic wind tunnel information on the potential behavior of experimental aeroplanes forced innovations in tunnel design to permit acquiring useful information. One of the most important was the 'sting' model support system (shown here with a model of the Douglas D-558-2) which eliminated the various struts and wires triggering airflow interference that contributed to unreliable data acquisition. Even so, such measures were not a satisfactory substitute for actual piloted research airplanes. *NASA*.

Plate 17: In the summer of 1946, convinced that supersonic flight was a long way off (if at all), Sir Ben Lockspeiser, Great Britain's Director General of Scientific Research at the Ministry of Aircraft Production, cancelled the Miles M.52 which had been under development since 1943. The M.52 constituted an ambitious and well-thought-out effort to develop a supersonic jet-propelled research aeroplane. Sir Roy Fedden, Great Britain's legendary aero engine designer, subsequently wrote 'No single act set back Britain's aircraft development quite so drastically, however, as the Government's decision in 1947 not to allow manned supersonic investigations'. *Hallion drawing*.

Plate 18: On 27 September 1946, Geoffrey de Havilland, one of Great Britain's most distinguished test pilots, perished over the Thames estuary when his experimental DH 108 Swallow pitched violently and broke up as it approached sonic speed. His death dramatically demonstrated that the risks associated with the 'sound barrier' were, tragically, far from theoretical. *British Aerospace.*

Plate 19: De Havilland's highly-polished Swallow, shown here several days before its fatal accident, was the second of three DH 108's the company built as technology demonstrators for a proposed commercial airliner. Aptly named, it had a futuristic sweptwing and tailless planform that belied serious stability and control deficiencies. *British Aerospace.*

Plate 20: Sobered by de Havilland's death, American researchers pressed on with their own supersonic research efforts. In 1942, under conditions of great secrecy and urgency, the Bell Aircraft Corporation had designed and flown the first American turbojet airplane, the XP-59A Airacomet, shown here at Muroc Dry Lake, California. With this record of accomplishment, the company was a natural choice to build America's first supersonic testbed, inspired by Ezra Kotcher's 'Mach 0.999' proposal of 1944 as well as indigenous company interest. *Bell Aircraft Corporation.*

Plate 21: Bell Aircraft Corporation's response to the Mach 1 challenge was to design a rocket-propelled, winged and instrumented bullet. Here a model of the XS-1 is readied for a wind tunnel test at the NACA Langley Memorial Aeronautical Laboratory. Ironically, it was the wind tunnel's inability to furnish reliable transonic information that forced development of research aeroplanes which, in turn, fed back useful data for future wind tunnel design. *NASA.*

Plate 22: The XS-1s were essentially flying fuel tanks in the shape of .50 calibre bullets, with a spherical oxygen tank ahead of the wing, a semi-spherical alcohol-water tank behind the wing and 12 nitrogen spheres to furnish pressure for the fuel-feed system and the cockpit. The instrument package nestled over the wing centre section and the rocket engine occupied the tail cone of the plane. *Bowers collection, USAF Museum.*

Plate 23: The first XS-1, completed in 1945, flew at Pinecastle Army Air Field, Florida, in early 1946 on glide tests to prove the plane's basic airworthiness. Here it is shown prior to being winched into its launch B-29's bomb bay. Note the steel pipe (below the word Bell on the side of the plane) temporarily added as a cautionary guide to prevent the XS-1 from drifting sideways and/or backwards during launch. *USAF AFFTC.*

Plate 24: Here the XS-1 is shown rolling out after its first landing, 25 January 1946. *NASA.*

Plate 25: With the first glide flight complete, technicians and observers gather around the saffron plane. After another nine flights, it returned to Bell in March for installation of its rocket engine and a thinner wing and horizontal tail. *NASA.*

Plate 26: To support the Pinecastle tests, the NACA sent an engineering and instrumentation test team from its Langley laboratory consisting of (*L – R*) Gerald Truszynski, John Householder, Walter Williams, Norm Hayes and Joel Baker. *NASA*.

Plate 27: Jack Woolams, Bell's project pilot for the XS-1 at the time of the Pinecastle tests, died tragically in a P-39 accident while practising for the 1946 Thompson Trophy air race. The company replaced him with Chalmers 'Slick' Goodlin. *Woolams collection, USAF AFFTC.*

Plate 28: From Florida, the XS-1 programme moved to Muroc Army Air Field, California on the edge of a large dry lake. Here is Muroc in October 1946 looking south-to-north. Note the old South Base complex, and visible at the top left of the lake is the North Base flight test facility used for the first American jet flights in 1942 and later for the Skystreak programme. Note the railroad tracks running completely across the lake, east–west, removed in the early 1950s, when the main base was relocated further north along the western lake shore. *USAF AFFTC.*

Plate 29: In early October 1946, the second XS-1 (the first completed with a rocket engine) arrived at Muroc Army Air Field for powered flight testing. *USAF AFFTC.*

Plate 30: Here the second XS-1 is shown at Muroc in front of its fuelling facilities. The loading pit where it was winched aboard the B-29 mothership is to the left. *USAF AFFTC.*

Plate 31: The simple 6,000lb-thrust four-chamber Reaction Motors Inc. 6000C4 rocket engine which, as the XLR-11 in the X-1 series and the XLR-8 in the Skyrockets, pushed both these designs past Mach 2. With each chamber fired individually, the 6000C4 used regenerative cooling, whereby its propellants were circulated in hollow walls around the combustion chamber and nozzle 'throat' to prevent the intensely hot rocket exhaust from melting the engine. *USAF AFFTC.*

Plate 32: Full-thrust engine ground test of the second XS-1. Note the frost forming on the outside of the supercold liquid oxygen tank, the open instrument bay, and the shock-diamonds visible in the exhaust stream, a tell-tale sign of its high velocity. *NASA.*

Plate 33: The Bell XS-1 test team at Muroc, 1946 (*L – R*): Dick Frost, project engineer and chase pilot; Harold 'Pappy' Dow, B-29 pilot; Mark Heaney, B-29 co-pilot; Al Bindig, instrument technician and B-29 aircrewman; Frank Nicolas, mechanic and B-29 aircrewman; Bill Miller, B-29 flight engineer; Chalmers 'Slick' Goodlin, XS-1 test pilot and Charles 'Mac' Hamilton, XS-1 crew chief. *USAF AFFTC.*

Plate 34: 9 December 1946: Locked in the belly of its launch B-29, the second XS-1 is carried aloft from Muroc Army Air Field for its first powered flight. *Ritland collection, USAF AFFTC.*

Plate 35: 9 December 1946: Slick Goodlin ignites the XLR-11 engine, beginning the first powered flight of the XS-1 series. *USAF AFFTC.*

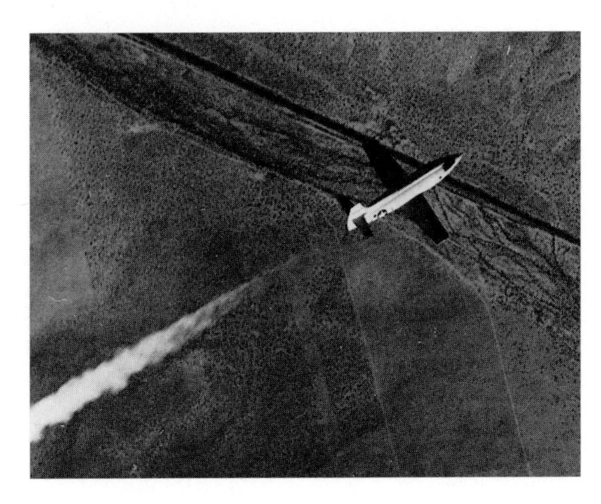

Plate 36: Bell completed its contractor flight test programme on the XS-1 in May 1947. Standing (*L – R*) Pappy Dow, Alvin 'Tex' Johnston and Slick Goodlin. The next month, the Air Force and the NACA assumed full control over the XS-1 flight test programme. *USAF AFFTC.*

Plate 37: That same month, allegedly, the initial attempt by the Soviet Union to exceed the speed of sound came to naught. This involved a Russian derivation of a Nazi design – the proposed DFS-346 rocket plane, fabricated in Russia as the *Samolyot* [Aircraft] 346-1. The 346-1 used XS-1-style air-launching from an interned Boeing B-29 that had suffered flak damage over Japan and force-landed in eastern Siberia. German test pilot Wolfgang Ziese reportedly had to abandon a high-speed flight attempt at Mach 0.93 when the 346-1 threatened to go violently out of control, probably because of a basically unsound aerodynamic design. With the 346-1 grounded for rework, the road was clear for the XS-1 to assault the 'sound barrier'. *Hallion drawing.*

Plate 38: In April, the Navy-sponsored Douglas D-558-1 Skystreak, more closely in line with NACA ideas than the XS-1, arrived at Muroc to begin its own test programme. Douglas test pilot Eugene F 'Gene' May began flying it from North Base. *US Navy.*

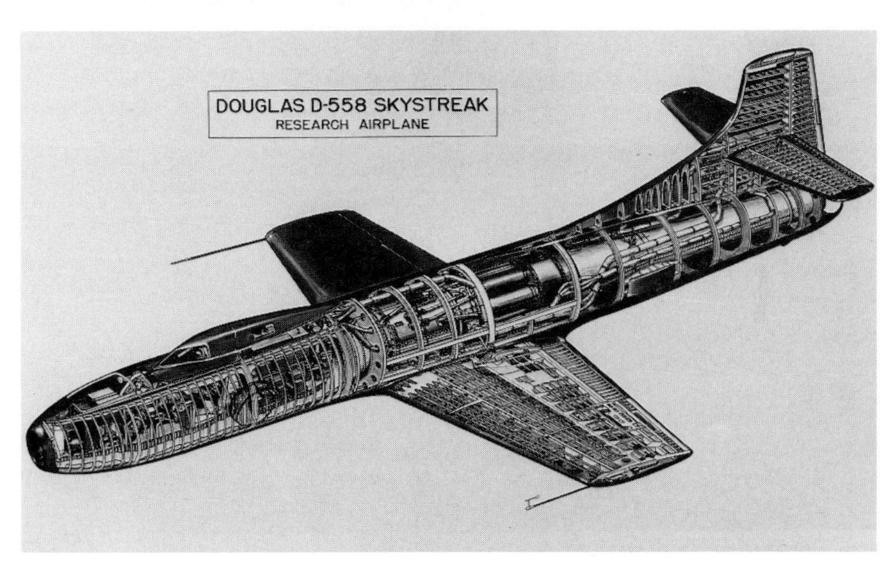

Plate 39: Overall, the D-558-1 represented a more conservative approach to building a transonic research aeroplane. It was, in essence, the smallest instrumented airframe that could be wrapped around an Allison J35 (TG-180) turbojet engine. The pilot sat in a jettisonable nose section. *US Navy.*

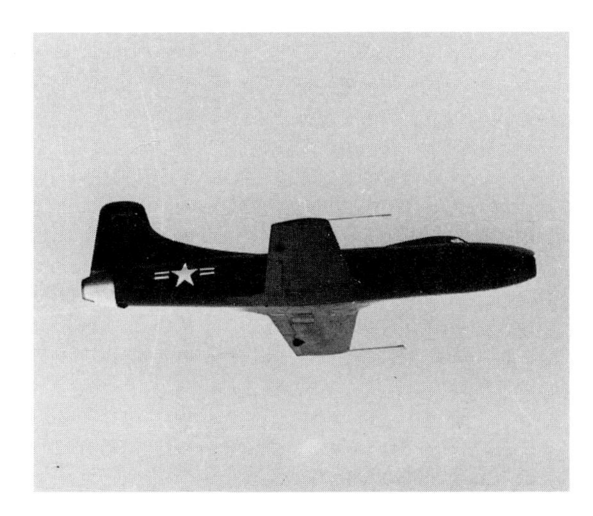

Plate 40: While the Air Force readied the first XS-1 for its initial Air Force powered flights, the Navy used the first two D-558-1 Skystreaks to set consecutive world's airspeed records. Here, flying the second Skystreak on 25 August 1947, Marine test pilot Major Marion Carl averages 650.796 mph during four low-altitude passes over the Muroc lakebed. (This plane subsequently claimed the life of Tick Lilly on 3 May 1948). *US Navy.*

Plate 41: Muroc's attention turned back to the XS-1 four days later when Air Materiel Command project pilot Captain Charles E 'Chuck' Yeager completed the plane's first Air Force powered flight. Here he poses by the first XS-1, named *Glamorous Glennis*, after his wife. *USAF AFFTC.*

Plate 42: Chuck Yeager standing beside his close friend and chief technical advisor, Captain Jack Ridley and *(L – R)* Merle Woods, Jack Russell and Garth Dill, the maintenance crew that kept the XS-1 in the air. *USAF AFFTC.*

Plate 43: Two of the key research administrators for the XS-1 effort were (left) Air Force Brigadier General Albert Boyd at Wright Field's Air Materiel Command, and (right) Walter C Williams, chief of the NACA's Muroc Flight Test Unit. Williams, in particular, worked with Ridley on planning XS-1 missions and supervised all data reduction from the XS-1s' comprehensive instrumentation, and from ground radar tracking. *NASA.*

Plate 44: 14 October 1947: Chuck Yeager reaches Mach 1.06, 700 mph at 43,000 feet, at approximately 10.28 am., the first piloted supersonic flight. This is the classic XS-1 photo, taken by chase pilot Bob Hoover from a photo-reconnaissance Lockheed FP-80 Shooting Star. (Discolouration on left outer wing panel of XS-1 is smudging from the B-29's left inboard engine exhaust). *USAF AFFTC*.

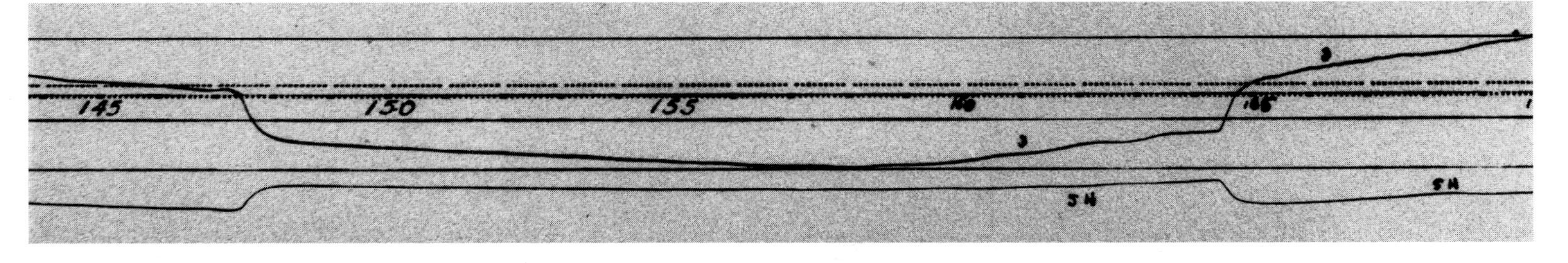

Plate 45: Inside the XS-1 as it goes supersonic, instrumentation records on oscillograph film reveal the tell-tale characteristic 'Mach jump' as the bow shockwave passed across the static pressure ports located on the side of the XS-1's nose, irrefutable proof that the plane had indeed, exceeded Mach 1. *NASA*.

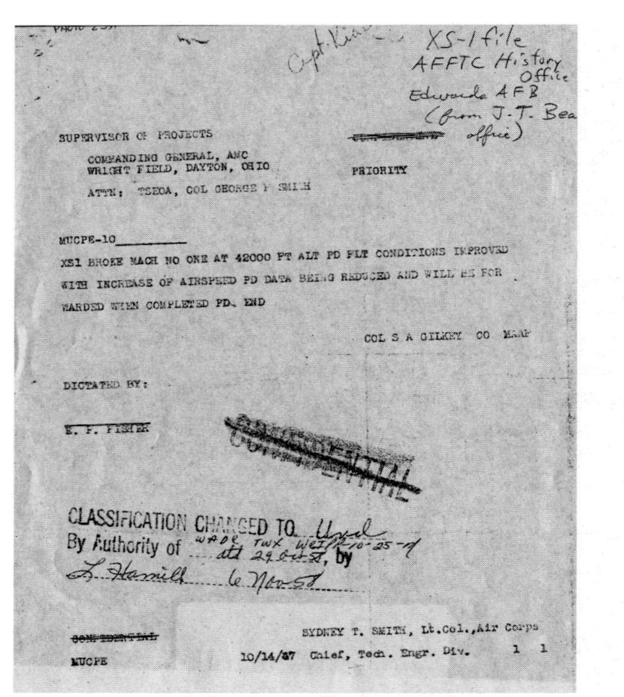

XS-1 file
AFFTC History Office
Edwards AFB
(from J.T. Bea office)

SUPERVISOR OF PROJECTS

COMMANDING GENERAL, AMC
WRIGHT FIELD, DAYTON, OHIO

PRIORITY

ATTN: TSEOA, COL GEORGE F SMITH

MUCPE-10

XS1 BROKE MACH NO ONE AT 42000 FT ALT PD FLT CONDITIONS IMPROVED WITH INCREASE OF AIRSPEED PD DATA BEING REDUCED AND WILL BE FORWARDED WHEN COMPLETED PD. END

COL S A GILKEY CO MAAF

DICTATED BY:

E. F. FISHER

CLASSIFICATION CHANGED TO Uncl
By Authority of WADC TWX ... dtd 29 Oct 58, by
L. Hamill 6 Nov 58

SYDNEY T. SMITH, Lt.Col.,Air Corps
Chief, Tech. Engr. Div.

MUCPE 10/14/47 1 1

Plate 46: Immediately after the flight, Colonel Signa A Gilkey, the Muroc base commander (and a test pilot who, six years previously, had narrowly escaped from a plunging YP-38 caught by compressibility) sent a priority message classified **Confidential** to Wright Field announcing the good news. *USAF AFFTC.*

Plate 47: Chuck Yeager's pilot report on the first supersonic flight; the Air Force immediately classified this one **Secret**, and it remained so for the next 11 years. *USAF AFFTC.*

Date: 14 October 1947

Pilot: Capt. Charles E. Yeager

Time: 14 Minutes

9th Powered Flight

1. After normal pilot entry and the subsequent climb, the XS-1 was dropped from the B-29 at 20,000' and at 250 MPH IAS. This was slower than desired.

2. Immediately after drop, all four cylinders were turned on in rapid sequence, their operation stabilizing at the chamber and line pressures reported in the last flight. The ensuing climb was made at .85-.88 Machi, and, as usual, it was necessary to change the stabilizer setting to 2 degrees nose down from its pre-drop setting of 1 degree nose down. Two cylinders were turned off between 35,000' and 40,000', but speed had increased to .92 Machi as the airplane was leveled off at 42,000'. Incidentally, during the slight push-over at this altitude, the lox line pressure dropped perhaps 40 psi and the resultant rich mixture caused the chamber pressures to decrease slightly. The effect was only momentary, occurring at .6 G's, and all pressures returned to normal at 1 G.

3. In anticipation of the decrease in elevator effectiveness at speeds above .93 Machi, longitudinal control by means of the stabilizer was tried during the climb at .83, .88, and .92 Machi. The stabilizer was moved in increments of 1/4 - 1/3 degree and proved to be very effective; also, no change in effectiveness was noticed at the different speeds.

4. At 42,000' in approximately level flight, a third cylinder was turned on. Acceleration was rapid and speed increased to .98 Machi. The needle of the machmeter fluctuated at this reading momentarily, then passed off the scale. Assuming that the off-scale reading remained linear, it is estimated that 1.05 Machi was attained at this time. Approximately 30% of fuel and lox remained when this speed was reached and the motor was turned off.

5. While the usual light buffet and instability characteristics were encountered in the .88-90 Machi range and elevator effectiveness was very greatly decreased at .94 Machi, stability about all three axes was good as speed increased and elevator effectiveness was regained above .97 Machi. As speed decreased after turning off the motor, the various phenomena occurred in reverse sequence at the usual speeds, and in addition, a slight longitudinal porpoising was noticed from 198-96 Machi which controllable by the elevators alone. Incidentally, the stabilizer setting was not changed from its 2 degrees nose down position after trial at .92 Machi.

6. After jettisoning the remaining fuel and lox a 1 G stall was performed at 45,000'. The flight was concluded by the subsequent glide and a normal landing on the lake bed.

CLASSIFICATION CHANGED TO Uncl
By Authority of WADC TWX ... dtd 29 Oct 58, by
L. Hamill 7 Nov 58

Charles E. Yeager
CHARLES E. YEAGER
Capt., Air Corps

Plate 48: Six days later, while meeting at Wright Field, key members of the XS-1 test project gathered for a private dinner in honour of the flight – and Yeager's second, secret, Distinguished Flying Cross, awarded earlier that day – at the Kitty Hawk Room in Dayton's Biltmore hotel. (*L – R*): Major Roberto 'Bob' Cardenas, Roy Sandstrom, Captain James Fitzgerald Jr, Lieutenant Bob Hoover, Dick Frost, Captain Chuck Yeager, Larry Bell and Captain Jack Ridley. *USAF AFFTC.*

Plate 49: As a co-recipient of the Collier Trophy for 1947, Larry Bell assembled key staff who had made the XS-1 a success with the trophy they had so rightly won. Standing (*L – R*): Roy Sandstrom, Jack Strickler, Bill Smith, Robert Woods, Paul Emmons and Dick Frost. In front (*L – R*) A J Marchese, Stan Smith and Ben Hamlin. *USAF AFFTC.*

EXTRA

RACE RESULTS — Los Angeles Times — Pictorial

MONDAY MORNING, DECEMBER 22, 1947 — DAILY, FIVE CENTS

U.S. MYSTERY PLANE TOPS SPEED OF SOUND

Suicide Pact Fatal to Man; Wife May Die

Test Pilots Pierce Dread Barrier at 70,000 Feet Altitude Above Desert

Three Americans have pierced the wall of sound!

The dread barrier to supersonic flight was first conquered at Muroc Air Base early last month when Capt. Charles Yeager, Air Force pilot, hurtled the XS-1 rocket plane through the wall at approximately 70,000 feet, The Times learned from reliable sources.

Man Struck by Auto Dies

Eight Killed by Gunfire and Knife in Palestine

Infant Abandoned on Porch Will Be 'Miss Christmas'

U.S. Strives to Prevent Western Union Strike

New White House Dog Received by President

Cold Wave Hits Italy

Announcement

Features Index

Plate 50: Two months after the epochal flight, word leaked out of Yeager's accomplishment. Here is the front page of the *Los Angeles Times* for 22 December 1947 with aviation reporter Marvin Miles' account of the flight. With the flight shrouded in secrecy, his article understandably differed in a number of particulars from what actually occurred. Public interest is evident in the size of type used for the banner headline: probably the largest typeface used by the *Los Angeles Times* since the surrender of Japan in 1945. *USAF AFFTC.*

Plate 51: Using the second thick-wing XS-1, the NACA flew a detailed series of flight investigations lasting into 1951 that conclusively demonstrated that wing thickness profoundly influenced aircraft performance. After being grounded for structural problems, this aircraft was rebuilt as the ultra-thin-wing Mach 2+ X-1E. Here it is shown in mid-1949. *NASA.*

Plate 52: Rocket research flights demanded thorough planning. Here NACA flight researchers map the 45th NACA X-1 research flight, a wing loads and aileron effectiveness sortie, subsequently flown by test pilot Scott Crossfield on 15 May 1951. (*L – R*) Joe Vensel, Crossfield (holding the mission test card for his kneeboard), and fellow test pilots Joe Walker and Walter Jones. *NASA*.

Plate 53: After being grounded for structural problems, NACA rebuilt the agency's X-1 as the ultra-thin-wing Mach 2+ X-1E. Here it is shown undergoing modification in the early 1950s for a new low-pressure turbopump feed system, an ultra-thin-wing and revised canopy design. *NASA*.

Plate 54: Flying the NACA XS-1, Herbert H Hoover (right) became the second pilot – and the first civilian – to exceed the speed of sound on 10 March 1948. He is standing next to Robert A Champine, who joined the NACA Muroc team after the death of Howard 'Tick' Lilly in the second D-558-1. Hoover himself subsequently perished in the crash of an experimental NACA North American B-45 Tornado bomber at Langley in 1952. *NASA*.

Plate 55: NACA began its Skystreak flight research programme in November 1947, with the second Douglas D-558-1. On 3 May 1948, this plane suffered catastrophic engine failure and crashed shortly after takeoff, killing Lilly, who was at too low an altitude to activate the plane's jettisonable nose and bail out. *NASA*.

Plate 56: Howard Clifton 'Tick' Lilly, like Chuck Yeager and 'Pete' Everest, a West Virginia native, was the first NACA pilot to die in the agency's long history of flight testing. Before his death, he had exceeded Mach 1 in the NACA XS-1, the second civilian (and third pilot) to do so. *NASA*.

Plate 57: Complementing the NACA XS-1 was the third Douglas D-558-1 Skystreak which NACA flew extensively until mid-1953. Here it is shown taking off from the dry lake for another research flight. *NASA*.

Plate 58: By mid-1948, the XS-1 was not the only aeroplane capable of breaking Mach 1. The elegant sweptwing North American XP-86 Sabre – shown over Muroc Dry Lake on 3 December 1947 piloted by Major Ken Chilstrom – became the second aeroplane type to exceed the speed of sound when, on 26 April 1948, North American test pilot George Welch dived it through the speed of sound. Less than a month later on 21 May, British test pilot Roland Beamont became the first British supersonic pilot when he dived the XP-86 past Mach 1 while visiting Muroc. *USAF AFFTC*.

Plate 59: In contrast to the smooth-flying XS-1 and XP-86, on 6 September 1948 test pilot John Derry, flying the third D H 108 Swallow – a more powerful and streamlined development of the design which had killed Geoffrey de Havilland Jr – reached Mach 1.04 in a courageous and wild out-of-control dive during which the vicious little aeroplane tried to tuck into a violent outside loop, the first supersonic flight in Great Britain. *British Aerospace.*

satisfactory low-speed handling characteristics. This was of paramount importance, for the D-558-2 design team recognised that the sweptwing could pose a danger to the pilot. While it offered performance increases at high speeds, it had undesirable low-speed characteristics, particularly at the stall. The Navy recognised the need for data on low-speed sweptwing behaviour by procuring, early in 1946, two highly modified Bell P-63 Kingcobras that featured sweptwings. Bell, Navy and NACA pilots flew the two planes, designated L-39s, in extensive low-speed investigations that ultimately benefited not only the D-558-2 programme, but the P-86 fighter programme and the future X-2 programme as well. In late October 1945, Douglas began initial low-speed investigations of the proposed sweptwing version of the D-558 using the Co-operative Wind Tunnel at Caltech. At the same time, the NACA at Langley Laboratory began testing models of swept D-558 configurations with both swept-aft wings and swept-forward wings. The Douglas design team, after several arrangement studies, found it impossible to fit a turbojet engine and fuel as well as a rocket engine and fuel into a D-558-1 fuselage. Clearly, the D-558-2 required an entirely new design. In the original D-558 proposal Douglas made to the Navy, the company included provisions for modifying two of the D-558 aircraft for rocket propulsion by installing rocket boosters and replacing the TG-180 turbojet with a smaller Westinghouse 24C turbojet. The design team now planned the new sweptwing aircraft around a Westinghouse 24C engine for takeoff and landing and a Reaction Motors, Inc. A 6000C4 engine for high-speed research, the same four-cylinder liquid oxygen-and-alcohol-fuelled rocket engine used in the XS-1. Though the Navy designated the RMI engine the LR-8, there was no significant difference between it and the AAF LR-11, aside from Navy policy to give engines even numbers and AAF policy to give them odd ones.[16]

The D-558-2 posed considerable challenges, for while the team already had the experience gained in designing the D-558-1, the earlier programme acted in some ways almost to constrain the freedom of the design team in approaching its sweptwing derivative. (For example, reflecting the Navy's requirement, the team wanted to keep its low-speed performance characteristics as near as possible to those of the straight-wing D-558-1). Responsibility for the aerodynamics of the new aircraft lay in the hands of Kermit E Van Every, for both Smith and Root had moved on to other activities. After returning to El Segundo, Smith became more involved with other company projects, though he did conduct performance investigations on inlet design and jettisonable nose design for the D-558-2. Likewise, Root also embarked upon other company ventures. Finally, in mid-1946, Douglas vice-president for Engineering, Arthur

Raymond, asked Root to join Project RAND, thus removing him from the D-558-2 project. Both men could leave the D-558 programme secure in the knowledge that their activities had contributed much to making the programme a reality. In designing the new sweptwing aeroplane, Van Every and his assistants had to arrive at a satisfactory aerodynamic configuration for both low and high-speed flight, which was a serious challenge in the 1940s. Douglas retained a circular cross-section fuselage, but employed side inlets for the Westinghouse 24C turbojet rather than a nose intake, thus giving the plane an ogival shape resembling that of the rival XS-1. Van Every decided to keep the fuselage of straight cylindrical section in the vicinity of the wing-fuselage junction, in order to minimise velocity increases in the airflow at that point which could act as a source of transonic drag rise. Further, because of a lack of specific information, Douglas did not design a wing-fuselage fillet for the D-558-2. Van Every had substituted for the wing itself a NACA 63 series airfoil section in place of the NACA 65 series that had been used on the D-558, because the 63 series had notably superior low-speed characteristics that would give the plane good stalling characteristics, while minimising degradation of its high-speed performance at the same time. Further, he reverse-tapered the section from a 10% thickness:chord ratio at the wing root to a 12% thickness:chord ratio at the tip, in order to give the plane the best possible flaps-up stall behaviour by delaying the onset of tip stall. He increased the wing area from 150 square feet – as on the D-558-1 – to 175 square feet on the D-558-2, thus ensuring that the sweptwing plane would have the same stall speed as its straight-wing predecessor. Finally, to improve behaviour at the stall, the design team incorporated wing fences and automatic leading edge slats, as well as simple flaps. As a compromise to achieve both acceptable high-and-low-speed performance, the team selected a wing sweep angle of 35° at the 30% wing chord line. The team retained the same wing span as the D-558-1 (25 feet) which caused a reduction in the aspect ratio from 4.17, as on the D-558-1, to 3.57, more desirable for a supersonic aeroplane. Both the XS-1 and D-558-1 teams incorporated thinner airfoil sections on the horizontal stabilisers in order to give the stabiliser a higher critical Mach number than the wing, but the D-558-2 team raised the critical Mach number of the stabiliser even further by giving it a greater sweepback angle of 40°. As on the other planes, the D-558-2 design team placed the horizontal stabiliser high on the vertical fin to keep it out of the wing wake and also made the stabiliser adjustable in flight. Both the D-558-1 and -2 had escape capsules for the pilot. Unlike the earlier aeroplane, however, the D-558-2 Skyrocket had a lower load factor. Instead of an ultimate 18g load factor as on the

D-558-1, the D-558-2 had an ultimate load factor of 12g at 11,250lbs and a limit load of 7.33g at 13,500lbs. Like the earlier D-558-1, the D-558-2 team expected it to attain Mach 1 at sea level, if required.[17]

On 29 January 1946, the Bureau of Aeronautics granted Douglas preliminary authorisation to develop a sweptwing version of the D-558, and by March, the design team had chosen the basic planform of the aeroplane, which did not differ appreciably from the final configuration actually built.[18] On 18 and 19 March 1946, Douglas held a mockup inspection and visiting NACA and Navy representatives approved the design, freeing the company to undertake detail design. Due to the sweptwing design and structural difficulties in assembling it, the design team did not attempt to locate either the landing gear or the turbojet fuel supply in the wing. This necessitated locating the gear and fuel in the fuselage and required a considerable increase in fuselage diameter over that of the earlier D-558-1. The team located the Westinghouse 24C turbojet in the fuselage below the wing with two tanks immediately above the wing centre section containing 250 gallons total of gasoline – not the more usual kerosene – for the turbojet. The jet exhaust passed out and slightly downwards – about 8° – from a ventral tail pipe. Douglas located the rocket engine's 195 gallon alcohol tank above the tail pipe in the aft fuselage. The Douglas team placed the 180 gallon spherical liquid oxygen tank in front of the wing above the turbojet intake ducts. Ahead of the liquid oxygen tank, Douglas positioned an 11-gallon tank to supply 90%-strength hydrogen peroxide to a turbine pump for, unlike the XS-1's, the Skyrocket could not make do with a pressurised fuel system and had to use a turbine pump to feed the RMI rocket engine. The turbine pump nestled between the alcohol tank and the aft gasoline tank for the turbojet and obtained its power from the decomposition of the hydrogen peroxide over a catalyst. Two separate centrifugal pumps connected to the turbine shaft fed the alcohol and liquid oxygen to the rocket engine. To ensure that the fuel tanks would have proper inlet pressures for the pumps prior to the operation of the turbine pump, the design team incorporated three spherical and four cylindrical 2,000 psi helium tanks to pressurise the alcohol tank to 30 psi, the liquid oxygen tank to 50 psi and the peroxide tank to 400 psi. (Ever conscious about weight, the design team chose helium over nitrogen because calculations showed that the helium would result in a weight saving of 30lbs.) The main consideration dictating the location of the propellant tanks was avoiding any shift in the plane's centre of gravity when the rocket engine ate its fuel.

Ahead of the liquid oxygen and hydrogen peroxide tank was the D-558-2's instrumentation compartment. The D-558-2 Skyrocket contained the same basic instrumentation package as the D-558-1 Skystreak, built around the 30-channel Miller oscillograph that Douglas had specially built for the D-558 programme. Its strain gauge installation was, however, far more extensive than that of the earlier D558-1 Skystreak. During the design of the Skystreak, Douglas engineers had succeeded in isolating the fuselage from the wing by using special links equipped with strain gauges. Due to structural differences in the Skyrocket's wing attachment, the design team could not do this, necessitating installation of a large number of strain gauges in the wings and tail surfaces to record wing loads. Douglas distributed 226 channels of four strain gauges each – a total of 904 strain gauges – throughout the wing for acquisition of stress information. More than four miles of strain-gauge wiring and over three miles of pressure tubing ran from wing and surface stations to the instrumentation compartment. Like the earlier Skystreak, the Skyrocket also had 400 pressure orifices cut into the wing and tail connected to the Miller oscillograph for pressure distribution surveys.

By October 1946, at which time the three D-558-1 Skystreaks were nearing completion, Douglas had released over 95% of the D-558-2's engineering drawings to the shops, completed about 80% of the tooling and 70% of the fabrication. The company had begun assembly of the first two wings for the aircraft. The structure of the Skyrocket followed closely that of the Skystreak. Of semi-monocoque design, the fuselage was of heavy-gauge magnesium sheet over heavy aluminium frame construction with no stringers. The wings and tail surfaces were of more conventional high-strength 75S aluminium alloy rib and stringer construction. As on the Skystreak, the Skyrocket employed tungsten alloy counterweights to balance the ailerons, and the flight controls lacked aerodynamic balance, power boost and tabs. As in the Skystreak, the Skyrocket also featured a control wheel for the pilot. With the D-558-2 well into the hardware phase, Douglas and the Navy agreed it was time to formalise the deletion of the last three Skystreak aeroplanes, together with the alternate wings for the aeroplanes and the provision for rocket propulsion in the D-558-1 which Douglas had set forth in their April 1945 engineering proposal to the Navy. On 27 January 1947, the Navy revised accordingly the Douglas contract deleting three Skystreaks and replacing them with three Skyrockets, issuing contract amendment 5A – the first formal authorisation to proceed with the Phase Two Skyrocket programme on 1 March – slightly over a month later. While Douglas constructed the three D-558-2 aeroplanes, NACA continued to furnish needed data on aeroplane performance derived from tests in the Langley 8-foot high-speed tunnel, the Langley spin tunnel and from firings of telemetry-equipped rocket-propelled models from Wallops Island. All of these tests generally confirmed the design choices that Heinemann and his team had made for the aircraft. In early November 1947, less than a month after Chuck Yeager's historic Mach 1 flight, Douglas completed the first D-558-2 Skyrocket. As with the trio of earlier Skystreaks, the three Skyrockets received a batch of Navy BuAer serials: 37973, 37974 and 37975. Like the first XS-1, the first D-558-2 did not have its RMI engine when completed. Though Douglas had frozen the design of the rocket installation in July, the company did not consider the installation flight-worthy, pending a full ground check of the entire rocket system. Douglas, the Navy and the NACA did not expect the RMI engines to be ready before August 1948 and research managers decided to conduct initial flight trials on the first two aircraft with just their turbojet engines. Douglas would retain the third D-558-2 until it could complete the aircraft with its LR-8 rocket engine. The NACA was willing to accept the second D-558-2 without its rocket engine to begin stability and control research as quickly as possible. At a later date, when the LR-8 became available, the NACA could return the plane to Douglas for rocket engine installation.[19]

The Skyrocket impressed most observers by its graceful and futuristic appearance for, from nose to tail, it consisted of a series of unbroken curves flowing smoothly together. Though later modified to a hooded, stepped design, the original cockpit of the D-558-2, like that of the XS-1, did not disrupt the plane's evocative lines, being flush with the fuselage's ogival shape. The nose itself ended in a long, pointed instrumented pitot boom vaguely reminiscent of a swordfish. As the plane did not yet have its rocket engine, the Skyrocket had a pointed tail cone fairing that added to its exotic looks and, unlike the crimson Skystreak or the saffron XS-1, the D-558-2 featured a glossy white finish overall. (Douglas had not chosen the colour for its aesthetic appeal, but from research necessity. Late in the summer of 1947, Walter C Williams, head of the NACA's Muroc Flight Test Unit at Muroc Dry Lake, had informed Douglas that red had proven undesirable from a flight safety and photographic standpoint and that white would stand out better both against the rich Mojave sky and tawny High Desert. Douglas complied to make the Skyrocket more easily trackable and, eventually the Skystreaks and XS-1s were painted white as well.) On 10 December 1947, Douglas trucked the first D-558-2, with 'Skyrocket' emblazoned on its nose, to Muroc Dry Lake where it joined the growing stable of research aeroplanes while undergoing final preparations for its initial contractor flights.[20] Given that, a mere decade before, the Navy's highest performance aeroplane had been the stubby 250 mph Grumman

F3F biplane, the sleek, sweptwing Skyrocket (which in time, would become the first aeroplane to exceed Mach 2) epitomised how far aviation science had come in 10 years and – given the challenge of its planned flight test programme – how much more remained to be done before the supersonic domain could be considered the stuff of routine flight.

PART TWO
Through the 'Sound Barrier'

4

Prelude to Mach 1:

The Bell XS-1 Contractor Programme

When Bell completed the first XS-1 46-062 in late December 1945, the company already had its initial flight test programme planned, and the AAF had delivered a Boeing B-29A Superfortress (45-21800) for the company to use as a launch aircraft. Technicians removed the bomb bay doors and cut out a portion of the lower fuselage in front of the aft pressure bulkhead of the B-29 for clearance of the XS-1's horizontal stabiliser. They added an extendible ladder from the forward right side of the bomb bay for the pilot to use for entering and, if necessary, leaving the rocket plane while it remained shackled to the launch B-29. Bell, NACA and the AAF hoped that Bell would eventually obtain a satisfactory turbine pump for the fuel system of the XS-1, so that the XS-1 could be converted from its 'temporary' high-pressure fuel system to a turbopump fuel system, which would permit attainment of its original performance estimates, including a possible 1,700 mph top speed. Langley personnel considered the air launching of the XS-1 a cumbersome method of operation, and some still hoped that some method of safe ground launching might emerge. By the autumn of 1945, the AAF and NACA had not selected the actual test site for the aeroplanes. NACA had expected the XS-1 to fly from Langley Laboratory where a large military airfield existed, but regarded with some concern a growing desire within Bell and the AAF to transfer its flight operations to some larger, more secluded site, such as Muroc Dry Lake in California. Before the XS-1 entered its research programme, the AAF and NACA determined to investigate the air-launch method of operation, and secondly, the feasibility of operating the exotic rocket aeroplane from a conventional flying field. Bell had completed the first XS-1 without its rocket engine and the company, together with the AAF and NACA, felt that preliminary glide tests of the aeroplane could answer both questions. The AAF selected Pinecastle Field, Orlando, Florida as the site of the XS-1 glide trials. During the trials, the NACA would supervise all details of the flight testing, and analyse all gathered data. Bell would furnish the B-29 launch crew and the XS-1 pilot.[1]

The Pinecastle Tests

Both the NACA and Bell teams consisted of supremely gifted and capable individuals, many of whom were not far removed from the college environment. The NACA assembled a small team to participate in the Pinecastle tests, consisting of representatives from Langley's Flight Research and Instrument Research Divisions. The director of the group was a young and energetic engineer, Walter C Williams. Williams held a BS in aeronautical engineering from Louisiana State University. After he graduated in 1939 he joined the Glenn L Martin company at Baltimore, Maryland but being research-minded, soon left for NACA's Langley Laboratory, researching aircraft stability and control. While at Langley, Williams became interested in the transonic research aeroplane concept, and had worked with John Stack, Milton Davidson and Harold Turner on research aircraft studies. When the XS-1 progressed to the flight stage, Williams came over from the Flight Research Division to be head of the flight-test programme on the aeroplane. Together with Williams was another young Langley engineer, Gerald M Truszynski. Truszynski held a BS in electrical engineering from Rutgers University and, after graduating in 1944, joined the Langley Instrument Research Division working on the design and development of radar and telemetry equipment. Truszynski had charge of the radar tracking equipment used on the Pinecastle tests, which consisted of a SCR-584 anti-aircraft gun-laying radar equipped with a camera for furnishing more accurate flight-path data. Both Williams and Truszynski later assumed key roles in directing the NACA flight research programme at Muroc Dry Lake and later, in the American manned space programme. The Bell XS-1 test team included both a launch plane crew and an XS-1 pilot. The B-29 launch crew consisted of pilots Harold Dow and Joseph Cannon and flight crewmen Ivan Hauptmann, William Means and Herman Schneider. The XS-1 pilot was Bell's chief test pilot Jack Woolams. Commissioned as an AAF pilot in 1938, Woolams subsequently served as a Curtiss P-36 fighter pilot with the 79th Pursuit Squadron. He left the service to return to college, graduated from the University of Chicago and joined Bell as a test pilot. While with Bell, Woolams participated in the P-59 programme, demonstrating notable airmanship on several occasions. He attained an altitude of 45,765 feet on 14 July 1943 in the second XP-59A, reaching a height of 47,600 feet on 15 December of that year. During P-59 dive tests at Niagara Falls, Woolams landed a YP-59A on its belly when air pressure changes sucked the gear down, breaking its retracting links. He had first-hand experience with the dangers of compressibility for, during the same series of tests, a P-59 he flew lost its tail in a high-speed dive. The cockpit canopy jammed, and Woolams only escaped after twisting in the cockpit and kicking the canopy loose with his feet. While the Bell design team built the XS-1, Woolams gave suggestions on the design from a pilot's point of view, particularly cockpit layout. Woolams was a consummate airmen and flight researcher and, like many of the breed, approached his work with a mixture of pride, professionalism and good humour that endeared him to those with whom he worked.[2]

One question that bothered the Bell engineers was whether the XS-1 would separate cleanly from the B-29 at launch or not. Unless positive separation forces existed, the XS-1 might snag the B-29, possibly resulting in the loss of both aeroplanes. Company engineers decided to make captive flights with the XS-1 shackled in the B-29's bomb bay and record pressure distributions between the two aeroplanes, completing the first captive flight of the XS-1 on 10 January 1946 which was a 2 hour 40 minute flight over Buffalo. After recording pressure distribution from the XS-1's pressure pickups in the instrumented wing, the company concluded that a positive separating force did exist. Cautious about relying solely upon recorded data, technicians modified one of the bomb bay's bomb releases to eject the XS-1 with a nose-down pitch to ensure a clean separation. They installed a steel tube forward of the XS-1's propellant tanks so that it projected from the fuselage for six inches on either side. They then added angled wooden guide posts in the B-29, so that if the XS-1 tended to yaw and drift at launch, the steel tube and guide posts would steer it clear of the mother aeroplane. On 18 January the B-29, with the XS-1 shackled in its bomb bay, flew nonstop from Buffalo to Pinecastle, droning south for 4½ hours; preparations for the all-important first glide flight consumed the next week. On Friday 25 January 1946, the silver and black B-29 piloted by Joseph 'Joe' Cannon and Harold 'Pappy' Dow rumbled down the runway at Pinecastle Field, lifted into the air and strained upwards, the orange XS-1 locked firmly in its belly. A Boeing B-17 Flying Fortress was already up for photo chase and Bob Stanley followed in a North American P-51H Mustang for safety chase. In order to see if the XS-1 needed the steel pipe and guideposts to prevent drifting, the Bell launch crew had daubed the tube and guide rails with red paint. On the first flights, Bell planned to launch the rocket plane with the inboard engines of the B-29 shut down and feathered with the B-29 flaps in the takeoff position and at a launch speed of 150 mph. Gradually, the company would then go to high power with flaps in cruise position and higher launching speeds. The company tracked the launches with cameras mounted on the B-29's wing and stabiliser tips. At 11.15am, 27,000 feet above central Florida, with both the Atlantic to the east and the Gulf of Mexico to the west clearly visible, Pappy Dow released Woolams and the XS-1. The little saffron plane dropped quickly and cleanly away from its mothership and, while the NACA test group monitored the glide flight, Woolams familiarised himself with the plane's handling characteristics. He found the plane very easy to fly, and became so preoccupied with the rocket plane that he almost undershot the landing. Post-flight inspection showed the XS-1 dropped clean without brushing the guideposts, and no traces of red paint showed from the pipe or guideposts. (Subsequent flights proved the guide rails and ejector unnecessary.) The XS-1 had passed another critical milestone: the first flight.[3]

In all, Bell completed 10 flights on the aeroplane at Pinecastle; one in January, eight in February and one in March. The XS-1 proved extremely responsive to the pilot, and Woolams considered the aeroplane the best he had ever flown. After several flights to familiarise himself with the general handling characteristics of the plane, Woolams made detailed data-gathering flights on the XS-1's longitudinal and directional stability and its rate of roll. The XS-1 suffered damage on two glide flights. On the fourth flight on 8 February 1946, the main gear retracted on landing and the plane damaged its left wing. On the next flight on 19 February just a week later, the nosewheel retracted during the landing runout and damaged the underside of the fuselage nose. Annoying as these accidents were, however, they did not delay the flight test programme. Highly pleased with the results of the glide flights at Pinecastle, Bell flew the XS-1 via its B-29 mother aeroplane back to Buffalo. The company now began preparations for the powered flight portion of the contractor test programme.[4]

Passing the Torch

Back at Buffalo, changes took place in the XS-1 design and flight test team. When the AAF contracted with Bell for the development of the X-2 supersonic sweptwing research aeroplane, XS-1 project engineer Stanley Smith left the programme to assume leadership of the X-2 programme and engineer Richard Frost replaced him on the XS-1 effort. Though an engineer, Frost was a test pilot by training and vocation. He had joined Bell as a test pilot in mid-1943, serving in this capacity until a serious accident in February 1945. He had to bail out of a burning P-63 Kingcobra at high altitude, and severe burns on his hands kept him grounded for six months. During that time he developed instrumentation to make radio-controlled dive tests in P-59's, one of Robert Stanley's ideas to acquire dive information without risking the life of the pilot in the attempt. Radio-control appeared so promising as a means of alleviating some of the dangers of dive testing that the Navy asked Frost to run a series of remote-control tests on the Grumman F7F Tigercat, and the AAF had him do the same to a Lockheed P-80 Shooting Star, important and precedent-setting examples in their own right on the path to unmanned air vehicles. When Frost moved up to take Stanley Smith's place on the XS-1 programme, he found his greatest challenges were firstly; getting the RMI powerplant installed and checked out; and then getting the XS-1 through its contractor-powered flight programme. While the first XS-1 was at Pinecastle, Bell completed fabrication of the 8% wing and 6% tail originally intended for the aeroplane. When the plane arrived back at the Bell plant, therefore, Bell technicians installed the thinner wing and tail, removing the thicker 10% wing and 8% tail for use on the second XS-1 serial 46-063, which was then being completed at the Bell plant. The first available RMI XLR-11 engine went into the latter aeroplane.[5] At the same time that Bell readied the second XS-1 for powered-flight trials, the company had to replace its chief test pilot as a result of a tragic accident.

An air racing aficionado, Woolams, together with Alvin M 'Tex' Johnston and Chalmers H 'Slick' Goodlin, two other Bell test pilots, had purchased two wartime P-39 Airacobras, under the auspices of a charter company that Goodlin operated, 'Skylanes Unlimited'. They named them *Cobra I* and *Cobra II* and replaced the drab wartime camouflage with striking colour schemes, scarlet and black for *Cobra I* and a yellow and black colour scheme for *Cobra II*. The three men readied the planes for the ill-fated 1946 Thompson Trophy air race; since only two pilots could fly, they held a coin toss, which Goodlin lost.[6] On Friday 30 August, the day before Labor Day, Jack Woolams took *Cobra I* out over Lake Ontario for an engine test. Skimming low over the water, something – perhaps a blown-in windshield – went wrong. The converted fighter plunged into the lake, killing Woolams instantly. At the time of his death, Woolams had little doubt he would be the first man to fly faster than sound. He did not consider the possibility particularly sensational. In an article, ironically published after his death, he wrote that the flights

> . . . will be the logical result of a great deal of hard, cooperative work. There will, of course, be some danger, as there must be in flying at new altitudes and speeds.

> But we are well prepared. I have complete confidence in the airplane and the men who have worked so long and hard on the project. And I am completely confident that the research we are conducting will more than justify the time and expense when the results are made available to the aeronautical industry.[7]

On Monday 2 September, Woolams' friend Tex Johnston took off from Cleveland Airport with 11 other competitors, determined to win the race for his friend. For the better part of an hour his powerful *Cobra II* led a mixed pack of modified fighters around the 30-mile closed course, and when Johnston landed, he had captured the Thompson Trophy at an average speed of 373.9 mph, 90 mph faster than the previous record set by the flamboyant Roscoe Turner in the 1938 Thompson race. After landing, sticking to their pre-race agreement, Goodlin and Johnston turned over what would have been Woolam's share of the $19,400 purse to his widow.[8]

At the Thompson Trophy race, Larry Bell introduced Goodlin to some of his friends as Jack Woolams' replacement as XS-1 project pilot, which was the first word the young pilot had of the decision. The next day, after Goodlin had ferried *Cobra II* back from Cleveland to Niagara Falls, Bob Stanley called Goodlin to his office and confirmed that the 23 year-old pilot would fly the contractor tests on the XS-1 including, as then planned, both its subsonic and supersonic flights. Stanley and Goodlin agreed on a compensation plan whereby Goodlin would receive a $10,000 bonus for the subsonic flights, with $2,500 of this amount to be paid to Jack Woolam's widow because of Woolam's singular contributions to the programme at Pinecastle. For the hazardous supersonic flights, Goodlin would receive $150,000 to be paid to a corporation of his choice. Final contract details were to be worked out between lawyers representing Goodlin and Bell, but negotiations dragged on into 1947 and, in fact, through the end of the contractor programme.[9]

Despite his youth, Goodlin had a considerable flying background; he soloed in several aircraft – a Piper J-3 Cub, Aeronca C-3 and Waco F-2, among others – before his 17th birthday. In 1941, he enlisted in the Royal Canadian Air Force for flight training, won a commission as a Pilot Officer and served as a flight instructor in Ontario. He transferred to the Royal Air Force, and flew Spitfires until he left the RAF in December 1942 for the US Navy. After carrier qualification, he flew as a Navy test and ferry pilot, then joined Bell in December 1943 as a test pilot. While with Bell, Goodlin had had a fair share of close calls. Shortly after joining Bell, a P-39 he was flying on a production test flight caught fire and the young pilot barely escaped from the smoke-and flame-filled cockpit in time to use his parachute. Besides making some remote control flights with Frost's modified P-59 and F7F, Goodlin flew as test pilot on the Bell L-39 low-speed sweptwing research aeroplane and on the Bell XP-83, a prototype long-range jet fighter. During one XP-83 test flight with an experimental ramjet installation, the plane caught fire in midair and the right wing began to burn away. Goodlin stayed with the burning jet until the second crewman, a Bell flight-test observer, could bail out, before himself parachuting from the stricken plane. In sum, Goodlin was an excellent pilot of the 'stick and rudder' school whose lack of formal engineering training did not detract in the least from his ability as a research pilot.[10]

On to Muroc

Though the Pinecastle trials had proven the feasibility of air launching the XS-1 and the good low-speed handling qualities of the little rocket aeroplane, they also demonstrated the inadequacy of attempting to operate the plane from a conventional airfield. Clearly, the XS-1 required a more suitable secluded location where its flight testing could take place without worry of overlying populated areas, and where the rocket plane could have a large emergency landing area on which to set down. Only one location fitted these qualifications: the AAF airfield at Muroc Dry Lake, California.

In September 1946 Hartley A Soule, chief of the Stability Research Division at Langley, designated Walter Williams as project engineer for the XS-1. Later that month, he detailed a group of 13 engineers, instrument technicians and technical observers from Langley Laboratory, under the direction of Williams, to Muroc Dry Lake. The small group arrived at the dry lake on 30 September. At first, NACA considered the group a unit of Langley Laboratory on temporary assignment; only later were they permanently assigned to the High Desert. The NACA personnel arrived at Muroc at a time of greatly increased awareness of the problems of transonic flight, largely because of gloomy predictions by Sunday supplement writers, but mostly from the recent death of Geoffrey de Havilland in the DH 108, which had given teeth to the threat posed by supersonic flight.[11]

Almost fittingly, the dry lake was nearly as exotic as the research aeroplanes themselves. In keeping with its strange environment, the Mojave desert houses many flat, dusty plains devoid of ridges, dunes or brush. These are dry lakes: Cuddeback, Searles and Koehn, to name a few. Early in the 20th century, a silver and gold mining firm set up an encampment on one such lake and dubbed it Rodriguez Dry Lake, after the company name. It is now known as Rogers Dry Lake, the largest dry lake in the world. A vast expanse of parched clay and silt, the lake is dry for ten months of the year. When dry, it can support up to 250 psi of pressure on its surface.

Shaped roughly like a lopsided figure '8', it has a surface area of 65 square miles. Over the year, winds erode its surface and heat chips and cracks the clay so that, by the autumn, the lake resembles a huge quilt or puzzle. Then come winter rains. The wind blows the water back and forth across the lakebed, smoothing and filling in the cracks. The rising temperatures of the new year then evaporate the water off, and the lakebed is once more virtually glass-smooth. For these reasons, automobiles and aeroplanes can move smoothly across its surface, and even the heaviest of aircraft can take off and land without fear of sinking into the lakebed. Rogers Dry Lake is the largest natural landing field in the world.

In 1910, Clifford and Effie Corum, together with Clifford's brother Ralph, settled on a 160-acre plot of land on the edge of Rodriguez Dry Lake. They recruited other settlers to join them, and soon a tiny community developed along the Santa Fe Railway where it edged across the dry lake. The brothers opened a small general store, dug wells for water and held Sunday church services in their home. They decided to name the community Corum, but post office authorities objected because California already had a township named Coram and the similarity in spelling would cause confusion. The Corums then suggested Muroc, created by spelling the name backwards, but the Santa Fe Railway objected because of a rail stop named Murdock. Instead, the railroad suggested Dorado, Ophir, Yermo or Istar. The Corum brothers persisted, however, and the little settlement became known as Muroc. The settlers soon applied the name to the dry lake as well. By the 1930s, Muroc represented little more than a spot on the map. North of the lake ran Highway 58 over which, during Depression years, 'Okies' streamed, entering California at Needles and then wending their way along through the desert communities of Essex, Ludlow, Barstow, Boron, Mojave and on into Bakersfield, in their search for a livelihood. (This is a path immortalised by John Steinbeck in his book *The Grapes of Wrath*.) At around the same time, Muroc gained notoriety as the site of supposedly the largest moonshine distillery in southern California. At night, liquor runners and prohibition agents attempted evasion and interception on the clay lakebed. In early 1933, then-Colonel Henry 'Hap' Arnold, commanding officer of March Field, Riverside, California desperately needed a bombing and gunnery range for his bomber and pursuit pilots to hone their deadly skills. The Navy refused the Army Air Corps the use of the Pacific Ocean. Just north of March Field, over the San Diego mountains, lay the Mojave, the most logical choice for such a range, for it was largely unpopulated and desolate to begin with. Arnold decided to scout the Mojave to determine its suitability. Arriving at Muroc one morning at 6am, he liked the look of the expansive dry lake; an ideal range and landing area. After his return to March Field, proceedings were initiated to gain title to the land. Most of the land already belonged to the government, and it was on these portions that the Air Corps immediately began operations. In September 1933 a cadre of soldiers from March Field, under the direction of Sergeant Harvey Fogelman, established a camp on the eastern shore of the lake and laid out the bombing and gunnery range. Except for an occasional forced landing or some limited private testing of aircraft from the lakebed, Muroc Dry Lake remained relatively tranquil until Pearl Harbor.[12]

The war brought a building boom – an influx of 40,000 people – and in 1942, designation of the bombing and gunnery range as Muroc Army Air Base. In November 1943, it became Muroc Army Air Field, complete with barracks, a control tower and a concrete runway, located on the southwestern lake shore. P-38 fighter pilots flew practice missions from the base and B-24 and B-25 crews departed for bombing runs over the range. To train bomber crews for anti-shipping strikes, engineers constructed a 650-foot replica of a Japanese *Mogami* class heavy cruiser in the middle of the dry lake, which pilots promptly dubbed the 'Muroc Maru'. (In 1950, when Army demolitions personnel disassembled the 'vessel', they found numerous unexploded bombs in its midst.)

In 1942, the Materiel Center at Wright Field designated the north end of the dry lake as the Materiel Center Flight Test Site. That autumn the nation's first turbojet, Bell's XP-59A, arrived at Muroc for its initial flight tests which were held in great secrecy. In 1944, the AAF redesignated the north end of the lake as the Muroc Flight Test Base. In October 1946, both the crew-training and flight-test activities at Muroc merged into a single test facility at the main Muroc Army Air Field, under the direction of the Air Materiel Command at Wright Field. Both during and after the war, Muroc witnessed a steady procession of exotic aeroplanes undergo tests from its lakebed. In July 1946, for example, a captured German Heinkel He 162 from Freeman Field made two flights. Later, two B-17s landed on the lakebed under remote control after completing a pilotless flight from Hawaii to California. It was into this environment of strange and unique aeroplanes that the XS-1 came.[13]

Meanwhile, Across the Atlantic

The XS-1 was not the only aircraft programme moving places in the late summer and early autumn of 1946. In Great Britain, Geoffrey de Havilland's tragic death in the fickle DH 108 Swallow had not stopped that programme, but had placed it on hold pending completion of a more refined model of the little jet. The long-awaited Miles M.52 programme had collapsed even before de Havilland's accident. With a fuselage inspired by a bullet, as with the XS-1, and wings and tail surfaces like razors, the M.52 possessed the right look, but had always proceeded at a slower pace than the XS-1, in great part due to the complexity of its planned engine, a technologically challenging centrifugal-flow ducted-fan afterburning turbojet. In February 1946, at which point the M.52's detail design was about 90% complete, the Director General of Scientific Research at the Ministry of Aircraft Production, Sir Ben Lockspeiser, had directed Miles to cancel the programme immediately, citing Britain's postwar poor economic situation but also, according to company founder F G Miles, the belief 'that aeroplanes would not fly supersonically for many years and perhaps not for ever'. Later, in a public announcement, Lockspeiser also mentioned the dangers of supersonic research as another reason for abandoning manned supersonic testbeds, preferring air-launched rocket-propelled models instead. The Vickers-Armstrong company later made rocket models of the M.52, equipped with telemetry that were dropped from a Mosquito bomber over the Scilly Isles. Though it is by no means certain that a piloted, jet-propelled M.52 would have been a success, (and it certainly could not have flown supersonically prior to the XS-1 in any case), it is worth noting that on one of these tests in October 1948, a M.52 model reached Mach 1.38 at 35,000 feet, approximately 930 mph, before exhausting its fuel and dropping into the sea, proof enough that Miles' aeroplane was indeed a genuine supersonic shape. Thus passed one of the XS-1's three major contenders for the supersonic crown.[14]

... *and in Stalin's Russia*

Whatever satisfaction the AAF, Bell and NACA might have taken from the fact that the supersonic race had narrowed to a match between the XS-1 and the lagging Skystreak-Skyrocket programme would have vanished had they known the status of a third contender – the Soviet Union. During the Second World War, the Soviets had maintained an active interest in high-speed flight, but with little real accomplishment. Researchers had flown modified propeller-driven fighters equipped with ramjets as a means of boosting their performance (though with disappointing results) as well as an abortive rocket-propelled interceptor that had killed its pilot. Unlike Germany, Great Britain, the United States and Japan, an indigenous turbojet engine eluded wartime Soviet aeronautical science. Such problems however, if anything, spurred the Soviet technological and intelligence communities to greater effort. The technical investigators of the TsAGI – the Central Aerodynamics and Hydrodynamics Institute, the Soviet equivalent of the NACA – were well aware from their own subsonic tunnel tests that as flow speed approached the speed of sound, the drag of a wing increased appreciably, as much as seven to ten times over its purely subsonic value, and that creative solutions had to be found lest the speed of sound really prove to be a barrier, at least for Soviet aeronautics.[15]

It is highly likely that Soviet interest in supersonic flight was accelerated by active espionage directed against the NACA, for the Soviets kept very close watch on their allies in the 'American TsAGI', even as both countries fought the menace of Nazism. Julius Rosenberg, head of a notorious spy ring most closely associated with Soviet espionage directed against the atomic bomb (but which also, for example, stole the radar-cued proximity fuse, one of the war's great technical innovations), arranged the recruitment of a brilliant young NACA engineer to join his growing spy stable: William Perl.[16] Perl, one of four engineering classmates of the City College of New York sharing strong Communist sympathies who joined the Rosenberg ring (the others were Joel Barr, Morton Sobell, and Alfred Sarant), subsequently proved a superb find. Perl's Soviet handlers assigned him the cover name *Gnome* (sometimes spelled *Gnom*), and later the name *YaKov*. (Rosenberg was first assigned the cover name *Antenna*, then *Liberal*; New York City, where many of the meetings took place between Soviet agents – including Perl – and their contacts, was *Tyre*.)

Intercepts of communications between Perl's Soviet contacts and *Viktor* – Lieutenant General Pavel M Fitin, the chief of the foreign intelligence directorate of the NKVD, the predecessor of the postwar KGB – reveal both the depth and zeal of Perl's espionage activities against his own country.[17] Perl had worked at Langley and at NACA Headquarters in Washington, then transferred to Cleveland when the agency created the engine laboratory that is now the NASA Lewis Research Center. At Cleveland, he established a reputation as the laboratory's 'resident genius', and 'a brilliant theoretician'.[18] A patriot he was not: by mid-1944 he was already passing along secrets of Westinghouse turbojets, thanks to a courier system set up by fellow agent Alexander Feklisov. So good was Perl's material that he earned a $500 performance bonus (a not inconsiderable sum in 1944) in September from Fitin. By the end of December, Rosenberg was travelling to 'YaKov's town' [i.e., Cleveland] to introduce other Soviet agents to his star find.

Perl was more than an engine man; he had a strong interest in supersonic flight, designed an early supersonic wind tunnel for tests of engine ducts at Lewis and, after the war, took a doctorate studying with no less than Theodore von Karman at Caltech. All this made him a genuine supersonic pioneer, and as such, extraordinarily valuable to the Soviets. Indeed, the official NASA history of Lewis laboratory states, 'By the early 1950s, papers by Leroy Turner, Herbert Ribner, William Perl [and others] established Lewis Laboratory's reputation for basic work in *supersonic aerodynamics*' (emphasis added).[19] As a key and highly respected NACA researcher in the critical years of 1944 onwards, Perl would have had routine access to the most up-to-date transonic and supersonic studies and reports as well as the latest in turbojet and rocket propulsion. These would have been available not only from Lewis, but from Langley and Ames as well (and from Muroc, once the X-1 and D-558 began flying) and from military and industry laboratories, too, particularly Wright Field and the Buffalo complex of Bell, Curtiss-Wright and, at a slightly later date, the Cornell Aeronautical Laboratory.

More than this, he would have been privy to all the shoptalk and gossip: what Stack, Diehl, and Kotcher were up to, what the Air Force and Navy were doing, what the status was of British transonic studies and, once the research aeroplanes were flying, what their progress results were. His close postwar association with von Karman could not have come at a worse time. Von Karman – always inclined to involve his students in his own activities – was busily mapping out the technological future for the US Air Force (and was thus privy to many new sensitive programmes and initiatives) as he headed the service's recently formed Scientific Advisory Board. (Intriguingly, another of von Karman's Caltech colleagues, the Chinese aerodynamicist Hsue-shen Tsien, had his own brush with charges of espionage, lost his security clearances and left via Canada for the People's Republic of China where he subsequently supervised creation of that nation's ballistic missile and atomic weapons programmes. It is unknown what ties existed between Perl and Tsien.)

In short, this little-known and even-less-appreciated agent may, in fact, be a key reason why the Soviet Union moved so quickly from a relatively disorganised supersonic research posture to a highly confident one by the mid-1950s. The Soviets placed their own first-generation supersonic jet fighter, the MiG-19 in service contemporaneously with the American F-100 of the early 1950s. This aircraft constituted a marked departure from its predecessors, incorporating many advanced design features either proven out or recommended by various NACA and industry aircraft and engine programmes and formed a departure point for studies leading to the Mach 2 MiG-21 and other designs.

From the TsAGI's standpoint, the next best thing to having a Soviet X-series programme of one's own would have been to have had an agent of Perl's qualifications embedded within the American X-series programme. With Perl, they had just the man. In August 1950, Perl's name came up in connection with the Rosenbergs, and he was arrested, charged and convicted of perjury (for having denied knowing Julius Rosenberg and Morton Sobell), subsequently serving a five-year jail term. Like Alger Hiss (saved by a statute of limitations), Perl was another very lucky Soviet agent. The extreme sensitivity of the *Venona* intercepts – a window on the inner workings of Soviet espionage, classified Top Secret and only available by special access – seems to have inoculated Perl against spy charges. In the absence of a confession from Perl or another agent as to his activities, only *Venona* material could have convicted him – and revelation of *Venona's* existence would likely have resulted in effective termination of the programme. Julius Rosenberg and his wife Ethel were not so fortunate. They were betrayed by a family member and went to the electric chair rather than jail, a controversial fate. Up until his death, Perl always denied being a Soviet spy, but the intercepts are damning and, as a result of their release, his denials – as with those of Alger Hiss and others – must be seen as, at best, crude attempts at disinformation.

So perhaps the acceleration of Soviet interest in supersonic flight had less to do with discovering what the Nazis had been up to and more to do with having a spy in the heart of the NACA just as America was structuring its planned postwar supersonic research programme. In any case, as one Soviet commentator has written,[20]

> After the Second World War . . . [the] Communist Party gave Soviet scientists a complex and responsible task: to work out questions in the design of jet aircraft, breaking the sound barrier and putting into production improved technology based on the use of all the achievements of related sciences.
>
> *The first requirement was the mastery of supersonic flight. From this, it was necessary first of all to establish the appropriate aerodynamic shape of the aircraft*. [Emphasis added.]

To this end, the Soviets pressed ahead boldly (if somewhat haphazardly) to exploit the rubble-filled magician's shop that was the former Nazi state.The Soviets took over the DFS 346 project and made it their own together with another German effort, the Junkers Ju 248, born as the Messerschmitt Me 263 which was itself a growth version of the earlier Me 163, but with considerably greater design refinement. The latter project served as the basis for a hurriedly built experimental rocket-propelled interceptor, the Mikoyan-Guryevich (MiG) I-270, with straight wings, a fuselage lifted from the Ju 248, an engine fuelled by nitric acid and kerosene and a T-tail having moderately swept horizontal tail surfaces as a means of giving it a higher critical Mach number than the straight wing. It is doubtful that the I-270 could ever have been more than a Mach 0.85-0.9 aeroplane, but its possibility as a rival to the XS-1 cannot be totally discounted; in any case, the MiG bureau manufactured two prototypes which were both apparently lost by the end of 1946 during ground and flight testing respectively. The DFS 346 was another matter entirely.[21]

As with the German rocket programme and remnants of the German atomic weapons effort, the Soviets moved large numbers of German aeronautical engineers and design staffs into the Soviet Union following the assembling of huge lists by NKVD state security and GRU military intelligence *apparatchiks*. The Nazi air ministry, the *Reichsluftministerium*, had entrusted development of the DFS 346 to the Siebel firm. The DFS 346 looked much like the Douglas Skyrocket, different in that it was entirely rocket powered, with a swept T-tail, had a jettisonable clear glass nose cone for the pilot (who would lie prone, on his stomach, looking ahead) and a skid landing gear. By the end of the war, a glider configuration of the plane was ready for testing and the actual aircraft was well underway. All this, plus the Siebel engineering staff, fell into Soviet hands. In October 1946, the NKVD moved the entire DFS 346 project and its personnel to the Soviet Union, relocating the design team to Podberezhye, approximately 75 miles east of Moscow, under the leadership of Semyon Alekseyev. The DFS 346, now known simply as *Samolyot* 346 (Aircraft 346), underwent extensive wind tunnel trials by the TsAGI together with drop tests of its nose cone. By the late autumn of 1946, the Soviets were aggressively pursuing the DFS 346, and intended to fly both a glider variant and a powered model by mid-1947. Thus, with the M.52 cancelled, the major foreign threat to the XS-1's primacy was a rocket plane conceived under one totalitarian state, built by a second and assisted by an American-delivered B-29 launch aeroplane, and it is likely that key information was furnished by an American engineer in the service of Soviet spymasters!

The First Powered Flights

In preparation for the XS-1, technicians at Muroc erected two huge tanks, one holding 15,000 gallons of liquid oxygen, the other 3,300 gallons of liquid nitrogen. They dug a loading pit for the XS-1 to roll into before being hoisted into the B-29 above it and modified a standard AAF fuel trailer to function as a mixing tank for the XS-1's alcohol-water fuel. On 7 October 1946, the second XS-1, joined to its B-29 launch aeroplane, departed from Bell's Niagara Falls plant accompanied by a P-51D Mustang chase aeroplane and a C-47 Skytrain for personnel transportation, arriving at Muroc on 8 and 9 October. Getting promptly to work, Richard Frost and the Bell test team readied the XS-1 for a glide flight the next day on

10 October. That first attempt at a flight met with failure. During the climb launch altitude, Goodlin noticed an increase in cabin pressure and actuated the cabin-pressure dump valve, which failed to function. The emergency pressure relief valve also failed to operate properly and, as a last resort, Goodlin released the cabin door lock. With internal pressure now nine psi above outside atmospheric pressure, the door shot out, bending the entry ladder from the B-29 and jamming between the ladder and XS-1. Goodlin could not return to the B-29 and remained in the little rocket plane while Harold Dow brought the B-29 in for a safe landing. After repairing the door and ladder damage and checking the operation of the dump valves, Bell attempted another glide flight the next day, 11 October. The Muroc Army Air Field historian reported laconically,

> The B-29 carrying the XS-1 took off at 1413 the 11th of October. The Rocket was released between 1455 and 1456, landing at 1503. The mission was completed successfully.[22]

On 14 October, a year to the day from when Chuck Yeager would exceed Mach 1 for the first time, Goodlin completed the XS-1's second glide flight. He followed it with the third three days later, on 17 October. For about the next month, Bell performed routine maintenance on the Mustang chase aeroplane, installed more efficient filters in the alcohol-water mixing unit and modified the propellant pressurisation system of the second XS-1 so that the pilot could regulate it from the cockpit. On 2 December, Bell checked the XS-1's fuel jettison system on a glide flight during which Goodlin vented 1,930lbs of alcohol-water ballast. Four days later, the company attempted the first powered flight. It almost ended in disaster. During the climb to altitude, a nitrogen valve froze in the intense cold, preventing nitrogen from entering the liquid oxygen tank. Therefore, the rocket engine could not be fired, nor could Goodlin jettison the propellants. The B-29 had to land with the XS-1, heavy with the full fuel load, still shackled in the bomb bay. The nose landing gear of the XS-1 then inadvertently extended. If the B-29 had landed with the XS-1's gear extended, the obstruction might have caused both planes to crash, since even with the XS-1's gear retracted, the separation between the XS-1's belly and the ground was less than a foot. Company president Lawrence Bell had brought a group of Bell company directors to watch the first powered flight of the rocket plane. The officials were now worried over the launch crew's predicament. Richard Frost, in the chase P-51D, suggested that Goodlin retract the XS-1's nose gear during the B-29's landing approach. As the bomber swept down towards the Muroc runway, Goodlin raised the nose wheel. It remained up as the B-29 landed. Relieved, Bell said tersely, 'Keep trying,' and returned to Buffalo. On Monday 9 December, Bell again tried to obtain a successful powered flight. After climbing to altitude, all aircraft systems indicated satisfactory operation. A Muroc observer reported later,

> The Rocket was dropped at 1154 from an altitude of 27,000 feet. When the first unit was turned on, a streak of flame came from the tail of the rocket and continued to glow until the unit was shut off. The second unit was fired with results coincidental with those of the first. A FP-80 was following the Rocket to observe its reactions and also to take photographs. When the pilot of the XS-1, Chalmers Goodlin, fired the four units simultaneously, the FP-80 could not maintain a speed consistent with that of the Supersonic.[23]

On this flight the XS-1 grossed 12,012lbs and, to Goodlin, the rate of drop away from the B-29 seemed twice as fast with the full fuel load as the glide-flight drops. Ten seconds after drop he ignited the XLR-11's first chamber, then fired the second. Under a 3,000-lb thrust, the XS-1 accelerated rapidly as it burned off its fuel load, outpacing the P-51D and setting a tough standard for the Lockheed FP-80 Shooting Star photographic chase plane. Goodlin shut off the second chamber and began a slow climb at 330 mph indicated air speed on the power of the first chamber alone. Sitting in the cockpit, Goodlin could barely hear the rocket's operation. At 35,000 feet he relit the second chamber and sped to Mach 0.79, finding the handling characteristics of the rocket plane very good. After shutting down both the first and second chambers, Goodlin dived to 15,000 feet. During the dive, the XS-1 began a slight snaking motion caused by lateral shifting of the fuel load. Though Bell had installed transverse fuel tank baffles to prevent longitudinal (i.e. fore and aft) shifting of the fuel load in flight, the baffles did not prevent fuel from sloshing from side to side. Goodlin did not consider the snaking serious, however, for he could easily control it. At 15,000 feet, he pulled out into straight and level flight and ignited all four chambers. As Goodlin reported, the orange rocket plane shot forward with 'Terrific acceleration . . . comparable to the application of water injection power in a conventional fighter from a standing start'. After a few seconds of full power, the liquid oxygen line pressure dropped and Goodlin shut the engine off. Just then, a fire-warning light came on in the cockpit of the plane, and Goodlin radioed Richard Frost in the chase Mustang to ask if Frost could see any signs of fire. Frost could not, but the fire-warning light remained on until Goodlin landed on the lakebed. The test team found that a small fire had occurred in the engine bay of the aeroplane, burning some wiring but not causing further damage. Replacement of wiring kept the plane grounded for slightly over a week. The XS-1 had passed another milestone: its first powered flight.[24]

Extending the Envelope

Though the XS-1 had outdistanced the slower FP-80 photo jet, the rocket-research aircraft completely outclassed any other conceivable chase plane, particularly Frost's piston-engine P-51D. Frost recommended that he be given a P-80 in place of the P-51D for the remainder of the programme and, as a result, the P-80 became the XS-1's standard chase plane. On 20 December, Goodlin completed the rocket plane's second powered flight, and on 8 January 1947 welcomed the new year by attaining Mach 0.8 at 35,000 feet. As part of its contractor requirements, Bell had to demonstrate satisfactory stability and control characteristics to Mach 0.8. At this speed, Goodlin reported the plane under 'perfect control,' with only a 'very slight shuddering' indicating compressibility phenomenon. The NACA strain-gauge installation in the plane indicated the compressibility buffet to be of negligible magnitude. By the end of January, Goodlin had completed 10 contractor flights in the second XS-1. In February, he began a series of flights to demonstrate the XS-1's structural strength and to gather buffet-boundary data. He made pullups to 8.7g at Mach numbers from 0.4 to 0.8 without damaging the XS-1's robust structure. At the same time, the RMI XLR-11 engine proved equally satisfactory. Though the RMI company only guaranteed the engine for one hour of service, the engine in the second XS-1 continued to function excellently beyond this time limit. On the last day of the month, Bell flew the second XS-1 via B-29 back to Buffalo for minor modifications to the propulsion system to improve ventilation of the engine compartment and also to perform a required 50-hour maintenance check on the launch B-29. The company planned to complete the check and modifications of the second XS-1 by early May so that it could complete its final contractor test flights.[25]

Early in March 1947, Bell completed the first XS-1, serial 46-062. The chief difference between this aeroplane and the second XS-1 lay in the wing configuration. After its glide trials at Pinecastle with the 10%-thick wing and 8%-thick tail, in early 1946 Bell installed the XLR-11 rocket engine in the plane and replaced the wing and tail with the newly fabricated 8%-thick wing and 6%-thick tail. The first XS-1 also had a slightly higher gross weight than the second XS-1. Therefore, because of the increased weight and thinner wing section, Bell predicted that the first XS-1 would have a higher stalling speed than its sister aircraft. On 5 April 1947, Bell's B-

29 arrived back at Muroc, the first XS-1 nestling in the bomb bay. After a weather delay, Goodlin made a glide flight in the first XS-1 on 10 April, noticing little difference in the handling qualities of the plane in comparison with those of the second XS-1 which had thicker wing and tail surfaces. The next day, Bell made the aeroplane's first powered flight. All went well until the landing. Upon touchdown, the first XS-1 skipped back above the surface, coasted for about 100 feet, then landed hard. The nosewheel and its locking structure absorbed the landing load, the locking structure sheared and the nosewheel retracted. The XS-1 skidded to a halt on its main landing gear and nose under-surface. Repairs took two weeks, and to make the best use of the delay, NACA engineers from Walter Williams' little unit completed instrumentation installation in the rocket plane. Late in April, Bell technicians at Muroc completed repairs to the first XS-1 and on 28 and 30 April, Goodlin completed the plane's second and third powered flights. After two unsuccessful flight attempts, he completed the fourth powered flight of the first XS-1 on 5 May. Bell completed modifications to the second XS-1 at Buffalo on 7 May 1947. That same day, the launch B-29 left Muroc to ferry it back to California, returning with the rocket plane two days later. At last, the end of the planned 20 powered flight programme was now in sight, and Bell concentrated on final demonstrations of airworthiness before delivery of the aeroplanes to the AAF Air Materiel Command. By 21 May, Goodlin completed two buffet-boundary investigations in the first XS-1. Concerns about the pace of the programme, which caused AAF personnel at Wright Field to complain directly to Bell test pilot Tex Johnston, led Johnston to venture to Muroc to check on the progress of the demonstration tests. Johnston took advantage of the trip to check out in the second XS-1 (the first flight of this aeroplane since its return to Muroc), flying to Mach 0.72 and executing an 8g pullout. He came away convinced that, aside from some concern regarding the plane's longitudinal trim characteristics, it could exceed Mach 1 in three or four more flights. After a one-week weather delay, Slick Goodlin performed an airspeed calibration flight in the second XS-1 on 29 May 1947, bringing the Bell contractor programme to a close. Robert Stanley, now Bell's vice-president for Engineering, offered project engineer Richard Frost the last flight in the XS-1 before Bell turned the planes over to the Air Materiel Command, but Frost, ever the professional, declined. He knew the importance of the aeroplane and he did not want to risk the programme's progress just to say he flew it. During the almost eight-month contractor programme on the aeroplane, Bell met all the requirements. The planes completed a total of 20 powered flights, pulled the requisite 8g loadings, and demonstrated satisfactory handling characteristics to Mach 0.8.[26]

Close Calls

Two close calls nearly upset the entire XS-1 programme and, while both were serious, the first was of profoundly greater significance than the other. In each case, fate more than skill preserved the XS-1's opportunity to pioneer the sonic skies. At long last by early 1947, the *Samolyot* 346 programme had at last resulted in 'real' aeroplanes. One of these, a glider designated the 346P, furnished useful flight test information in support of the rocket-powered 346, air-launched from beneath the wing of a Boeing B-29 that had first arrived in the Soviet Union, after sustaining damage on an air raid over Japan. In the spring, the Soviets began tests of the first rocket-powered aircraft (the 346-1) air-launched from the B-29 as the 346P was, nestling between the inboard and outboard engines under the right wing with a slight semicircle cut in the B-29's landing flap so that it could extend without slicing into the spine of the rocket plane. Captain Wolfgang Ziese, one of the Germans transported to Russia, served as project pilot for the powered tests which were conducted at Tyoplystan. On one of these tests in May 1947, he allegedly attained Mach 0.93, approximately 620 mph, but 'violent vibrations' caused him to decelerate to safer speeds. Given that the 346 family had a controversial aileron design, as well as a fixed horizontal stabiliser and an obviously too-small vertical fin, any combination of lateral – longitudinal – directional motions may have been to blame. It was 'back to the drawing board' for the 346, which would appear in hardly more flight-worthy configurations – the 346-2 and 346-3 – in 1949 and 1951. But by any measure, only a narrow thread of fate had preserved the XS-1's chances to be first through the 'sound barrier'.[27]

The second close call could not destroy the programme, but it was serious enough. Shortly after Bell completed the contractor programme, an incident occurred that nearly claimed the second XS-1. Ironically, it happened not in the air but on the ground. On 5 June five C-54s, loaded with representatives of the Aviation Writers' Association, landed at Muroc to observe a flying show at the field. The tremendous display numbered many of the latest military aeroplanes, including experimental bombers such as the AAF's Douglas XB-43 Mixmaster, North American XB-45 Tornado and Convair XB-46, the Northrop N9M flying wing demonstrator (a subscale XB-35), new fighters including the North American P-82 Twin Mustang and Republic P-84 Thunderjet as well as the Navy's first two jet fighters: the North American XFJ-1 Fury and Vought XF6U-1 Pirate. Douglas test pilot Gene May showed off the red D-558-1 Skystreak, Slick Goodlin made a powered flight from the B-29 in the first XS-1 and, on the ground, Richard Frost gave a demonstration of the second XS-1's rocket engine. He fired the first cylinder, then lit off the third which exploded. Amid the roar of the engine, the explosion went unnoticed until a fire-warning light flashed in the cockpit. Frost, leaning into the cockpit on a ladder, quickly got down and ran along the fuselage to see what was wrong. The paint on the rear fuselage blistered before his eyes in the invisible alcohol flame. Frost's shouts for assistance brought a quick response from the Bell ground crew, who removed access doors so that the Muroc fire department could put out the fire. The fire department doused the flames quickly, but the fire severely damaged the engine bay and aft fuselage of the plane, and Bell had no choice but to ferry it via B-29 back to Buffalo for repairs, which occupied the next month: a sobering warning that nothing about rocket aeroplanes could be taken for granted.[28]

Programme Redirection

Originally, when Bell had the XS-1 under development, the AAF planned to place a separate contract with the company for research flights after the company completed acceptance tests on the airplane; the NACA would get one aeroplane for its own flight tests. At a joint AAF Air Materiel Command-NACA conference at NACA Headquarters in February 1947, the NACA agreed to furnish fuel, maintenance and a flight crew for the XS-1 aeroplanes, and the AAF agreed to furnish fuel spare parts, maintenance and a flight crew for the launch B-29. In anticipation of the NACA XS-1 programme, De Elroy Beeler joined the NACA Muroc unit in January from Langley Laboratory as engineer-in-charge of the XS-1 loads programme, Gerald M Truszynski rejoined the project as instrumentation engineer in March 1947 and Joseph R Vensel, a former test pilot, arrived from NACA's Lewis Flight Propulsion Laboratory in April to head up NACA flight operations at Muroc.

In early May, more significantly, the AAF decided to take over the first XS-1's flight test programme completely. Colonel Albert Boyd, chief of the AAF's Air Materiel Command Flight Test Division, telephoned Larry Bell and broke the news that Bell test pilots would not be making the first supersonic flights in the aeroplane after all. For reasons that are unclear, this decision was not conveyed to the X-1 test team and, in particular, to Slick Goodlin, for whom it had special meaning. For several weeks the issue went back-and-forth between the service and Bell. When it was finally announced in early June, it shocked and disheartened the Bell team, already wracked by the resignation of Slick Goodlin. Goodlin had believed his contract terms firm, based on his conversation with Stanley after the Thompson Trophy race. Disturbingly, negotiations among the legal representatives of Goodlin and Bell had dragged on inconclusively into May

1947, leading Goodlin's lawyer to believe the company was stalling. (By this time, of course, the company's leadership already knew that, in all likelihood, no Bell pilot would ever fly the X-1 beyond Mach 1, but still it kept Goodlin and the X-1 test team in the dark.) In early June, Stanley informed Goodlin that the contract terms were unacceptable. The company could not pay the agreed-upon supersonic test bonus to a corporation for tax reasons; it could only pay to an individual. Stunned, Goodlin considered the turnaround 'unethical and unprincipled', and dictated a letter of resignation. When he received it, Stanley apparently considered the resignation a threat: an attempt at improving the pilot's negotiating position. Goodlin was serious, however, and 'told him he was quite mistaken'. He left the programme, having ably shepherded it through its subsonic demonstration phase. After deductions for taxes and payment of Jack Woolam's portion of the subsonic bonus, he had earned just over $5,000, less than his share of the Thompson Trophy purse, for his part in the XS-1 effort.[29] Ironically, Stanley himself soon resigned from Bell in a corporate dispute, leaving to form his own aviation company that would, in time, help transform the second X-1 into the Mach 2+ X-1E.

So Bell was out, and the Air Force and NACA were in. Following extensive discussions, including discussions with Bell, on 30 June 1947, NACA and AAF Air Materiel Command representatives met in conference at Wright Field to discuss the future programme of the Bell XS-1. The conference marked the initiation of a two-pronged transonic research programme. The representatives decided to let the AMC's Flight Test Division conduct an accelerated test programme to investigate transonic flight phenomena to a Mach number of 1.1 using the first XS-1; the AMC would seek to attain this speed as rapidly as was consistent with safety, in as few flights as possible. The NACA, on the other hand, would conduct a slower and more detailed research programme, making thorough examinations of transonic stability and control and flight loads, using the second XS-1. Due to the different emphasis in each phase of the flight programme, the two XS-1s would have different instrumentation packages. The first XS-1, the AAF aircraft, would have detailed instrumentation for speed measurement; measurement of normal, longitudinal and transverse acceleration; measurement of elevator-stick force and recording of aileron, elevator, and rudder positions. The NACA's second XS-1 would have more detailed instrumentation to record all of the quantities measured on the AAF XS-1, plus measurement of angular velocity in roll or pitch; sideslip angle; and aileron, elevator, and rudder forces. Most significantly, both had a 6-channel telemeter installed to transmit airspeed, control surface position, altitude, and normal acceleration to the ground, so that, as Walter Williams later recalled, 'if we lost the airplane, we could at least find out a little about what had happened'.[30] Now would come the most crucial test, the real challenge: flight to and through the 'sound barrier'.

5
Beyond Mach I:
The Air Force and NACA XS-I Programme

In midsummer 1947, the AAF began preparations to fly the XS-1 faster than sound. Overall responsibility for directing the AAF XS-1 programme rested upon James H Voyles, AMC XS-1 project engineer; Paul F Bikle, chief of the AMC Flight Test Division's Performance Engineering Branch and Colonel Albert Boyd, chief of the AMC's Flight Test Division. An AAF pilot since 1929, Boyd specialised in flight testing military aircraft and was highly respected as a pilot by other service test pilots. On the XS-1 programme, he had responsibility for selecting the AMC test crew that would fly the aeroplane at Muroc. Most recently, Boyd had flown from Wright Field to Muroc to win back the world's air-speed record for the United States. At Muroc, he flew a specially modified Lockheed Shooting Star, designated the P-80R, over a 3-kilometre course and during four low-altitude speed runs, averaged 523.738 mph, returning the speed record to the United States for the first time in 24 years. In searching for a pilot, Boyd turned to the Fighter Test Section of the AMC's Flight Test Division. One day, Major Kenneth Chilstrom, chief of the Fighter Test Section, called a meeting of the fighter test pilots at Wright Field. He announced the AAF was going to fly the XS-1 and he wanted the names of those pilots interested in flying the rocket aeroplane. Among the many pilots who expressed an interest was Captain Charles E 'Chuck' Yeager, a 24-year-old fighter ace from Hamlin, West Virginia.

Chuck Yeager had enlisted in the AAF in September 1941 and, after a brief stint as a mechanic and crew chief on a Beech AT-11 twin-engine trainer, left for pilot training. He proved an adept pupil, with all the aggressiveness and skill of a born fighter pilot. In March 1943, he won his wings as a Flight Officer and began training as a fighter pilot with the 357th Fighter Squadron based at Tonopah, Nevada. During a practice mission against some B-24's, the supercharger on his P-39 disintegrated, setting the plane on fire. Yeager jumped clear and pulled his ripcord. The opening parachute knocked him unconscious, and when he landed he cracked a vertebra. After recuperating in hospital, he left for England in January 1944 as a replacement P-51 pilot. He set an impressive record in the European Theatre of Operations. On 4 March 1944, while on a bomber-escort mission he shot down a Messerschmitt Bf 109. The next day, however, three Focke Wulf FW 190s teamed up to shoot down his Mustang. He parachuted safely to earth and managed to contact a French underground group. Over three weeks later, he slipped across the Spanish border, carrying a wounded fellow escapee to safety. Back in England, after promoting Yeager to captain, the AAF restricted him to non-combat flying for security reasons to prevent his possible recapture on future flights. The restriction lasted until after D-Day. He then began a meteoric combat record. On 12 October 1944, he shot down five Bf 109s; on 6 November he destroyed a Me 262 jet fighter and damaged two others and on 27 November, four FW 190s fell before his guns. By the time of his return from combat, he had shot down 11 German warplanes, shared in the destruction of another and damaged three more. He held the Distinguished Flying Cross, the Bronze Star, the Silver Star with oak leaf cluster, the Air Medal with six oak leaf clusters and the Purple Heart. Back in the United States, Yeager served briefly as a flight instructor at Perrin Field, but the AAF posted him to Wright Field as a pilot in the Fighter Test Section. He checked out in the jet-propelled P-59, then into the higher performance P-80. Shortly after Major Kenneth Chilstrom asked for the names of pilots interested in flying the XS-1, Colonel Boyd called Yeager into his office. He filled the young pilot in on the details of the project, and again asked him if he would be interested in flying the plane. Yeager replied affirmatively. Boyd then asked Yeager who he would recommend for the project, excluding himself. Yeager recommended 1st Lieutenant Robert A Hoover and Captain Jack Ridley.[1]

Like Yeager, Robert Hoover had seen combat in Europe during the war. He was likewise a superlative airman, best known for extraordinary aerobatic demonstrations in aircraft ranging from small biplanes to piston-engine and jet-powered fighters and business aircraft. He joined the AAF upon the outbreak of hostilities as a flight cadet and received training in twin-engine aircraft, though he preferred flying fighters. After arriving in England, he test-flew aircraft at a repair depot, flying aircraft ranging from fighters to four-engine bombers. He requested combat duty and the AAF transferred him to the 52nd Fighter Group, flying British-designed

Spitfire Mk.V fighters. On his 60th mission, a FW 190 shot his Spitfire down. Not as lucky as Yeager had been, Hoover spent the remainder of the war in *Stalag Luft I*, a Nazi POW camp for captured Allied airmen. After release, Hoover joined the Flight Test Division at Wright Field. Captain Jack Ridley held a BS in mechanical engineering from the University of Oklahoma. During the war he served as engineering liaison officer at the Consolidated Vultee plant in Fort Worth, Texas on the B-32 and B-36 programme. He had completed pilot training in May 1942. In March 1944 he attended the AAF School of Engineering at Wright Field. He then received assignment to Caltech where he studied under Theodore von Karman, earning an MS in aeronautical engineering in July 1945. The AAF assigned him to the AMC Flight Test Division at Wright Field where he underwent test-pilot training in the spring of 1946. Ridley was an excellent blend of engineer and pilot, and Yeager trusted him implicitly on all technical matters.[2]

The Air Force–NACA–Bell Team Gets Ready

Yeager's choices coincided with those of Colonel Boyd. Boyd selected Charles Yeager as project pilot, Robert Hoover as alternate and Jack Ridley as flight-test project engineer. He notified General Benjamin W Chidlaw, deputy commander of the Air Materiel Command, of his decision. He then sent Yeager, Hoover and Ridley to Buffalo to familiarise themselves with the plane. At Bell, the three pilots met with the XS-1 design team and observed a test firing of the second XS-1's rocket engine. By mid-July, Bell had completed repairs to the aircraft following its near-disastrous fire at Muroc on 5 June. Rather than waste time trying to patch the damaged aeroplane, Bell simply replaced the entire tail assembly of the second XS-1 with that of the third still under construction. As an additional safety measure, Bell replaced all the aluminium lines containing alcohol, oxygen and nitrogen in the engine compartment with lines of stainless steel. The AMC Flight Test Division chief selected Major Roberto L Cardenas, an experienced bomber test pilot (and close relative of a former president of Mexico) as B-29 launch aircraft pilot and Lieutenant Edward L Swindell as B-29 flight engineer. Late in July, Bell technicians attached the second XS-1 beneath the B-29 Superfortress for ferrying back to Muroc. The AMC test team then departed from the Bell plant for California. They arrived at the desert test centre on 27 July 1947, one day after President Harry S Truman signed the Armed Forces Unification Act creating a separate Department of the Air Force. When, and if, the XS-1 exceeded Mach 1, it would be the first major accomplishment of the nation's newly independent United States Air Force.[3]

Bell had closed down the company test facility at Muroc in mid-July, but the AMC borrowed Richard Frost to run a little ground school for the Air Materiel Command test team. For the remainder of July and into early August, Frost briefed the Air Force crew on the aeroplane systems and the intricacies of launching it in flight. On Wednesday 6 August, Yeager made his first glide flight in the first XS-1. At 12,000 feet while the B-29 climbed to launch altitude, Yeager clambered down the access ladder into the XS-1's cockpit and wriggled inside. At 18,000 feet, Bob Cardenas dropped the rocket plane as Frost and Hoover watched from two chase P-80's. Yeager was on his own. Like Woolams, Goodlin and Johnston before him, he found the plane's control characteristics highly satisfactory and, as the plane glided towards the lakebed, he made three slow rolls. The next day, Yeager completed another glide flight with equal success. On 8 August, Yeager dropped away from the B-29 on his final familiarisation glide flight and engaged Hoover's chase P-80 in a mock dogfight all the way down. The AMC test crew now concentrated on preparing for the first Air Force powered flights.[4]

Walter Williams' NACA Muroc Flight Test Unit also moved forward with plans for its own part in the two-pronged AF-NACA transonic research programme. In August, the unit's first two research pilots arrived at Muroc, Herbert H Hoover and Howard C Lilly. A former Navy pilot, Lilly came to Muroc from NACA's Lewis Flight Propulsion Laboratory at Cleveland. During the 1946 Thompson Trophy Race at Cleveland where Woolams had died and Johnston had triumphed, Lilly was placed seventh in a P-39. Herbert Hoover, a NACA pilot since 1940, had been one of the first pilots deliberately to fly through thunderstorms on bad-weather flying research. He already had a reputation as a cool pilot in tight situations; once, during an instrumentation calibration flight in a NACA SB2C Helldiver, the plane's cockpit canopy hood came loose in flight smashing Hoover across the forehead and inflicting a deep cut that bled profusely. Though stunned by the blow and blinded by blood flowing into his eyes, Hoover instinctively retained control of the dive bomber, cleared his eyes and, despite his injuries, brought the plane back to Langley for an emergency landing. On another occasion, while firing a rocket-propelled model from a P-51 in a Mach 0.7 dive, the model disintegrated, showering the Mustang with wreckage. The wreckage punctured the plane's coolant tank, but again Hoover brought the plane in for a successful forced landing. After arriving at Muroc in August, Hoover studied the XS-1 under Richard Frost's careful tutelage and also kept abreast of the Douglas contractor programme then underway on the D-558-1 Skystreak. Howard Lilly had no background in compressibility research, so Hoover fitted a P-51 with a Machmeter and accelerometer and had Lilly fly it into the region of trim changes and buffeting to familiarise him with the characteristics of compressibility. Step-by-step, Hoover and Lilly were preparing themselves for their own sonic assault.[5]

Yeager's First Flights

Late in August, after checking out the rocket engine of the first XS-1 on the ground, Yeager readied himself for the first Air Force powered flight. Continuing a tradition he had begun in the Second World War with his P-51s, Yeager named the XS-1 in honour of his wife, painting *Glamorous Glennis* in red and white letters on the plane's nose. For the first flight, Yeager had strict instructions not to exceed Mach 0.8 in the rocket plane; the flight was for pilot familiarisation rather than for acquisition of research data. On 29 August, Cardenas lifted the heavy bomber off the runway at Muroc and droned upwards to launch altitude. At 7,000 feet pressure altitude, Yeager entered the first XS-1. During the final minutes before drop he pressurised both the fuel and liquid oxygen tanks, checked the fuel jettison system and turned on the data recording switch that automatically activated the NACA recording instrumentation. Then, as Cardenas finished intoning the countdown, Yeager dropped away from the B-29 in the XS-1 at 255 mph and 21,000 feet above the desert.[6]

Heavy with fuel, the rocket plane fell more than 500 feet before Yeager ignited the first rocket chamber. Then, as the plane accelerated, he shut down the first chamber and ignited the second then shut it down and ignited the third, following this with a slow roll. During the roll the plane attained zero g and engine shut down due to a drop in lox tank pressure, but he reignited the third chamber when the plane regained a nose-down attitude. He then shut it off and fired the fourth, shut down the engine and nosed into a dive, completely powerless. During the dive he vented some of the propellants to check the XS-1's jettison system, and noted the same snaking oscillation due to fuel sloshing that Goodlin had noticed on his first powered flight. The full fuel load, combined with a dive entry speed of Mach 0.7, caused the XS-1 to pick up speed rapidly in the dive. Richard Frost, flying a chase FP-80, followed Yeager down and suddenly found he was travelling slightly over Mach 0.8 with the plane buffeting and shaking from compressibility. At 5,000 feet Yeager recovered from the dive, fired the first chamber and initiated a shallow climb and then fired the remaining chambers in sequence. Climbing steeply in a vain attempt to keep below Mach 0.8, the pilot reached 35,000 feet and began recovery into level flight, finally dropping the nose and rolling out to a normal attitude at 30,000 feet, having attained Mach 0.85 during the recovery. He then shut off the engine, jettisoned his remaining propellants and glided down to the lakebed. Chuck Yeager's flight to Mach 0.85 brought a rapid note from Colonel Albert Boyd that neither the XS-1 pilot nor aeroplane was expendable, 'so please approach higher speeds progressively and safely to the limit of your best judgement'. Yeager and Jack Ridley responded that discussions between the AMC test team, NACA and Bell project engineer Richard Frost suggested that no trouble would be encountered to Mach 0.85 and that when Yeager attained Mach 0.83 without difficulty, he felt safe in proceeding to Mach 0.85. The reply reiterated that the AMC test team regarded safety as of paramount importance. With Yeager now fully checked out in the aeroplane, the Air Force could begin research-flying the first XS-1. The Air Force desired to attain supersonic speed as quickly as was safely possible. The NACA favoured a more cautious incremental approach to Mach 1. Writing about the programme a decade later, Walter Williams stated:

> We were enthusiastic, there is little question. The Air Force group – Yeager, Ridley – were very, very enthusiastic. We were just beginning to know each other, to work together. There had to be a balance between complete enthusiasm and the hard, cold facts. We knew and felt that if this programme should fail the whole research airplane program would fail, the whole aeronautical effort would be set back.[7]

The successful flights of the XS-1 during the contractor programme had not served to mute the voices of Cassandras who predicted the rocket would explode or that it would disintegrate from compressibility effects. In the background lurked the image of Geoffrey de Havilland's DH 108 coming apart over the Thames less than a year before. Frost was more optimistic; he knew Bell had designed the plane according to the best aerodynamic knowledge available with thin wings, a bullet fuselage and a high tail. Beyond that, some educated guesswork went into the design. 'If it could be done,' he reflected later, 'we knew we could do it.' NACA and the Air Force did have some information to go on in predicting the transonic behaviour of the XS-1. The Bell acceptance tests of the aeroplane agreed with low-speed wind tunnel test results. Wind tunnel data ended at about Mach 0.85, but predicted satisfactory behaviour at those speeds. Wing-flow data ended around Mach 0.93 with inconclusive results. With the test information disappearing at the point where scientists expected critical changes to occur in stability and control characteristics, NACA engineers 'developed a very lonely feeling as we began to run out of data'.[8]

Yeager completed his second powered flight in the Air Force XS-1 on 4

September, attaining approximately Mach 0.89. On the flight, however, the telemeter failed, preventing acquisition of stability and control data, so Yeager successfully repeated it on 8 September and followed with a fourth flight to Mach 0.91 on 10 September. The AMC team then decided to probe gingerly around Mach 0.9 to evaluate elevator and stabiliser effectiveness and buffeting. Yeager reached Mach 0.92 on 12 September, the fifth powered Air Force flight of the XS-1 but the AMC temporarily suspended flights of the rocket plane after this flight, pending installation of a faster stabiliser actuator. Yeager departed for Wright Field for final checks in a T-1 partial pressure suit (itself a relatively new development from aviation medical research). At the high altitude the XS-1 would make its top-speed runs (50,000 feet and above) the young pilot would need a pressure suit to survive should the rocket plane lose its cockpit pressurisation. While at Wright Field, Yeager and Jack Ridley reported to Colonel Boyd on the progress of the flight programme from whom they received cautions to proceed carefully and then flew back to Muroc. Meanwhile, the NACA Muroc Flight Test Unit, in addition to monitoring the Air Force flights and performing data reduction and technical advising, had readied their second XS-1 with its 10%-thick wing and 8%-thick tail for its acceptance flight. Since neither Herbert Hoover nor Howard Lilly had yet checked out in the plane, the task fell to Yeager, who made its proving flight on September 25.[9]

The Assault on the 'Sound Barrier'

If the programme had dodged a bullet with the Soviet 346-1's problems, a new presence at Muroc nevertheless added its own pressures for the team to get on with the business of supersonic flight. In early September, a flatbed truck had wended its way over the Los Angeles hills, delivering the first prototype North American XP-86 Sabre sweptwing jet fighter from its Inglewood manufacturing plant to Muroc. The XP-86, sire of a long line of what would arguably be the most graceful jet fighters ever flown, exemplified the ugly duckling that evolves into a beautiful swan.[10] Originally conceived with a straight wing and tail, it benefited from the tremendous interest in sweptwings that accompanied both Robert Jones' research and the discovery of Nazi interest in them. Hastily but thoroughly redesigned in the autumn of 1945, the XP-86 had a 35° swept wing, similarly swept tail surfaces and, like the XS-1, an adjustable horizontal stabiliser. Lithe, light and aerodynamically clean, the Sabre looked like it was going Mach 1 just sitting on the ground; while none of the Sabre family ever were supersonic 'on the level', they all could dive to just past the speed of sound, the XP-86 included. North American test pilot George Welch made the first flight in the XP-86 on 1 October 1947 and shop-talk around Muroc suggested strongly that he was not the kind to wait for the XS-1 to be first. While prudence would dictate a cautious approach – for North American could not afford to risk a costly prototype on a casual stunt – it would not be too long (a few months at most) before the XP-86 would be nudging the sonic frontier.*

It was in this atmosphere that Yeager and the XS-1 returned to the air. Though flying the XS-1 could hardly be considered routine, by now the AMC flights had settled into a standard pattern of operation. Frost flew low chase for the XS-1, and Robert Hoover flew high chase, piloting a FP-80 photo plane. Frost flew in formation about 1,000 feet above and behind the B-29 and when Bob Cardenas began the countdown, Frost would nose over in a dive, pass below the B-29 and be off Yeager's wing as the XS-1 dropped from the bomb bay. Frost would pull into level flight and then into a slight climb as Yeager ignited the rocket engine. The accelerating rocket plane invariably left Frost's P-80 straggling far behind. Ten miles ahead of the B-29 and at about 48,000 feet, Bob Hoover flew along in his FP-80 on high photo chase. After the drop, Yeager would climb towards Hoover's plane so that at the climax of the mission he would pass Hoover at close range to the Shooting Star's powerful cameras. By now Yeager was flying into the region of unknowns, extending each flight a little bit faster, encountering shock wave buffeting on every flight that reminded him of 'driving on bad shock absorbers over uneven paving stones'. The broad clear contrail generated by the engine furnished Frost with a big marker to follow. After Yeager passed Hoover and shut off the engine, if it had not

---

* There have been persistent rumours since 1947 that Welch broke the speed of sound before Chuck Yeager, but that 'the Air Force hushed it up'. While Welch was reportedly an individual who might well have attempted to do so, an event of this sort would, in all likelihood, have generated more substantial proof than mere rumour, and, for its part, North American (now Rockwell) has always stated that the first supersonic foray of the XP-86 was on 26 April 1948, which is consistent with the Sabre's planned flight test programme. Until – if ever – substantial evidence is produced, rumours that the Sabre was first seemingly fall into the same mythic category as rumoured flights before the Wrights, or propeller-driven aeroplanes that exceeded the speed of sound. And in any case (as with the Wrights), it was the XS-1 that had the impact upon both aeronautical science and the popular imagination alike. What is worth noting is that the pace of aeronautical development in the 1940s was so explosive that a prototype jet fighter could arrive for flight testing with potential performance believed limited only to specialised research aeroplanes only two to three years before.

already starved itself of propellants, the contrail ended. Hoover and Frost (catching up from launch) then had to keep the little rocket plane in sight as it descended powerless to the lakebed which was an extremely difficult job, for the plane was small and hard to see against the tawny earth and, if the chase pilots momentarily shifted their glance, they could lose sight of the plane.[11]

After completing an airspeed calibration flight to Mach 0.925 on 8 October, the AMC test crew readied the aeroplane for another stability and control investigation on 10 October, one more flight in the incremental step-by-cautious-step towards the first supersonic forays. That day, Cardenas dropped the XS-1 from the launch B-29 above the yellow-brown Mojave. Chuck Yeager fired all four chambers of the rocket engine, nosed the XS-1 upwards in a climb to 38,000 feet and then shut down the chambers one by one. After performing an accelerated stall, he dropped 2,000 feet, tripped two of the rocket chamber switches on the cockpit panel and accelerated in a climb to 40,000 feet. The rocket plane picked up speed rapidly at the high altitude and Yeager reached an indicated Mach number of 0.94 at 45,000 feet. He pulled back on the control column and – nothing. The elevator had completely lost control effectiveness because at that speed the shockwave on the horizontal tail had shifted aft and was now standing right on the elevator's hingeline. 'The control column,' as he wrote later in his flight test report, 'could be moved to the limits of travel each way with little force and very slow response in airplane attitude.' Then, as the last of the liquid oxygen and water-alcohol fed into the combustion chambers, the rocket engine sputtered for lack of fuel and the young pilot switched it off. The XS-1 coasted upwards to 45,000 feet before arcing over for its return to earth. On the long glide downwards, frost formed on the inside of the canopy which persisted despite Yeager's efforts to scrape it off. Robert Hoover and Richard Frost, riding alongside the powerless plane in their chase P-80's talked the XS-1 pilot down to a risky but safe 'blind' landing on the lakebed. In the back of Yeager's mind, though, was the thought that 'We had had it. There was no way I was going faster than .94 Mach without an elevator'.[12]

Discussions with Ridley, the NACA team and with Colonel Boyd led to a suggestion by Ridley that the flights should now proceed using the adjustable horizontal stabiliser as the primary means of control up to Mach 1. NACA urged more caution, but Yeager, Ridley and Bell representative Richard Frost wanted to press on. The NACA's reduction of the data recorded by the plane's internal instrumentation, combined with ground radar tracking, offered its own surprise to the waiting NACA and AMC test teams. It revealed that the XS-1 had flown faster than ever before, perhaps even as fast as the speed of sound. Although the cockpit Machmeter indicated only 0.94, corrections for the readings pushed the true Mach number up to 0.997 at a pressure altitude of 37,000 feet, approximately 658 mph. That morning, high above Muroc, Chuck Yeager had possibly become the first man to exceed the speed of sound. To Richard Frost, the message was clear: repeat the flight, but at a higher Mach number to ensure a clear-cut case of supersonic flight.[13]

Early on the morning of Tuesday 14 October, long before the sun cast its light upon the dry lake, the participants in the continuing drama of Chuck Yeager and the *Glamorous Glennis* rubbed sleep from their eyes and set about making final preparations for the rocket plane's ninth powered flight. Yeager himself was not in the best of condition that day. Over the intervening weekend, he had taken his wife to aviatrix Florence 'Pancho' Barnes' Fly-Inn, a popular dude-ranch and gathering-spot located southwest of the base between Muroc and Rosamond Dry Lake. He took a horse out for a night ride in the desert but on returning to the ranch, failed to see a gate locked across the corral. The horse ran into the gate, bolted and threw Yeager off. Though he made little of the accident, he was in considerable pain. Realising the goal of the last few months to be just within reach and unwilling to risk losing the chance to make the first supersonic flight, Yeager visited a civilian physician in nearby Rosamond rather than go to the base hospital. The doctor's inspection revealed two broken ribs, but Yeager decided to keep quiet about his accident. If the pain prevented him from crawling from the mother B-29 into the cockpit of the XS-1 in flight, he would cancel the mission before the drop.[14]

Muroc came alive with activity at 6 o'clock on the morning of the flight. The saffron XS-1 lay bare, its insides and plumbing exposed, as ground crewmen readied it and its rocket engine for flight as the sun rose over the ridgeline east of the lakebed and cast its first tentative rays upon the base. As Yeager gathered with Jack Ridley and the rest of the B-29 launch crew, the talk centred on the upcoming flight: that they should check out the stabiliser setting for optimum control effectiveness before initiating the high-speed run. The test crew already had heard of Yeager's weekend accident (though only his long-time friend Jack Ridley knew about the injury) and to relieve tension, Ridley, Frost, Hoover and Cardenas presented the test pilot with glasses, a rope and a carrot. After coffee in the service club, Yeager gave the XS-1 a preflight check and then suited up. He conferred with Williams and De E Beeler of the NACA Muroc Flight Test Unit. The NACA engineers stressed caution and warned him not to exceed Mach 0.96 unless absolutely certain that he could safely do so. Then Yeager entered the B-29. At 10.02am the B-29's four Wright R-3350 piston engines began to clatter and the huge four-bladed propellers slowly revolved, with ever increasing speed. From the Muroc control tower, a skeleton structure, came clearance for takeoff. Cardenas began the takeoff roll; the bomber, with its orange parasite, trundling down the concrete runway and finally lifting into the clear blue sky.

As the Superfortress began straining for altitude, a complex series of operations took place on the ground. The Air Force closed Muroc Dry Lake and its environs to unauthorised aircraft Richard Frost and Robert Hoover prepared to take off in their chase P-80s. At the NACA facility, now grown to 27 from the original 13, Gerald Truszynski and his instrumentation engineers readied their telemetry reception equipment and two SCR-584 radar sets to track the flight. Others remained close to the flight communications monitors to keep informed on the progress of the mission. After three years of intensive research and development the final trial would come within a mere 30 minutes, eight miles above the desert floor,

At 5,000 feet, Yeager left the relative security of the B-29's bomb bay for the transfer ladder and, with Ridley helping him, he squirmed painfully into the open cockpit hatch of the XS-1. Ridley closed the hatch in place, sealing Yeager into the rocket plane and the plucky pilot locked it using a sawn-off broom-handle to give him leverage and make reaching (painful due to his broken ribs) unnecessary. The B-29 continued its climb; Hoover and Frost joined up for chase. On the ground below, Truszynski's radar tracked the spiralling bomber as it climbed upwards. At the five minute warning before launch, Yeager pressurised the fuel and liquid oxygen tanks and checked the fuel jettison system. Then, with nothing else to do, he waited for the drop. At the 1-minute warning, NACA radar cleared the B-29 to launch the rocket plane and Jack Ridley asked Yeager, 'You all set?' Hunched in the cockpit, tense but ready for flight, the young pilot replied, 'Hell, yes, let's get it over with'. Bob Cardenas then voiced the countdown. At 10.26am, at a pressure altitude of 20,000 feet and an indicated airspeed of 250 mph, the first XS-1 dropped free from the dark bomb bay of the B-29, out into the bright sunlight.

As the plane dropped earthwards like a winged bomb, Yeager fired the fourth engine chamber, then the second, shut-down the fourth, fired the third, then shut down the second and fired the first. All four chambers had checked out. Under the combined thrust of the first and third chambers, the XS-1 began to climb, racing away from Richard Frost in his P-80 and the ponderous B-29. Yeager reignited the second and fourth chambers and, under a full 6,000-lb thrust the rocket plane accelerated for altitude, trailing a cone of fire and streaming shock-diamonds in the exhaust as it rapidly expended its propellants, becoming lighter and faster by the second. From Mach 0.83 to 0.92, Yeager cautiously tested stabiliser effectiveness. Though the rudder and conventional elevator lost effectiveness, the adjustable stabiliser still proved effective, even as speed increased to Mach 0.95. At 35,000 feet, he shut down two chambers, continuing the climb on the remaining two. Yeager began to level off around 40,000 feet and fired one of the two shutdown chambers. Describing the flight later, Yeager said,

> With the stabiliser setting at 2° the speed was allowed to increase to approximately .98 to .99 Mach number where elevator and rudder effectiveness were regained and the airplane seemed to smooth out to normal flying characteristics. This development lent added confidence and the airplane was allowed to continue to accelerate until an indication of 1.02 on the cockpit Mach meter was obtained. At this indication the meter momentarily stopped and then jumped to 1.06 and this hesitation was assumed to be caused by the effect of shock waves on the static source. At this time the power units were cut and the airplane allowed to decelerate back to the subsonic flight condition. When decelerating through approximately .98 Mach number a single sharp impulse was experienced which can best be described by comparing it to a sharp turbulence bump.[15]

Yeager had attained an airspeed of 700 mph, Mach 1.06, at approximately 43,000 feet, making him the first pilot to exceed the speed of sound and Hoover, in his FP-80, had caught the moment on camera. Inside the aeroplane, the recording instrumentation captured the classic 'Mach jump' pressure trace as the sonic shockwave passed over the plane's static pressure port as it accelerated through the speed of sound. Aside from the Machmeter readings, Yeager did not note anything spectacular about breaching the 'sound barrier'. There had been no violent buffeting, no wrenching of the plane; instead, as Yeager described it later, 'it was as smooth as lemonade. Grandma could be sitting up there sipping lemonade'. He shut down the XLR-11 engine for good, though 30% of the propellants remained. The first XS-1 coasted up to 45,000 feet, and Yeager performed a 1g stall in clean configuration. The nose of the plane then dropped downwards towards Muroc Dry Lake. Before the flight, ground crewmen had bathed the cockpit glass with Drene shampoo to prevent frosting and Yeager did not encounter any on the trip earthwards. During the glide back, Yeager could hear the ticking of the cockpit clock in the virtually soundless and chilly cockpit, cold from the residual liquid oxygen in the tank. Frost's and Hoover's P-80s joined up, and 14 minutes after dropping away from the B-29, the tyres of the XS-1's landing gear brushed the clay of Muroc Dry Lake. The little plane, travelling at 160 mph, rolled approximately 2.2 miles across the lakebed before coming to a stop. Yeager got stiffly out, his ribs paining him, fatigued having completed the first manned supersonic flight in aviation history.

The Air Force did not rest on its accomplishment. Though word of Yeager's achievement immediately flashed out to those with a need to know within the NACA, industry, the Air Force, and the Navy, it remained highly classified and officials did not release it immediately to the public. Not everyone at Muroc knew of the feat – Yeager's reward was to have been a free steak dinner that night at Pancho's – but he had to settle instead for a martini-fuelled party with friends at his house. After the flight, Yeager and key members of the test team left for Wright Field to brief Colonel Boyd and the Air Materiel Command's senior staff. On 20 October at a secret ceremony, Yeager received a second Distinguished Flying Cross. That night, key team members gathered with Larry Bell in the Kitty Hawk Room of the Biltmore Hotel for a celebratory dinner. To the world at large however, the events of 14 October had not occurred; it did not even appear in the Muroc semiannual historical report. To outsiders, then, the XS-1 remained just another exotic aeroplane flying at the lake. After returning from Wright Field, Yeager made four more flights in the XS-1 before the end of October, on one of which he experienced a full-blown electrical failure, forcing him to jettison propellants (using a jury-rigged nitrogen expulsion system installed by Richard Frost) and land. On 6 November, he dropped away from the B-29, fired up the XLR-11 engine and streaked to Mach 1.35 at 48,600 feet, slightly over 890 mph – twice as fast as a conventional P-51D Mustang fighter. By early November, the myth of a 'sound barrier' had been laid in its grave. Many questions still remained to be answered about aircraft stability and control and loads in the transonic region, but the barrier syndrome disappeared. The XS-1 had proven that a properly designed aeroplane flown by a skilled pilot, could pass from subsonic to supersonic flight without damage.

The Air Force and NACA Explore the Supersonic Domain

One week after Chuck Yeager became the first pilot to exceed the speed of sound, Herbert Hoover initiated the NACA programme on the second XS-1 with a familiarisation glide flight. Hoover thought the plane had pleasant stall characteristics with adequate warning and mild motions but on landing, the little rocket plane touched down hard on its nosewheel, and the landing strut collapsed. Both XS-1 aircraft had peculiar behaviour on landing, that could – as in Hoover's case – trigger nose wheel failure. In part, it stemmed from pilot technique and training. The XS-1 would touch down at about 130 mph, sometimes bouncing back into the air. During the bounce, its speed would drop off to 125 mph or less and at these low velocities, the elevator lost control effectiveness. Simultaneously, the nose would begin to drop. After the bounce, most pilots would be reluctant to pull the control column all the way back – exactly the corrective action required on the XS-1 – for fear of stalling out. With the XS-1's low elevator effectiveness at this speed – around 125 mph – the plane required almost full up elevator to keep the nose from dropping through. Consequently, most pilots would handle the control column gingerly and by the time the nose dipped, it was too late: the nose wheel would hit hard and collapse. Repairs and subsequent maintenance work, along with the annual winter rains, kept the XS-1 grounded until 16 December when Hoover made the first rocket flight ever performed by a NACA pilot, attaining Mach 0.71. He flew again on another checkout flight the next day reaching Mach 0.8.[16]

By this time, word was leaking out about Yeager's accomplishment two months before, and Secretary of the Air Force Stuart Symington directed that initial data from the programme be prepared for presentation to the aircraft industry for use by contractors. On 22 December, the industry trade paper *Aviation Week* disclosed Yeager's supersonic flight in a story subsequently picked up by newspapers coast-to-coast. The Air Force and NACA neither confirmed nor denied the story and the Justice Department under Attorney General Tom C Clark began an investigation to see if the government could prosecute the magazine for violation of national security. (Department attorneys subsequently concluded that the magazine had not broken any laws.) On 9 January 1948, senior officials of the Air Force, the Navy, the NACA and private industry met at Wright Field for an historic conference on the XS-1's research results. Among those who attended were John Stack, Walter Williams, Hartley Soule, Ira Abbott and Hugh Dryden of the NACA; Ezra Kotcher, General Lawrence C Craigie, General Franklin O Carroll, Colonel Albert Boyd and Dr Hsue-shen Tsien of the Air Force and Larry Bell, Bob Stanley and Dick Frost of Bell, as well as Ridley and Yeager of the Air Force test team. The seminar covered the history of the XS-1 programme, the operation of the aeroplane, its tests and test results. Speakers pointed out that while the XS-1 generally experienced most of the difficulties predicted previously by engineers, the magnitude of such problems had not been as serious as had been expected. In speaking of the future programme of the XS-1 aeroplane. Charles L Hall of the AMC's Aircraft Projects Section spoke of possibly using the XS-1 for applied research and equipment testing, such as research on aerodynamic beating, weapons systems for supersonic aircraft and problems of crew safety. He announced that for these and other reasons the Air Force had already contracted with Bell for four advanced XS-1 aircraft.[17]

As events turned out, however, both XS-1's continued to fly as aerodynamic, control and structural research vehicles and neither the Air Force nor NACA made tests of aircraft equipment or systems in the two rocket planes. During January 1948, Yeager made three supersonic flights in *Glamorous Glennis*, by now showing an alarming predilection for minor but unnerving engine fires. Faithful chase pilot Bob Hoover, sadly, never got his chance to fly in the plane. In addition to their special duties, the AMC XS-1 test team participated in test programmes on other military aircraft. At the end of September 1947, the first Republic P-84B Thunderjets arrived at Muroc for accelerated service tests. Late in November, while flying one of the planes, Hoover had experienced an inflight fire that burned through his control cables, forcing him to eject in a high-speed dive, but the seat would not fire, and he had to bail out, striking the horizontal stabiliser and breaking both legs. The accident grounded him for three months, taking him away from the XS-1 programme. The test team replaced him with Captain James T Fitzgerald Jr. Fitzgerald attempted a checkout flight in *Glamorous Glennis* on 24 February 1948, but after launch the plane experienced an inflight fire and he had to jettison the remaining propellants, landing on the lakebed. On 26 March, Yeager attained the highest speed ever made by an XS-1 aeroplane. After launching from the B-29 Superfortress at 31,000 feet and 240 mph, he climbed on three chambers at Mach 0.85 to 50,000 feet, levelled off and, under full power, began a shallow dive. *Glamorous Glennis* reached Mach 1.45 at 40,130 feet, as determined later by NACA analysis of radar tracking data and instrumentation readings, an airspeed of approximately 957 mph. After performing two stall checks, he continued down to land on the lakebed. During April and May, Captain Fitzgerald and fellow Air Force test pilot Major Gustav E Lundquist performed a series of pressure distribution surveys and stability and control investigations around Mach 1.1. Air Force officials grounded the aeroplane in May for installation of strain-gauge instrumentation and also for engine work. On 25 May, Fitzgerald returned the plane to the air on a buffet-boundary investigation and on 3 June, during a low-altitude transonic investigation by Lundquist to Mach 0.95, the left main gear door of the first XS-1 opened in flight. Despite the trim problem caused by the opening, Lundquist succeeded in landing on the lakebed. The nose wheel collapsed upon landing and the B-29 ferried the aeroplane back to Wright Field – recently redesignated Wright-Patterson Air Force Base – for repair.[18]

Less spectacularly, the NACA programme went somewhat more smoothly, as Williams' Muroc Flight Test Unit concentrated on making the agency's first supersonic flight. In January 1948, NACA pilots completed seven subsonic flights in their XS-1. In January, Howard Lilly checked out in the plane and by the end of the month, Hoover had worked up to Mach 0.925. On March 4, Hoover attained Mach 0.943 at 40,000 feet. Then six days later, Hoover dropped from the B-29 on a stability and loads investigation, fired three of the four rocket chamber and began to climb, levelling out at 42,000 feet. The second XS-1 rapidly accelerated to Mach 0.93 and Hoover fired the fourth chamber. Under full thrust, the rocket research aircraft shot to Mach 1.065, approximately 703 mph. Herbert Hoover had become the first civilian pilot to exceed the speed of sound.* After engine burnout he coasted to the dry lake. Despite emergency efforts, Hoover found he could not extend the nose landing gear. He held off the nose as long as possible, and even though the plane skidded to a stop on the underside of the nose, the damage was slight. NACA had the plane back on flight status within 10 days and later that month Lilly also exceeded Mach 1. In April, Lilly continued the NACA stability and control and aerodynamic loads programme on its XS-1, but on 16 April the NACA plane came to grief. As the plane touched down, it began a skipping motion, and on the third contact the nose landing gear collapsed, tripping the landing gear operating mechanism and causing the left main gear partially to retract. The rocket plane skidded to a halt resting on the nose wheel, right main wheel, left wingtip and tail cone. Repairs to the NACA XS-1 occupied the better part of the summer and modifications to the plane kept it grounded into the autumn. By September 1948, the Muroc Flight Test Unit had prepared a test programme for approximately 50 flights to investigate aerodynamic loads and stability characteristics over a range from Mach 0.6 to Mach 1.2 at an altitude of approximately 40,000 feet. Early in October the Muroc engineers painted the plane white overall to facilitate tracking and in November, it returned to flying status. In October 1948, NACA had sent another research pilot from Langley, Robert A Champine, to Muroc to replace Howard Lilly who had died the previous May in the crash of the second Douglas D-558-1 Skystreak. A former Navy fighter-bomber pilot, Champine had joined the Langley

---

* In January 1950, President Truman awarded Hoover the Air Medal for 'meritorious achievement while participating in aerial flight on 10 March 1948 . . . in piloting an experimental aircraft faster than the speed of sound, thereby providing valuable scientific data for research in the supersonic field'.

research-pilot staff in December 1947 and had flown the Bell L-39 low-speed sweptwing research aeroplane. After Lilly's death, Herbert Hoover remained on at Muroc to train Champine in the intricacies of transonic test flying. Hoover made a check flight in the NACA rocket plane, now designated simply the X-1, on 1 November. After a second check of the plane on 15 November, Hoover turned it over to Champine, who made his first flight in the aircraft on 23 November. Over the next two weeks, Champine completed a further three flights for pilot familiarisation. NACA technicians installed handling qualities instrumentation in the aeroplane for further research work.[19]

The Mackay and Collier Trophies

Despite the revelation of the first supersonic flight by *Aviation Week*, the Air Force and NACA waited until June 1948 before confirming the flight. On 15 June 1948 in a joint press conference, General Hoyt S Vandenberg, Chief of Staff of the Air Force and Dr Hugh L Dryden, NACA's newly appointed Director of Aeronautical Research, confirmed that the XS-1 had repeatedly exceeded the speed of sound while flown by Air Force pilots Yeager, Fitzgerald and Lundquist, and by NACA pilots Herbert Hoover and Howard Lilly. Secrecy still prevented disclosure of particular facets of the supersonic flight and Yeager responded to more inquisitive questioning by replying, 'If you had a gold mine, you wouldn't tell the world where it was, would you?' Asked how it felt to fly through the 'sound barrier', the young pilot replied, 'It was pretty nice, I thought'. At the ceremony, Yeager received another oak leaf cluster to his Distinguished Flying Cross as well as the Mackay Trophy for 1947, presented annually for the most meritorious flight of the year. With shrewd insight, Hugh Dryden, one of the original propeller researchers whose work had triggered the first study of transonic flight conditions, noted,

> The achievement of Captain Charles E Yeager as the first man to attain sustained horizontal supersonic flight in a piloted aircraft brings to public attention the power of a new tool, the research airplane, in obtaining the basic aeronautical knowledge essential to the design of military aircraft of outstanding performance . . .
>
> It is not easy for us so close to the event to appraise correctly the significance of the dawn of the supersonic age. There is danger on the one hand of seeming to minimize the accomplishments of a pioneer who ventured where none had traveled before and on the other hand there is danger of arousing undue expectations on the part of the public as to the performance to be expected of tactical military

> aircraft. The XS-1 is a small research airplane, flown at high speed at high altitude where the air loads on the structure are small. Between it and tactically useful military aircraft of larger size flying at lower altitude where the air loads are much greater there remains much research and development on many difficult problems.

Some doubt still remained in the public mind about the origins of the XS-1. Shortly after completion, many sources incorrectly stated it drew on German wartime technology and some suggested its engine was merely an American modification of the German Walter HWK 109-509 that had powered the Messerschmitt Me 163 *Komet* interceptor. An editorial written after the Air Force-NACA disclosure stated bluntly, 'Moreover, since the XS-1 is admittedly merely an improved copy of a German wartime fighter, it is not impossible that the Germans also knew a thing or two about supersonic flight.' Not unexpectedly, this brought a swift and sharp denial from the Air Force and NACA. In fact, the only foreign country that yet had any experience with manned supersonic flight was Great Britain, courtesy of British test pilot Roland Beamont, who had flown the experimental North American XP-86 Sabre at Muroc on 21 May 1948, exceeding Mach 1 in a dive remarkable for its tranquillity. In contrast, on 6 September 1948, test pilot John Derry, flying the third DH 108 Swallow – a more refined, powerful and streamlined development of the second aircraft which had killed Geoffrey de Havilland – reached approximately Mach 1.04 in the course of an essentially out-of-control dive during which the vicious little aeroplane tried to bunt – i.e. tuck under in a violent outside loop. This was the first supersonic flight in Great Britain. (Later, in 1949, it actually did bunt in excess of Mach 1, fortunately without killing its pilot.) On 26 December 1948, the Soviet Union became the third nation to exceed Mach 1 when test pilot I V Federov, flying an experimental Lavochkin La-176 jet fighter, dived through the speed of sound reaching Mach 1.02.[20]

While the USAF and NACA still laboured on the X-1 programme and foreign nations attempted to equal it, the time had come for richly deserved public honour: the award of the Robert J Collier Trophy for 1947, aviation's most prestigious award. At 12.30pm on 17 December 1948, the 45th anniversary of the first successful flight at Kitty Hawk, John Stack, Chuck Yeager and Larry Bell gathered in the office of President Truman at the White House to receive the trophy, awarded annually for 'the greatest achievement in aviation in America, the value of which has been demonstrated by actual use during the preceding year'. The citation to the award read:

> To John Stack, Research Scientist, NACA, for pioneering research to determine the physical laws affecting supersonic flight, and for his conception of transonic research airplanes; to Lawrence D Bell, President, Bell Aircraft Corporation, for the design and construction of the special research airplane X-1; and to Captain Charles E Yeager, US Air Force, who, with that airplane, on October 14, 1947, first achieved human flight faster than sound.

Along with the citation, the description of the project aptly summarised the impact of the saffron speedster upon aviation history:

> This is an epochal achievement in the history of world aviation – the greatest since the first successful flight of the original Wright Brothers' aeroplane, forty-five years ago. It was not the achievement of any single individual or organization, but was the result of a sound aeronautical research and development policy involving fine teamwork and cooperation between research scientists, industry, and the military – a factor essential to keeping America first in the air. Without this teamwork the achievement would have been impossible.

In retrospect, there was one notable omission from the Collier trophy that deserves correction. Undoubtedly, given his strong advocacy and personal persistency that resulted in the XS-1 being the aircraft it was – an air-launched, supersonic, rocket-propelled, research aeroplane – Ezra Kotcher deserved, as much as any of the other three awardees, to have been named a recipient of the 1947 Collier trophy. Always a gentleman, and modest to a fault (John Stack referred to him emphatically as 'a nice man – a very nice man') – Kotcher never complained. If it hurt him deeply, he never let it show, never let others know. In the opinion of this author, however, the omission of Ezra Kotcher from the Collier trophy will always constitute a blemish that should be corrected.[21]

Ground Takeoff, at Last and but Once

Shortly after the award ceremony, the Air Force called Richard Frost back from Bell to arrange a ground-rocket takeoff in the X-1. Though Bell had designed the aircraft with a conventional retractable landing gear, the Air Force and NACA air-launched the aeroplane for safety and added performance reasons. However, interest in rocket takeoffs from the ground lingered on, not least of which was that Douglas and the Navy had the first jet-and-rocket-propelled ground-launched Skyrocket at Muroc awaiting its first rocket flight. At the annual 1948 Wright Brother's dinner, Lawrence

Bell broached the subject unofficially to Yeager and General Vandenberg. To put it mildly, both men expressed a keen willingness to perform a ground-launched takeoff using *Glamorous Glennis*. Accordingly, Bell engineer Richard Frost prepared a list of recommendations for the Air Force to follow in planning the flight attempt.

To overcome any danger to the landing gear, Frost recommended that new tyres, chambers and brake biscuits be installed, and that the Air Force fill the propellant tank to no more than ¾ capacity in order to limit the takeoff weight to 10,600 lbs. He suggested further that the takeoff be made as early as possible in the day before the sun heated the runway and that Yeager avoid braking so as not to put undue strain on the nose wheel. To minimise the shifting of the centre of gravity, Frost recommended that the Air Force conduct a thorough weight and balance analysis prior to the flight, and make the flight as soon as possible after the analysis so that liquid oxygen boil-off would not affect materially the centre of gravity's location. Frost left the choice of stabiliser settings up to Yeager and recommended that the pilot should reduce accelerations of less than 1g – which might result in engine shutdown if they interfered with the fuel flow – if he made an early takeoff when atmospheric turbulence was at a minimum and also made a spiral climb, which would ensure that normal accelerations did not fall below 1g. Frost also advised that Yeager retract the landing gear before the aeroplane exceeded 250 mph, that he keep climbing speed below Mach 0.85 in order to retain control effectiveness and that he fire the rocket engine with brakes locked so as to ascertain proper engine ignition before the plane began its takeoff roll. At Edwards Air Force Base, Jack Ridley transformed the recommendations into reality and added some of his own.

On 5 January 1949, Yeager strapped himself into the X-1, completed a pre-flight check, fired chambers two and four, ignited chambers one and three and released the brakes. With a shattering and angry roar, *Glamorous Glennis* lunged forward and darted into the air, spiralling upwards in a flight Yeager recollected nearly four decades later as 'fabulous . . . There was no ride ever in the world like that one!' Eighty seconds after engine ignition, Yeager attained 23,000 feet and Mach 1.03. As the powerless X-1 began its glide earthwards, Yeager jettisoned the plane's remaining liquid oxygen and alcohol. The rocket plane touched down on the lakebed and rolled to a stop in a cloud of dust 2½ minutes after takeoff as a chase F-86 Sabre winged victoriously overhead. Too bad for Gene May, Douglas and the Navy – the Air Force claimed a first (and only) – in the postwar rocket research aircraft programme. (The Skyrocket used both its jet and rocket engine to become airborne and, though it could have taken off solely on the power of its rocket engine, no such attempt was ever made.) The next American manned rocket aeroplane to take off under its own power would be the NASA Space Shuttle *Columbia* in 1981.[22]

High-Altitude Flight

Following its unique takeoff, the Air Force X-1 remained on the ground because of winter rains that flooded the lakebed. Flight operations resumed in March 1949 and Jack Ridley, recently returned from a stint as pilot and project officer on the sweptwing Boeing XB-47 Stratojet six-engine bomber at Moses Lake, Washington, checked out in the aeroplane. Though he noticed no difficulties, post-flight inspection revealed that a loose engine wire had caused a small engine fire. Tragically Captain James Fitzgerald, who entered the X-1 programme as Yeager's alternate following Robert Hoover's P-84 accident, subsequently becoming the fourth man to fly faster than sound, had by now himself been killed. Early in September 1948, Fitzgerald was severely injured in a TF-80 crash at Van Nuys, California dying from his injuries 11 days later. His place was taken by a young test pilot who would, in his own right, establish himself as one of the most significant of the supersonic pioneers: Major Frank K Everest, Jr.[23]

Like Yeager, Everest also hailed from West Virginia. He had joined the Civilian Pilot Training Program at West Virginia University in 1940 and, after receiving a private pilot's licence in February 1941, had applied for AAF flight training. He received his commission and wings as a fighter pilot on 3 July 1942, and arrived in the Middle East as a Curtiss P-40 Warhawk pilot in early 1943 attached to the 314th Fighter Squadron. He saw extensive combat against Nazi Bf 109 and FW 190 fighters and it is a measure of his piloting ability that he succeeded in scoring against both these first-rate fighters, even though the P-40 was decidedly inferior to both. He intercepted and shot down Ju 52 transports desperately attempting to both resupply and evacuate Field Marshal Erwin Rommel's *Afrika Korps* and, after the collapse of 'The Desert Fox's' forces in North Africa, flew close air support missions during the Italian campaign. He transferred back to the United States in 1944 as a P-40 instructor in Venice, Florida and used the opportunity to fly a wide range of aircraft, including the A-20, A-24 (AAF SBD) and A-25 (AAF SB2C) attack bombers; the C-78 utility transport; and the powerful P-47 and P-51 fighters. Everest soon requested a return to combat flying and transferred to Chihkiang, China as a replacement pilot. The AAF soon assigned him as acting squadron commander of the Mustang-equipped 29th Fighter Squadron. Japanese flak shot down his P-51 in May 1945 while he flew an anti-shipping mission over the Yangtze River and he spent the remainder of the war as a POW, enduring the characteristic brutality of Japanese captivity at Nanking and Fengtai prison camps until release in August 1945. After arrival back in the US, Everest joined the test pilot staff at Wright Field, beginning formal test pilot training in May 1946. He checked out in the P-59 and P-80, and later flew a variety of acceptance tests on the P-80, P-84 and F-86 Sabre. In January 1949, he became assistant chief of the Fighter Test Section at Wright Field.

Early in January 1949, Colonel Albert Boyd called Everest into his office at Wright Field for a discussion on the Air Force X-1 programme. Boyd stated that the Air Force no longer planned any more high-speed record runs in the aircraft, but that the service was planning high altitude investigations to determine the behaviour characteristics of manned high-speed aircraft at altitudes above 50,000 feet. Further, such flights could lead to breaking the world's altitude record of 72,395 feet set by the AAF Air Corps-National Geographic Society balloon *Explorer II* on 11 November 1935, piloted by Captains Orvil A Anderson and A W Stevens. Boyd then asked if Everest would be interested in making the altitude flights. Everest replied affirmatively and, after being fitted for a high altitude T-1 partial pressure suit, drove west to Muroc. On 21 March 1949, he completed a check-out flight in *Glamorous Glennis*, ambitiously reaching Mach 1.22 at 40,000 feet. During Everest's second flight in the X-1 on 25 March, the aeroplane suffered a rocket fire and the automatic engine shut down. Despite the disappointing flight, the Air Force X-1 test team and Bell agreed to attempt a maximum altitude investigation on Everest's next flight. On this flight, 19 April 1949, Everest could only get two of the four chambers firing and he had to turn back at approximately 60,000 feet He was wearing an early-model pressure helmet with a fixed visor that could not be opened and, as the aeroplane glided earthwards, he felt a pressure in his eardrums. Due to the design of the helmet, he could not lift the visor, hold his nose and blow, nor could he remove the helmet as the cockpit contained only nitrogen from the pressurised system and no oxygen. Inevitably, one eardrum ruptured and, despite tremendous pain, he managed to land safely. Fortunately the injury was slight and, after letting the rupture heal, Everest resumed research flights in the rocket plane. Serious engine difficulties returned to plague Everest on his next flight on 2 May. After launching from the B-29, he attempted to ignite all four chambers but only three fired. As he pulled the X-1 into a climb, the plane shuddered from an internal explosion, the XLR-11 engine abruptly quit and, at the same time, Everest lost rudder control. As the X-1 dropped down, chase pilot Richard Frost came alongside in his chase F-80 and observed that the number one chamber had exploded, jamming the rudder. Everest retained sufficient control to land on the lakebed and the Air Force ferried the X-1 to Wright Field for repairs. Ground inspection of the aircraft revealed that the exhaust of the other chambers had ignited vapour streaming from the chamber which had not fired. The fire moved inside the chamber, igniting the heavy concentration of liquid oxygen and alcohol inside with a resulting excess pressure that blew the chamber out of the aeroplane. Understandably, the Air Force did not like the apparently continuing problem of engine fires in the X-1 series and re-employed Richard Frost from Bell, together with Bell engineer William Smith, to investigate the cause of the recurring fires. Smith and Frost traced the engine trouble to the spark plugs used in the engine. They found that the regular plugs, with gaskets made of asbestos covered with a thin sheet of copper, could not withstand the high internal chamber pressures and would often blow out. Smith solved the problem by installing solid gaskets made of copper rather than the copper-covered asbestos.[24]

Repairs and engine modifications kept the X-1 at Wright-Patterson until mid-June 1949. The Air Force still planned to continue the high-altitude research programme, and hoped eventually to reach 80,000 feet during the course of a further four flights. (Aside from simply exceeding the altitude record of *Explorer II* the Air Force wanted to investigate flight at high lift coefficients.) During the remainder of June and July, Air Force and NACA technicians readied the X-1 for flight, reinstalling research instrumentation removed for engine repairs. After a ground run to ensure proper rocket operation, Everest went aloft on 25 July for his fifth X-1 flight, his third attempt to attain high altitude. Everest only attained 66,846 feet, but it was higher than the rocket aeroplane had ever flown before. On 8 August, Everest launched from the B-29 on his fourth altitude attempt. *Glamorous Glennis* separated cleanly from the B-29 and Everest fired three of the four rocket chambers, then pulled into a steep climb. The rapidly accelerating X-1, growing lighter as it burned off its heavy load of propellants, quickly outdistanced Chuck Yeager's chase F-80. Everest used the stabiliser trim for transonic control and at around 45,000 feet, nosed over into level flight prior to going supersonic. The little rocket plane quickly accelerated from Mach 0.9 to over Mach 1 and Everest again used the stabiliser trim to re-enter a climb. He turned off the Machmeter which could not register accurately at high altitudes, and as the X-1 climbed, the altimeter reading lagged by at least 5,000 feet For the final portion of the climb, Everest attempted to light the X-1's fourth rocket chamber, but the ignition failed. The X-1 continued climbing, still under only 75% power. The controls grew sluggish at the increasing altitude, and finally at around 65,000 feet, the three chambers exhausted their propellants and shut off from fuel starvation. The X-1 abruptly decelerated, hurling Everest forward in his shoulder harness. Still coasting powerless into the deep purple sky above it, the X-1 continued upwards in an arc for a few more thousand feet, until it attained the apex of its climb at 71,902 feet Then, momentum spent, it slowly began its drop back to earth. Everest performed several stall checks during abrupt pull-ups (to achieve the requisite high lift coefficient conditions) down to Mach 1.1 as he descended from 60,000 to 30,000 feet. At 30,000 feet, Yeager's chase F-80 picked up the orange rocket aeroplane and the two aeroplanes descended to the lake. Although he did not know it, it was to be the highest any of the original X-1 aeroplanes ever flew.[25]

On 25 August, Everest again went aloft in the X-1 for another attempt to reach extreme altitudes, resulting in another X-1 'first', but this time a strictly inadvertent one. As Jack Ridley closed *Glamorous Glennis*'s entrance hatch, Everest noticed a tiny crack in the Plexiglas canopy, measuring only about an inch long, and apparently confined to the inner shell of the canopy. After a smooth drop and light-off, Everest began his climb, levelling off to go supersonic before resuming the climb. As he passed 65,000 feet, a loud sighing sound filled the cockpit and Everest's T-1 partial pressure suit inflated automatically. Everest realised at once that he had lost cockpit pressurisation and, glancing up, he saw that the tiny one-inch crack had extended to six inches and broken through to the atmosphere. Recognising the danger he was in – for should the suit fail, he would die in seconds – Everest shut off the XLR-11 engine and nosed over into a dive. F-80 chase pilots Yeager and Captain Arthur 'Kit' Murray saw the vapour trail abruptly break off, and Yeager radioed Everest asking if anything was wrong. As the life-saving pressure of the suit severely constricted the body and made normal speech impossible; Everest could only reply to Yeager's anxious questions with a succession of grunts that further alarmed those listening in on their conversation. When the X-1 reached 30,000 feet, Everest eased out of his supersonic dive, and at 20,000 feet he opened the dump valve to depressurise the suit, reported the canopy failure and assured those listening that he was all right. After jettisoning the remaining propellants, he brought the X-1 in for a landing on the lakebed, having made the world's first operational emergency use of a pilot's pressure suit to save both the pilot and aircraft. The cracked canopy forced the Air Force to suspend the X-1's altitude flights until Bell could deliver a new canopy to Muroc. While waiting for the new canopy, however, the Air Materiel Command decided to cancel further high-altitude investigations with the rocket aeroplane. The flight experience indicated that the aircraft would probably not exceed Everest's 8 August record height by more than a few thousand feet and such an incremental increase would not warrant the preparation and added costs of further high-altitude investigations. Everest's X-1 height of 71,902 feet thus remained a record for the two original X-1's with their high-pressure fuel systems.[26]

Glamorous Glennis Retires with Honour

By late 1949, the first X-1 was approaching the end of its service life. The Air Force had already contracted for the more advanced X-1A, X-1B and X-lD aeroplanes, the potentially Mach 3+ Bell X-2, the hopefully Mach 2 Douglas X-3 (though it would achieve only Mach 1.21), the Northrop X-4 semi-tailless research aeroplane and the Bell X-5 variable wing-sweep test-bed. Though a true supersonic fighter still lay in the future one aeroplane, the North American F-86 Sabre, could safely exceed the speed of sound during slight dives and did so routinely. (In contrast, in the rest of the world, supersonic flight was a far from routine endeavour. In the Soviet Union, engineers still did not have the 346 programme at a point where it could fly past Mach 1 and authorities cancelled an indigenous programme to develop a rocket-propelled sweptwing supersonic test-bed – the Bisnovat B-5 which was an exceedingly unattractive aeroplane – before it had progressed beyond Mach 0.775.) Under a contract between the Air Materiel Command's Aircraft Laboratory and the Cornell Aeronautical Laboratory, the Cornell Aero Lab would get both the first X-1 and the new X-1A, and instrument the two aircraft for autopilot studies and structural loads measurements. However, in mid-November 1949, the AMC's Aircraft Laboratory decided to relinquish ownership of *Glamorous Glennis* and deliver it to the NACA. Since the NACA could not use the aircraft without a major overhaul, this decision virtually terminated the research programme on the first X-1 aeroplane. Forecasting ahead, Hartley A Soule, NACA's Research Aeroplane Projects Leader, accurately predicted that 'it is probable that the No. 1 airplane will be sent to the Smithsonian Institution'.[27]

On 7 September 1949, Colonel Albert Boyd assumed command of Muroc Air Force Base, replacing Colonel Signa A Gilkey who had commanded the desert base during the pioneering flights of Yeager two years previously. Boyd himself had flown *Glamorous Glennis* earlier in the year in a spectacular full-power flight to Mach 1.2 over the Pacific coast under the watchful tutelage of Chuck Yeager. When it returned to flying status in October, it served primarily as a checkout aircraft to give selected test pilots a taste of rocket flight past Mach 1. Colonel Patrick D Fleming, former chief of the AMC's Fighter Test Section, flew the plane on 6 October, attaining Mach 1.2. Three weeks later, Major Richard L Johnson, then-chief of the AMC's Fighter Test Section and holder of the world's air-speed record (for conventional aeroplanes) of 670.981 mph, set in a F-86A on 15 September 1948 at Muroc, also flew the lithe rocket plane. Early in the new year, *Glamorous Glennis* completed its final four research missions. The final flight, its 59th, came on 12 May 1950. For the last time, fittingly piloted by Chuck Yeager, *Glamorous Glennis* dropped away from a launch Boeing B-50 Superfortress (a modified and much more powerful development of the B-29) to pose for camera footage for Howard Hughes' RKO movie *Jet Pilot* starring John Wayne and Janet Leigh – playing a *Soviet* experimental fighter! The flight came off without difficulties and Yeager brought the rocket plane down on the lakebed for the last time. *Glamorous Glennis* remained at the desert test site, now named Edwards Air Force Base (in honour of Captain Glen W Edwards, killed in the crash of the experimental Northrop YB-49A Flying Wing on 5 June 1948) until August 1950. By this time America was again at war and before the end of the year, Air Force Sabres would be duelling above the Yalu with Soviet MiG-15s in the world's first transonic dogfights. While awaiting shipment to the Smithsonian, NACA technicians removed its instrumented horizontal stabiliser for use on the third Bell X-1 the turbopump-equipped *Queenie* which, at last was ready for flight. On 19 August, the Mach-busting first X-1 left Edwards nestled snugly under the B-29 from which it had dropped that historic October day almost two years previously. Fittingly piloted by Jack Ridley and Chuck Yeager, the Superfortress winged its way east to AMC headquarters at Wright-Patterson AFB where Air Force technicians repainted the X-1. Then Yeager and Ridley ferried the X-1 via B-29 to the Air Force Association convention at Boston, Massachusetts, landing at Logan International Airport. There, on 26 August, Air Force Chief of Staff General Vandenberg presented *Glamorous Glennis* to Dr Alexander Wetmore, Secretary of the Smithsonian Institution. The X-1, he stated, 'marked the end of the first great period of the air age, and the beginning of the second. In a few moments the subsonic period became history and the supersonic period was born'. After Ridley and Yeager ferried the X-1 to Washington, the Air Force delivered it to Paul E Garber, Head Curator of the National Air Museum on 28 August. Visitors could look at it and the 1903 Wright biplane – itself only delivered to the air museum less than two years before – and ponder on the progress of aviation in the intervening years.[28]

The NACA Concludes its Comprehensive Programme

While the Air Force X-1 embarked upon its new career as a Washington tourist attraction, the NACA still fired up its engine in the Mojave sky on transonic research. In September 1949, John H Griffith had joined the

NACA unit as a research pilot. In the autumn, NACA Headquarters changed the name of its desert establishment to the High-Speed Flight Research Station, a more accurate title reflecting both the growth of the unit and its expanded mission, which now included flight testing and assisting in tests on a number of research aircraft besides the original X-1s. Like their Air Force counterparts, NACA pilots participated actively at every step of mission planning. Langley Laboratory established the generalities of the aircraft programme through Hartley A Soule, the Research Aeroplane Projects Leader for the NACA, who chaired a Research Aeroplane Projects Panel. At Edwards, agency engineers planned the order and detail of every flight down to every minute. NACA found that the various research aircraft – particularly the rocket-propelled ones – were very difficult to keep on flying status. The research pilots kept their skills sharp by practice flying and flying chase for other projects in conventional aircraft, such as P-51s and F-80s. After completing a research flight, engineers analysed the records from the internal instrumentation, radar tracking and telemetry gear. They then prepared the material for publication as NACA Research Memorandums, circulating copies of the drafts to other NACA installations for comment and review. After publication, these documents became part of the general storehouse of aeronautical knowledge available to those with a need to know, particularly by the American aircraft industry. The characteristically precise methodical NACA operation did not preclude, however, eventful flights. Prior to one launch, Bob Champine experienced a radio failure. He wrote 'Secure the Drop' on a flight card, and passed it into the bomb bay of the B-29. None of the B-29 crewmen were Navy-trained, so they failed to realise that Champine had just cancelled the mission. Suddenly, Champine realised that he was about to be launched. He managed to get everything in working order before the launch crew dropped him, innocently unaware of his predicament and fortunately, the flight came off without further difficulties.[29]

Despite their highly experimental nature, both the X-1 and D-558 aeroplanes had good flying qualities, and pilots liked them. The pilots, however, were under no illusions as to their possible escape in the event of trouble. Jumping from the X-1, a pilot would almost certainly hit the sharp leading edge of the wing. If he jettisoned the nose of the D-558, he might not have time to use his parachute if the emergency occurred at low altitude. It was good to have the adjustable horizontal stabiliser as a means of ensuring adequate transonic control. Test pilot Robert Champine was particularly impressed by it and undertook an extensive programme to investigate stabiliser efficiency at transonic speed, pioneering work in the development of the genuine all-moving tail for transonic and supersonic aeroplanes that appeared on the North American F-100 Super Sabre, the first of the 'Century series' supersonic jet fighters.

NACA grounded the X-1 from the autumn of 1949 until May 1950 for installation of additional recording instrumentation to measure wing loads and pressure distributions. Due to its 10% wing and 8% tail, the X-1 was not as fast as the Air Force's thinner-winged X-1 and so, when John Griffith reached Mach 1.2 on 26 May, it was the fastest flight ever made by the second X-1.* The nose landing gear collapsed on landing and repairs plus subsequent instrumentation changes (including changing the instrumentation over from direct current to 400-cycle alternating current) grounded the plane until August 1950. By mid-October 1950, NACA had completed nine successful pressure-distribution surveys to determine aerodynamic characteristics at specific wing locations, before discontinuing the programme in favour of investigating the aeroplane's lateral and longitudinal stability and control. Then technicians discovered that the alcohol tank had severely rusted on the inside, forcing an immediate grounding for an overhaul and tank replacement. NACA used the delay to install new instrumentation, finishing the overhaul by mid-March. On 6 April 1951, Chuck Yeager flew the aircraft as part of a NACA cooperative agreement with the Air Force for Hughes' film *Jet Pilot*. During the flight it suffered a small flash fire in the engine bay caused by a slow-closing alcohol valve, but an automatic fire extinguisher quenched the blaze before it caused any serious damage or posed a threat to either pilot or plane.[30]

In June 1950, Scott Crossfield joined the NACA research pilot staff at Edwards, beginning an association with the X-series that would eventually take him from the X-1 through to the X-3, X-4, X-5 and on to the proving flights of the hypersonic North American X-15. A native Californian, Crossfield had received his private licence under the Civilian Pilot Training Program in 1941 and had completed three semesters as an engineering student at the University of Washington before the Japanese struck Pearl Harbor. In February 1942, fed up with inactivity, he resigned from an AAF cadet programme and joined the Navy as a flight student, graduating as a pilot that December. After serving as a flight instructor in advanced bombing and gunnery, he transitioned into SBD dive bombers and ended the war as a Grumman F6F Hellcat pilot with Air Group 37 on board the carrier

* In its original configuration, of course. The second X-1, *nee* XS-1, was eventually rebuilt to a completely new aerodynamic configuration and flown as the Mach 2+ X-1E, as is related subsequently.

USS *Langley*, preparing for the invasion of the Japanese home islands. After V-J Day, he returned to the University of Washington, received a BS in aeronautical engineering in June 1949, received his MS a year later and joined the NACA as a research pilot that same month. For nearly a year, he flew a variety of NACA aeroplanes on research and proficiency flights before being offered the chance to check-out in the second X-1 on 20 April 1951. Technicians normally adjusted the stabiliser setting on the X-1 on the ground before its flights. On this one it had been set 1° greater than normal, a change proving far from innocuous. When Crossfield and the X-1 dropped away from the launch B-29, as chase pilots Frank Everest and Jack Ridley watched in amazement, the X-1 pitched up, stalled and rolled over on to its back. With a mix of crisp reflexes and aplomb, Crossfield recovered nicely, rolling the little rocket plane back to level flight, adjusted the stabiliser, and ignited all four rocket chambers. He began to climb rapidly, the X-1 reaching Mach 1.07 at 41,000 feet before the engine exhausted the fuel. On his return, ice formed on the inside of the canopy: the dehumidifying equipment installed was too weak to rid the cockpit of all the moisture that built up, especially on humid days, and which then froze at the high altitude. Using his right sock – it was a very untypical flight – Crossfield cleared the windshield sufficiently to see out, with Jack Ridley flying chase close on his wing, the novice (but hardly unprofessional) rocket pilot brought the X-1 in for a safe landing.[31]

Following three more check flights, NACA grounded the X-1 for installation of a new rocket engine. This, coupled with a routine maintenance grounding of both the B-29 and B-50 used to drop the X-1's kept the programme in the hangar until July. That month, NACA began a wing-loads and aileron-effectiveness programme, taking data as Crossfield abruptly rolled the plane left and right at transonic speeds. This continued for the next six flights, until the agency switched to measuring fuselage loads during abrupt pull-ups using both the elevator and adjustable stabiliser. Late in August, yet another new NACA pilot, Joseph A Walker, initiated himself to rocket research, with a flight to Mach 1.16 at 44,000 feet. Walker, destined to be another Edwards legend – he eventually held the record for the highest flight of a winged aircraft (the X-15) prior to the Space Shuttle – held a BA in physics from Washington and Jefferson College. In 1942 he had entered the AAF flying Lightnings on reconnaissance missions over the Black Sea, Austria and southern France. He joined the Lewis Flight Propulsion Laboratory in 1945 as a physicist before transferring to the High-Speed Flight Research Station in 1951. Review of flight records indicates that NACA research priorities fluctuated back-and forth across a range of transonic issues in this time period; between pressure-distribution studies, stability and control, wing and fuselage loads and other specialised studies. On 5 September, Scott Crossfield completed the plane's 53rd flight, one of a planned series of fuselage pressure-distribution surveys; though one rocket chamber failed to fire, Crossfield nevertheless still reached Mach 1.07. At this point NACA suspended the programme temporarily to make way for a special programme to investigate whether wing-mounted vortex generators – spanwise rows of small little fin-like tabs set perpendicular to a wing's surface – could reduce buffet severity in the transonic region. Walker attempted the first of these flights on 23 October, but the rocket engine cut out and he had to finish it as a glide flight. Bad weather, together with unavailability of a launch aeroplane, kept the X-1 grounded into November. On 15 November, engineers detected the presence of battery acid in the midsection of the aeroplane and NACA grounded the craft for further inspection. To complete the picture, an investigation by the Bell Aircraft Corporation shortly afterward revealed that the nitrogen pressurisation spheres in the plane were now so old as to be likely to fail from fatigue. The X-1 was now grounded for good.[32]

It was a remarkable coincidence in timing for multiple reasons. First, the X-1 had outlasted and bettered all of its Mach 1 rivals. The Skystreak had been too slow and the Skyrocket too tardy. The M.52 had been cancelled, as had the Soviet's Bisnovat B-5. The DH 108 Swallows – every one – had crashed fatally. Its most serious rival, the DFS 346, had proven a tremendous disappointment.* Had conditions at Edwards at that time been other than they were, the NACA would have speedily retired the second X-1 to some museum or boneyard. Circumstances, however, forced NACA to take another look at the worn-out rocket plane. As will be discussed subsequently, in August 1951 the X-1D exploded in flight before launch and had to be jettisoned over the Mojave. Less than a month before NACA grounded the X-1, the third X-1 (the long-overdue and ill-fated *Queenie*) blew up on the ground. Both aircraft were sorely missed and as NACA engineers surveyed the tired but trim X-1 they concluded that, like the

---

* The fastest and best of them, the 346-3, had entered flight testing in the summer of 1951. On 2 September, test pilot Wolfgang Ziese dropped from his B-29 over the Tyoplystan test site, fired up the two HWK 109-509 engines and streaked upwards, reaching Mach 1.3 – approximately 840 mph – at 39,000 feet The plane's accumulated deficits then came to the fore, for it plunged into an uncontrollable dive and one wing tore off. Ziese, most fortunately, jettisoned the nose cone and safely parachuted from the wildly gyrating rocket plane, which tumbled to earth and exploded in a spectacular fireball.

Phoenix of old, a new high performance aeroplane just might be developed from the remains of an ancestor aircraft. That, though, is another story, the story of the X-1E. It is enough to say that at the time of the second X-1's retirement, both it and its more famous stablemate had securely earned their place in aviation history, fully validating the hopes, expectations and far-sightedness of those within the Air Force, Bell and NACA that had pressed for their development. Any more to come from a rebuilt X-1E could only add additional icing to an already very rich supersonic cake.

6
The D-558-1 Skystreak Flight Programme

As late as October 1946, considerable doubt existed as to whether Douglas would flight test the D-558-1 at Muroc Dry Lake or at China Lake (the naval station at Inyokern), further north in the Mojave. While Inyokern did not have the facilities and suitable qualities that the Muroc site possessed for flight testing, it did have better living conditions in contrast to the 'roughing it' atmosphere at Muroc. NACA did not intend to take an active role in the contractor flight-test programme on the Skystreak and informed Douglas of this early in 1947 while company technicians readied the aircraft for flight. The NACA, however, did assure Douglas that it would co-operate in whatever way it could to provide technical consultation or even instrument supply, if requested by the Navy or the contractor. Otherwise, NACA would simply look on as an interested observer. Early in April 1947, Douglas sent a test team to Muroc together with the first D-558-1 (BuNo 37970). Headed by Alden B Carder, the team included company test pilot Eugene F 'Gene' May, only recently designated as project pilot, and members of the company's Experimental Flight Test Division. May had long experience in testing Douglas company aircraft, from the prewar B-18 bomber to the wartime A-26 Invader attack bomber and the postwar AD-1 Skyraider naval dive bomber, among others. He flew the Douglas XC-53 (an experimental DC-3) in a bad-weather flying programme aimed at evaluating heated wings for de-icing purposes. His closest call came during the war, while testing a modified A-26. Instructed to perform a terminal velocity dive in the bomber, May nosed over to pick up speed, then found that the plane tended to tuck under. Unable to pull the nose up, May, in desperation, pushed the engine throttles forward to full power. The gambit worked and the A-26, far above its maximum speed limitations, pulled out low above the ground.[1]

First Flights and a World Airspeed Record

On 15 April, Gene May took the first D-558-1 aloft for its initial flight. Immediately after lifting off the lakebed, the scarlet plane experienced a partial power loss and May landed the Skystreak straight ahead. As he braked to a halt, the left brake completely disintegrated, necessitating judicious use of full left rudder to hold on course. May terminated another flight attempt on 21 April under similar circumstances, and the Skystreak did not return to the air until the end of May. Landing gear trouble plagued the next six flights, the gear failing to lock up or retract but, by early July, Douglas technicians had remedied the problem. Following the 12th contractor flight on 12 July, the Douglas test team felt confident that the aircraft had satisfactory general handling qualities. Accordingly, the company removed the clear canopy used on early low-speed flights, and replaced it with a hooded high-speed cockpit enclosure with flat 'V' glass panels. After an airspeed-calibration flight, the company team embarked on serious performance investigations to high Mach numbers beginning with the 14th flight on 17 July. By the 20th flight on 5 August, May had attained Mach 0.85. In August 1947, the second D-558-1 (BuNo 37971) joined the Douglas test programme. Though destined for NACA service, Douglas completed the aeroplane without its special research instrumentation, for the agency planned installation only after Douglas satisfactorily demonstrated the airline to Walt Williams' engineers at the Muroc Flight Test Unit. Gene May performed the first contractor flights on this aircraft on 15 August and the next day, it was flown no less than four times, twice by Navy Commander Turner F Caldwell Jr and twice by Marine Major Marion E Carl.[2]

Both of these pilots flew the second Skystreak in preparation for an assault on the world's airspeed record which, nearly a year before on 7 September 1946, RAF Group Captain E M Donaldson had established at 615.778 mph flying a Gloster Meteor. Geoffrey de Havilland had died in the quest to exceed Donaldson's record, which stood well into 1947. Accordingly, in March, Navy officials discussed the possibility of setting a new world airspeed record using the D-558-1 with Douglas personnel. Douglas and Navy representatives concluded that the D-558-1 programme should not be delayed to make a record flight, but felt that the attempt would permit good calibration of instrumentation during the speed trials and demonstrate the ability of skilled pilots to fly at high Mach numbers close to the ground while retaining complete control of the aeroplane. The

Plate 60: In February 1948, Douglas test pilots began flying the first D-558-2 Skyrocket, at this time a pure-jet (and seriously underpowered) aeroplane, but one with clear promise of supersonic speeds. Note tailcone fairing over unused rocket exhaust and original short vertical fin. *USAF AFFTC.*

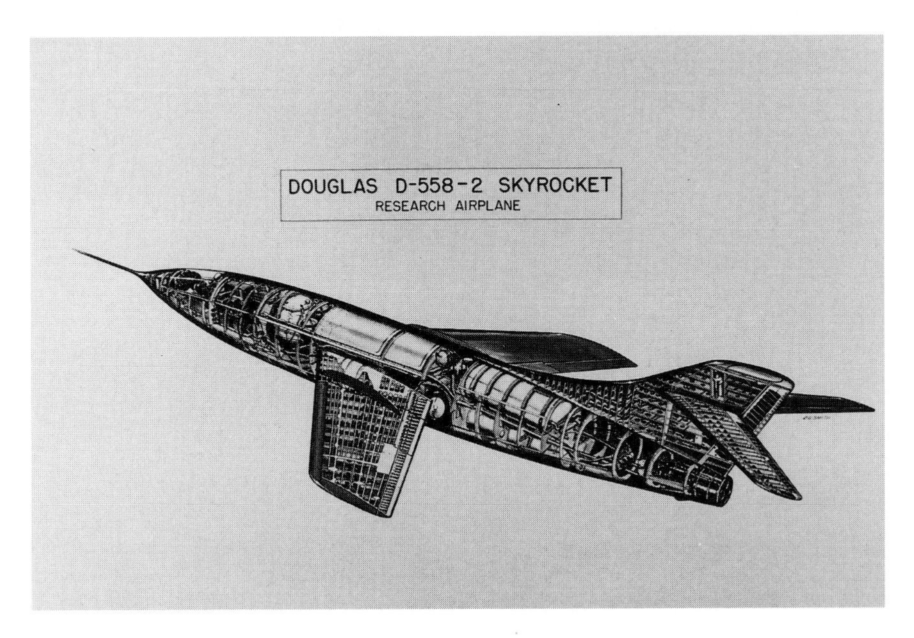

Plate 61: The Skyrocket differed greatly internally from its Skystreak predecessor, particularly since it required two very different propulsion systems for its jet and rocket engines. Note too, that the original Skyrocket configuration incorporated a flush canopy design, as with the XS-1. *Navy.*

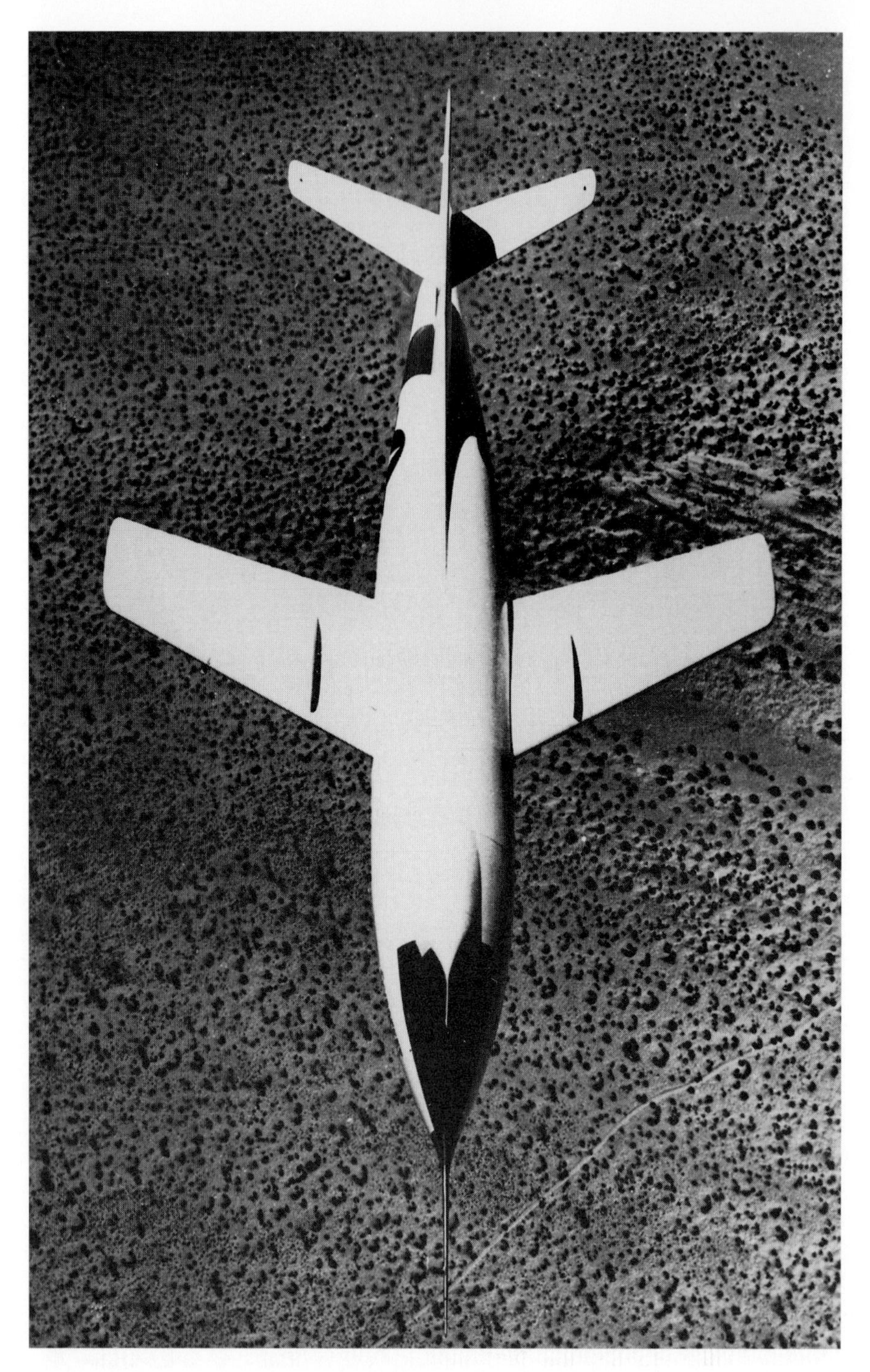

Plate 62: This classic photograph of the first Skyrocket in flight shows its extraordinarily clean lines by 1948 standards. A mere decade before, the US Navy's highest performing aeroplanes had still been metal-and-fabric biplanes; here now was a design suitable for Mach 2, over five times faster than their maximum speed. *USAF AFFTC*.

Plate 63: Skyrocket operations from the ground were extraordinarily dangerous, and when Gene May began flight testing the third D-558-2, equipped with both a J34 turbojet and LR-8 rocket, in 1949, the plane still required four jettisonable JATO (Jet-Assisted-Take-Off) boosters to get it off the ground reasonably quickly—a powerful argument for air-launching. USAF AFFTC.

Plate 64: Here the third Skyrocket (the first rocket-equipped Skyrocket) streaks past the camera at low-level, under full jet and rocket power. This was the first Skyrocket to exceed Mach 1 piloted by Gene May, 24 June 1949. *USAF AFFTC.*

Plate 65: Douglas and the Navy wisely modified the Skyrocket for air launching, dramatically increasing its performance. Here is the all-rocket second D-558-2 at the moment of launch from its Boeing P2B-1S (Navy B-29) mothership on 18 May 1951. On this particular flight, Douglas test pilot William 'Bill' Bridgeman reached Mach 1.7 at 62,000 feet. With this kind of performance, it was immediately a contender for the Mach 2 crown, but it encountered serious stability problems above Mach 1.5 that required extensive investigation before proceeding further . . . *USAF AFFTC.*

Plate 66: . . . then Bell's X-1D blew up while being readied for launch, fortunately without claiming both planes or any lives. All hopes now rested on the long-awaited X-1-3, *Queenie*, a turbopump-equipped X-1 potentially capable of Mach 2.44, shown here being readied for mating with its EB-50A Superfortress mothership. (This spectacular hydraulic lifting method for joining the two aircraft replaced the more awkward loading pit method used in the days of the XS-1). *NASA.*

Plate 67: Following only a single successful glide flight by company test pilot Joe Cannon, it too, blew up under its mothership, destroying both planes and seriously injuring its pilot. Despite its promise, therefore, 1951 ended as a year of shattered Mach 2 hopes. *USAF AFFTC.*

Plate 68: By 1953, the X-1 and D-558 programmes had spawned an entire 'X-series' which continues to the present day. Here are the 'X-series' under test at that time (clockwise from lower left): the advanced Bell X-1A; the third Douglas D-558-1 Skystreak; the Convair XF-92A, progenitor of all American delta aeroplanes; the first Bell X-5, the world's first variable-sweep wing aeroplane; the all-rocket second Douglas D-558-2; the second Northrop X-4 semi-tailless research aeroplane, and centre, the glamorous (and one-and-only) Douglas X-3 Stiletto, which anticipated the F-104. *NASA.*

Plate 69: That same year, aeronautical researchers were anticipating exceeding Mach 3 with Bell's glamorous X-2 sweptwing rocket research airplane. Early glide trials, however, revealed teething problems with its landing gear, and its complex two-chamber Curtiss-Wright rocket engine experienced serious delays. Then on 12 May 1953 – disastrously – the second airplane (shown here) exploded during a captive test flight over Lake Ontario, while locked in the bomb bay of its B-50 mothership. The accident killed Bell chief test pilot Jean 'Skip' Ziegler and aircrewman Frank Wolko, and seriously damaged the B-50 mothership whose crew in a feat of remarkable airmanship, managed to land safely. The surviving X-2 would not make its first powered flight until late 1955. *USAF.*

Plate 70: By 1953, new generations of military aircraft were appearing that could fly at speeds limited only a few years before to the rocket-research aeroplanes. Here is the prototype North American YF-100A Super Sabre, the first Western supersonic 'on the level' jet fighter. Some of its features – a thin, sharply swept wing and an all-moving horizontal tail, for example – reflected lessons learned from, or validated by, the X-series research aircraft program. *USAF.*

Plate 71: Research aircraft operations were complex and demanded the highest standards of professionalism and coordination. Here Scott Crossfield and the all-rocket D-558-2 pose with the ground crew, support staff, chase planes, launch aircraft, tracking radar and associated equipment needed to support a single flight. *NASA.*

Plate 72: Like the later X-1s, the Skyrocket was rolled under its P2B-1S mothership while the larger plane was hydraulically raised above it. The high vertical fin of the Skyrocket posed its own problems in mating the two machines. *NASA*.

Plate 73: Three outstanding Skyrocket pilots who probed the Mach 2, high-altitude frontier (*L* – *R*): Douglas pilot Bill Bridgeman reached Mach 1.88 and 79,494 feet in 1951; Marine pilot Marion Carl reached 83,235 feet and NACA's Scott Crossfield became the first pilot to reach Mach 2, in 1953. *NASA*.

Plate 74: Crossfield's Mach 2 flight, on 20 November 1953, combined rigorous preflight preparations and exacting piloting technique to extract the absolute maximum performance possible from the elegant second Skyrocket. Here, the plane's graceful, futuristic lines are shown to good advantage. *NASA*.

Plate 75: Crossfield's record lasted less than a month, as the Air Force's X-1A was back at Edwards Air Force Base with Major Chuck Yeager. Here, Yeager poses with the Bell-Air Force test team and equipment required to support flights of the advanced rocket plane.

Plate 76: In November and December 1953, Yeager made three exploratory flights in the X-1A before his all-out attempt to exceed Mach 2. Here, the silver X-1A glides downwards in seemingly solitary splendour over Edwards Air Force Base, west of the dry lake. Visible directly under the aeroplane is the new main base being constructed, with a 15,000-foot runway; in the lower left is the single runway of North Base and in the lower right is South Base, which was then still in use. *USAF AFFTC.*

Plate 77: Larry Bell and Chuck Yeager share a reflective moment beside the X-1A after one of its flights. *USAF AFFTC.*

Plate 78: On 12 December1953, Chuck Yeager launched from the B-29 reaching Mach 2.44 at approximately 76,000 feet and then nearly losing his life as the rocket plane tumbled wildly out of control. After a 50,000-foot drop, he managed to recover and land safely on the lakebed. *USAF AFFTC.*

Plate 79: Eight months after Yeager's wild flight, Major Arthur 'Kit' Murray cautiously took the X-1A to 90,440 feet on 26 August 1954. Here Yeager (on ladder) and Murray pose by the X-1A, having set records no other X-1 aeroplane ever exceeded. *USAF AFFTC.*

Plate 80: Delivered to the NACA, the X-1A was extensively instrumented in anticipation of a detailed high-Mach high-altitude research programme. Unfortunately, it too blew up under its mothership, and was jettisoned into the desert not quite a year after Murray's altitude flight, fortunately without injury, the third of the X-1 family to meet such a fate. *NASA.*

Plate 81: The NACA also received the X-1B, delivered to the agency in December 1954 following a harrowing flight to Mach 2.3 by Colonel Frank K 'Pete' Everest. After the loss of the X-1A, the agency used it for aerodynamic heating and high-altitude flight control studies using small rocket thrusters. *NASA.*

Plate 82: By the mid-1950s, a series of Mach 1.5+ fighter prototypes appeared, indicating the maturation of the supersonic revolution even beyond the F-100 experience. First was the Lockheed XF-104 Starfighter, a preproduction prototype of which became the world's first fighter to exceed Mach 2 in April 1956

Plate 83: . . . then the Republic YF-105A Thunderchief, progenitor of a nuclear strike fighter that evolved into a legendary 'dumb bomb' and Wild Weasel deep-strike workhorse of the Vietnam war, and . . .

Plate 84: . . . the Navy's Chance-Vought XF8U-1 Crusader, the first supersonic naval jet fighter and the finest pure dogfighter of its time. *USAF.*

Plate 85: In 1954 and 1956, two significant flights gave ample evidence that the technological gap between American and British aviation that had opened in the post-Second World War era (in large measure because of the misguided decision in 1946 not to pursue piloted research airplanes) was at last, closing. First, on 11 August 1954, Roland Beamont exceeded the speed of sound in Great Britain's first genuinely supersonic airplane, the English Electric P.1, reaching Mach 1.01. The P.1, with a distinctive wing and engine layout, evolved into Europe's first Mach 2 interceptor, the British Aerospace Lightning. *British Aerospace.*

Plate 86: On 10 March 1956, flying the shapely Fairey FD.2 Delta research aeroplane, test pilot Peter Twiss reached 1,132 mph. Alas, the next year Duncan Sandys' infamous Defence White Paper cancelled or imperiled a number of promising British high-speed initiatives, foreshadowing equally misguided cancellations in the 1960s, most notably termination of the TSR.2 and P.1154. (The FD.2 flew in the 1960s with a sinuous 'ogee' wing planform in tests supporting development of the magnificent Anglo-French Concorde SST). *The Fairey Company Ltd.*

Plate 87: After the momentous events of 1953 – 1954, the X-1 and D-558-2 programme wound down to retirement. The air-launched jet-and-rocket-powered third Skyrocket flew on an extensive NACA research programme to study means of preventing transonic pitch-up on sweptwing airlines. Here it is shown sporting a 'sawtooth' wing leading edge extension, one of several experimental configurations examined. *NASA.*

Plate 88: Other pitch-up fixes evaluated included small wing fences to minimise spanwise flow (a tendency of the airflow to stream approximately parallel to the wing's leading edge) . . . *NASA.*

Plate 89: . . . and fully-locked-open leading edge slats. The locked-open slat configuration worked better than other attempts at pitch-up fixes. *NASA*.

Plate 90: After its pitch-up studies, the third Skyrocket embarked on a research programme to assess how external stores (such as bomb and drop tank shapes) affected transonic and supersonic aircraft performance. Here, it is shown with two 150-gallon fuel tank shapes; this workhorse aeroplane eventually retired at the end of August 1956. *NASA*

Plate 91: In September 1956, the trouble-prone Bell X-2 program came to a tragic (if spectacular) end. Earlier in the summer, the surviving rocket plane had attained Mach 2.87 while piloted by Lt. Col. Frank Everest, and then Capt. Iven C. Kincheloe had set an altitude record of 126,200 feet, taking it so high that its controls lost aerodynamic effectiveness and the plane followed a ballistic flight path, like a missile. On September 27, the X-2 dropped away from its B-50 launch airplane for a maximum speed flight attempt . . . NASA photo.

Plate 92: . . . In the tiny cockpit, though new to the X-2, test pilot Captain Milburn 'Mel' Apt flew a picture-perfect mission profile, achieving Mach 3.2, 2,094 mph, the world's first Mach 3 piloted flight. As he turned back to Edwards, the plane coupled violently and tumbled out of control, reminiscent of Yeager's wild ride over two years previously. Highly experienced in inertial coupling testing, Apt courageously and vainly fought to control the plane. Already badly battered from the high g forces, he jettisoned its escape capsule; the separation knocked him unconscious. The capsule plunged to earth, carrying him to his death. *USAF*.

Plate 93: The last three of the X-1 and D-558 family, photographed in 1956, flown by the NACA High-Speed Flight Station (*L – R*): the X-1E, the second Skyrocket and the X-1B. The Skyrocket made its last flight on 20 December 1956 and the X-1B on 23 January 1958. *NASA*.

Plate 94: The little X-1E, *nee* the second XS-1, soldiered on, making its last flight on 6 November 1958. Radically rebuilt from its original XS-1 configuration, it featured a very thin low-aspect-ratio wing, a low-pressure fuel-feed system and a redesigned cockpit, as well as other changes. Fuel tank cracks ended hopes of possibly operating it to Mach 3 by using new high-energy fuel and boosted internal engine pressures, as well as ventral fins (shown here) to improve its directional stability above Mach 2. With its retirement, the saga of the first era of supersonic flight came to a close. *NASA*.

Plate 95: The legacy (1): In 1959, NASA began flight tests of the North American X-15, the world's first hypersonic (Mach 5+) research airplane. The X-15 was a far cry from the XS-1, but over its decade-long flight programme – which witnessed the first winged flights into space, and the first flights above Mach 6 – the three X-15's were flown by 12 research pilots, four of whom – Scott Crossfield, Jack McKay, Neil Armstrong and Joe Walker – had cut their teeth flying the X-1 and D-558-1 and -2 aircraft. *NASA*.

Plate 96: The legacy (2): Neil Armstrong would go even further, carrying the legacy of the X-1 and the X-15 to the dusty Sea of Tranquillity as mission commander of the Apollo 11 lunar landing in July 1969. *NASA*.

Navy decided to continue with the attempt after Colonel Albert Boyd, flying the P-80R, set a new world airspeed record of 623.738 mph on 19 June 1947. BuAer detailed Commander Caldwell and Major Carl, a distinguished Marine fighter ace, to make the flights. Both Caldwell and Carl had extensive flying backgrounds; Commander Turner Caldwell was the Navy D-558 project officer and Major Marion Carl, a Marine test pilot, had scored 18 victories over Japanese aircraft while stationed in the Pacific during the war, some while flying against overwhelming odds at the battle of Midway. Since 1945, Carl had served at the Patuxent River Naval Air Station, Maryland where he flew early jet aircraft including the Bell P-59 and a 'hooked' Lockheed P-80 Shooting Star, which he trapped aboard a carrier to prove jets could operate at sea, and even a captured German Messerschmitt Me 262. In addition, he became the Marines' first helicopter-rated pilot.

To capture Boyd's record, Carl and Caldwell had to fly the Skystreak below 75 metres (less than 250 feet) in altitude and make speed runs along a three kilometre (1.864 miles) measured course. After practising in the second Skystreak on 16 and 17 August, Caldwell took the first D-558-1 aloft on 20 August. Carl flew the second aeroplane that same day, but only Caldwell broke the record, averaging 640.663 mph in four passes over the measured course. Elated, BuAer sent a message of appreciation to the NACA, stating in part, 'A great measure of the credit for the success of the D-558 airplane speed record flight is due to the NACA. The highly important introductory research and investigation programme leading to recommendations on airplane configuration problems was essential in the development of this airplane.' Five days later, flying the second D-558-1, Carl made another attempt, this time to exceed Caldwell's record. He had noticed that the Skystreak's TG-180 (better known as the J35) engine indicated 100% rpm on the ground but once aloft, would decrease to 98%. He convinced Douglas technicians to increase rpm on the ground to 102% so that rpm in the air would be a full 100%. On 25 August during an 18-minute flight, Carl averaged 650.796 mph during four passes, exceeding Caldwell's record by a healthy 10 mph.[3]

Three days after Carl's record-breaking flight, Captain Frederick M Trapnell, USN, completed two brief evaluation flights in the first D-558-1, attaining Mach 0.86 at 7,000 feet (approximately 615 mph) on the second flight. Aside from finding lateral control 'undesirably coarse' due to the small control wheel, Trapnell felt the aircraft had exceptionally fine flying characteristics and concluded that, 'The model D-558 gives the impression of being an outstandingly excellent job of design and engineering, and a very sound airplane for research purposes . . .' Following the record flights, the first Skystreak returned to its contractor programme. The second completed a further 18 evaluation flights before being turned over to the NACA for flight research on 23 October 1947.[4]

Beginning the NACA Programme

On 4 November 1947, the Bureau of Aeronautics sent a letter to Douglas and NACA concerning delegation of responsibility during the programme. As set forth, Douglas would continue to fly a contractor programme for performance investigations using the first aeroplane. NACA would get the second and third aircraft, performing routine flight maintenance and inspection on them. Douglas would undertake major maintenance items and modifications with funds provided by the Bureau of Aeronautics. In addition, the Navy would assume financial responsibility for engine overhaul and replacement expenses and the NACA would procure fuel and oil for the aircraft from the Air Force. This established the pattern of future D-558 administration. The Navy provided administration and appropriations, the NACA coordinated and conducted the research programme as well as supplying consultation and research support, and Douglas assumed responsibility for the design and construction of the aircraft, as well as demonstrating their satisfactory handling qualities.

After receiving the second Skystreak from Douglas, the NACA-Muroc instrumentation section, headed by Gerald M. Truszynski, began installation of research instrumentation using a similar package to that being put in their Bell XS-1 at the same time. The payload included a 12-channel oscillograph for strain gauges; a 60-capsule manometer for recording pressure distribution; wheel force and pedal force transmitters; aileron-, elevator- and rudder-control position recorders; a three-component recording accelerometer; a four-channel telemeter and switch to transmit airspeed, altitude, normal acceleration and elevator and aileron positions; an airspeed-altitude recorder; a sideslip-angle transmitter and a gunsight camera to photograph the instrument readings on the pilot's control panel. A 1⁄10-second timer synchronised the instrumentation. By late November 1947, NACA technicians completed installation except for the rudder-control position recorder and the telemetering equipment. Research pilot Howard C 'Tick' Lilly completed the craft's first NACA flight on 25 November and made another the next day. Instrumentation malfunctions occurred on both flights and Lilly terminated the second one shortly after takeoff because the landing gear would not lock up. Otherwise, Lilly reported that the Skystreak handled satisfactorily. NACA cancelled further flight plans when winter rains flooded the lake, and engine modifications to comply with engine-maintenance technical orders kept the Skystreak grounded into the new year.[5]

During other Skystreak flights, Douglas and the NACA discovered that the scarlet aircraft had extremely poor visual characteristics from a tracking or photographic standpoint, being exceedingly difficult to pick out against the dark blue Muroc sky. Accordingly, Walt Williams requested that Douglas paint the aircraft white or yellow. Over the winter of 1947–48, NACA took advantage of time spent on the ground by painting the fuselage of their Skystreak white to facilitate optical tracking. Following final instrumentation checks, Lilly returned the Skystreak to the air on 16 February for an airspeed calibration flight, but failure of a ground radio receiver prevented the calibration from taking place. With the pilot-familiarisation flights drawing to a close, the NACA Muroc unit readied the aircraft for its actual flight-research programme. On 17 February 1948, William H Barlow, NACA engineer on the D-558-1 programme, outlined the proposed flight-research plans for the aircraft. The initial programme planned covered stability and control and aerodynamic loads measurements, and included:

- two flights for airspeed calibration at high-altitude (30,000 feet) and low altitude (during flybys);
- 13 flights to evaluate static directional stability (through slowly increasing sideslips), dynamic directional stability (through abrupt rudder kicks) and lateral control (through abrupt rudder-fixed aileron rolls) from Mach number range to 0.4 to 0.85 at 10,000 and 30,000 feet;
- 16 flights to measure longitudinal stability and aircraft loads in both straight and accelerated flight during straight-and-level speed runs to maximum speed at 10,000 and 30,000 feet (to determine static longitudinal control characteristics and wing and tail loads in straight flight);
- turns from 1 to 4g in ½g increments from Mach 0.4 to 0.85 (to obtain stick-force-per-g data, elevator position variation with normal acceleration and wing and tail loads in accelerated flight); and
- abrupt pull-ups from Mach 0.4 to 0.85 at 30,000 feet (to acquire variation of the coefficient of lift with the rate of pull-up, available control, buffet boundary, buffet loads data, elevator hinge moments and dynamic stability characteristics).

Engine and instrument checks, combined with a wet lake, kept the Skystreak grounded until March. When Lilly did resume flights in the red-and-white aircraft, landing gear difficulties marred the next four flights. The landing gear doors would not lock and, eventually, technicians traced the trouble to a door latch that closed before the landing gear came up. After increasing the clearance between the latch and the door's engaging roller, the landing gear difficulties disappeared. On 8 April 1948, Lilly completed two flights for airspeed calibration, repeating the first because of radar beacon failure. He completed a tower flyby calibration on 12 April and during the rest of the month, flew the Skystreak on its first pure research flights. By the end of April, the Skystreak had completed five directional stability investigations. On 29 April, he had attained Mach 0.88 (approximately 580 mph).[6]

Tragedy and Aftermath

On 3 May 1948, Lilly took off for the second Skystreak's 18th NACA flight. After lift-off, the landing gear would not retract and he brought it back down. Following minor maintenance work and an inspection, a ground service crew towed the Skystreak to the west end of the paved east-west runway, disconnected the tug and Lilly began his takeoff roll. The red-and-white Skystreak broke ground after about 5,000 feet of run, its landing gear came up normally and the research plane continued to accelerate in level flight over Muroc Dry Lake at about 100–150 feet, its J35 engine shrieking loudly. Approximately 2.2 miles east of the start of take-off, travelling about 250 mph, witnesses saw the Skystreak shed a large piece of external fuselage skin, followed by a gush of smoke mixed with flames. The plane held steady for a few seconds before slipping into a left yaw and roll and then almost inverted, striking the ground and breaking up, the fuel-filled wing immediately burst into flames. At low altitude, with no means of escape, Lilly had had no chance. He had become the first NACA research pilot killed in the line of duty since the creation of the agency in 1915.[7]

Howard Clifton Lilly's flying career spanned a scant seven years, but they had been years filled with accomplishment. Like Yeager and Everest, Tick Lilly came from West Virginia. After graduating from the Civilian Air Pilot Training Programme in 1941 he had earned navy wings. In September 1942, he had received an honourable discharge from the Navy so he could join the NACA's Langley Memorial Aeronautical Laboratory as a research pilot and he later transferred to NACA's Lewis Flight Propulsion Laboratory in Cleveland. He joined the Muroc team in August 1947, trained under Herbert Hoover and, while flying the agency's XS-1, had become the third man in the world to fly faster than sound. A highly experienced Skystreak pilot, he had flown its previous 18 flights, and had made one flight in the third aircraft as well. All indications – particularly the in-flight fire – pointed to a mechanical failure of the aircraft itself. Like Lilly, though, the Skystreak, with but minor exception, had an excellent record. Before NACA received the second D-558-1, Douglas and Navy pilots already had made a total of 27 flights in the aircraft. Taken as a whole, at the time of the accident the three D-558-1 aircraft had successfully completed 127 flights without serious incident, a highly satisfactory record of operations.

NACA formed an accident board to investigate the crash chaired by chief pilot Melvin Gough, and security personnel cordoned off the area of the crash and the flight path from runway to crash site. Searchers carefully located bits of wreckage and mapped their location on a grid. About 0.2 miles short of where searchers had found the big, section of white fuselage skin, other searchers located bits and pieces of the compressor case, blade fragments, paint chips and small pieces of fuselage skin. No evidence of burning or fire damage existed, and the findings indicated that the turbojet compressor blades and case had torn out of the fuselage. Close examination of the aircraft wreckage revealed that the disintegrating compressor had sheared the right rudder cable and up-elevator cable, impairing both longitudinal and directional control. The crash itself had broken the remaining control cables. Evidence, then, indicated an engine failure in which the compressor section had disintegrated, severing vital control lines and making a crash inevitable. A memorandum to the D-558 accident board written slightly over a week after the accident, expressed concern over the lack of protection around control lines and criticised the small size of the Skystreak's canopy, since a pilot could not wear a protective helmet. The writer also criticised the unproved jettisonable-nose method of escape and suggested that when the D-558 aerodynamic research programme ended, NACA could equip one of the two remaining aircraft with an autopilot and telemetry equipment and deliberately jettison the nose in flight to measure cockpit motions. (Interestingly, given the battles NACA had waged over air-launching research aeroplanes a mere three years previously, the author also recommended the air launching of all research aircraft because of its proven feasibility demonstrated by XS-1 operations and because it would provide altitude for the pilot to escape or take corrective action in case of difficulty. The Accident Board agreed that more consideration should be given to the potential of air launching for all research aircraft.)

In their report released later, the board agreed with most suggestions made in this memorandum. The board exonerated all personnel involved in any way in the ill-fated flight and placed the blame for the accident on the disintegration of the compressor section of the TG-180 (J35-C-3) engine. The compressor had severed the main fuel lines, elevator and rudder cables causing an in-flight fire and subsequent loss of controls. The board recommended that all control lines, fuel lines, pumps and electrical circuits be protected, and that the control cables be shielded, armoured or duplicated. Additionally, the board recommended that the cockpit canopy be redesigned to permit greater visibility and the pilot's wearing of a crash helmet and that only the latest-type engines complying with the most recent changes in design be used in research aircraft. In particular, the board recommended that the engine access for inspection should be improved in the remaining two D-558 aeroplanes. NACA quickly pointed out that the value of the Skystreak to high-speed flight research remained unaltered by the accident and the incorporation of these recommended changes.

Essentially the loss of the second D-558-1 meant that the third, BuNo. 37972, had to assume the research programme of the destroyed aeroplane. In September 1948, NACA returned the third D-558-1 to the Douglas plant for modifications recommended by the accident board. The company had already completed modifications to the first, and this aircraft continued flying in the Douglas contractor programme in a hazardous series of high-speed dives to obtain stability and control information up to the plane's maximum Mach number. On 29 September 1948, it became the only Skystreak to exceed Mach 1 in the course of a 35° dive by company test pilot Gene May. Originally, Douglas had delivered the third Skystreak on 4 November 1947 and, by the time of Lilly's accident, it had completed four flights. After the aircraft arrived back at the Douglas plant, Douglas technicians added duplicate control cables and ¼-inch stainless steel armour over the emergency fuel pump and fuel lines. They replaced the Skystreak's standard high-pressure fuel hoses with wire-wound fuel hoses following tests in which .22 calibre rifle bullets, simulating disintegrating engine fragments, easily pierced the standard hose. In addition to making safety modifications, Douglas also painted the aeroplane white.[8]

Recovery and Resumption

Early in November 1948, Douglas sent the third D-558-1 back to the Muroc facility. NACA engineers already knew before initiation of its flight-research programme that the aircraft experienced more severe longitudinal and lateral trim changes than those experienced with the XS-1 in the same speed range. Douglas had to complete a satisfactory demonstration flight in the aircraft before NACA would accept it. On 3 January 1949, Gene May made the demonstration flight which included a 6.8g pullout, constant maximum sideslips left and right at 580 mph and a high-speed low-level pass at 605 mph indicated airspeed. Post-flight inspection by a

NACA safety representative revealed a badly mangled brass safety-wire that had evidently passed through the engine. Rather than pull the engine for inspection, Douglas installed a new engine so as not to delay delivery. On 22 January 1949, the NACA took charge of the third D-558-1, giving it the call sign 'NACA 142', and spending until April readying it for flight. During April, Douglas concluded their performance investigation using the first Skystreak, subsequently delivering it to NACA for spares support. In total, the first Skystreak had completed 101 flights. During its demonstration programme, hopes that the Skystreak series might generate a tactical aircraft had been quashed. Although it attained Mach 1 in a dive, its behaviour characteristics at this speed deteriorated to such an extent that it could not possibly be used for tactical military flying. NACA's own test flights quickly highlighted these deficiencies. On 22 April, NACA pilot Robert A Champine performed the third Skystreak's first NACA flight since modification. Instrument malfunctions marred both this flight and a second on 28 April, but Champine did get some indication of the aircraft's behaviour characteristics at transonic speeds, and it was not encouraging: during one dive on the second flight, wheel force for trim rose from 5lbs push at Mach 0.82 to 30lbs push at Mach 0.87. Throughout the dive, the plane exhibited increasing tail heaviness and initiated a lateral oscillation with very low damping.[9]

NACA again grounded the aircraft for maintenance work, for the turbojet engine had to be removed and replaced. Replacing the Skystreak's engine required removal of flight instrumentation and this work kept the Skystreak grounded until August 1949. The plane then embarked on a comprehensive handling-qualities investigation programme flown by Champine and John H Griffith, a newly arrived NACA pilot. On these flights, instrumentation recorded data during aileron rolls, sideslips and abrupt rudder kicks around Mach 0.9. The Skystreak quickly exhausted its fuel supply on these flights, forcing NACA to install tip tanks for increased endurance. Additionally, during September and October 1949, NACA instrumentation engineers installed manometers to record pressure distribution from orifices in the right wing. The Skystreak resumed flying on 31 October, beginning an uninterrupted series of pressure-distribution research flights that continued over the next seven months . On the last of these flights, NACA flight 22, 13 June 1950, Griffith attained approximately Mach 0.99, about the Skystreak's limiting Mach number. In mid-June 1950, NACA engineers removed its horizontal stabiliser and replaced it with the instrumented stabiliser from the first D-558-1 then in 'dead' storage at the NACA facility to enable engineers to measure tail loads. At the same time, as they had with the agency's X-1, NACA technicians converted the electrical instrumentation system from direct current to 400-cycle alternating current, also installing a special picture manometer obtained from Langley Laboratory.[10]

The Transonic Picture

Transonic pressure-distribution measurements indicated that a shock wave first formed at the 45% chord position of the upper wing surface at about Mach 0.76. As the Mach number increased, this shock moved aft, together with the centre of pressure. At about Mach 0.82, a shock wave would form on the lower wing surface. As the Mach number increased further, the lower shock moved aft while the upper shock remained stationary, thus causing a forward shift in the location of the centre of pressure. With a further increase in Mach number, the two shock waves and the centre of pressure would again move aft. This behaviour corresponded to that experienced on the NACA XS-1, which also had a 10% wing section. At transonic speeds the Skystreak became laterally unstable and presented the pilot with considerable difficulty in maintaining wings-level control. 'Wing-dropping' appeared, for example, the right wing would drop suddenly and, as the pilot attempted to bring it up, the D-558-1 would roll rapidly left. If the pilot applied corrective aileron, the plane would roll rapidly right. Usually, the Skystreak displayed a marked tendency to roll right beginning at about Mach 0.84, perhaps from manufacturing irregularities. As Mach numbers increased, a trim change caused by a loss in elevator effectiveness appeared. NACA engineers were uncertain whether changes in pressure distribution or actual physical distortion of the stabiliser and elevators caused the loss in elevator effectiveness and the resulting trim changes. This uncertainty led to the decision to install the stabiliser from the first Skystreak on the NACA D-558-1. One interesting experiment NACA engineers undertook on the Skystreak – and also on the XS-1 – was the installation of wing-mounted vortex generators – rows of little airfoil-like tabs intended to create small swirling vortices that might help stabilise the positions of the shock waves on the wing – thus possibly reducing or eliminating the undesirable changes in lateral and longitudinal stability. Before the conclusion of the pressure-distribution research programme, engineers fitted the Skystreak with the vortex generators and John Griffith completed five vortex-generator research flights. Test results indicated that they had little effect on wing-pressure distribution until the aircraft reached Mach 0.85. Between Mach 0.85 and Mach 0.89, the generators caused the shock wave to displace more to the rear, improving pressure recovery near the wing trailing edge. Further, the generators prevented flow separation and delayed the onset of wing-dropping by 0.5

Mach. Encouraged by such results, Boeing was the first manufacturer to make widespread use of the vortex generator concept with rows on the B-47 and B-52 strategic bombers, the KC-135 tanker transport, and on to its commercial jetliners, though other manufacturers took advantage of the idea as well, notably Douglas itself with Ed Heinemann's small A-4 Skyhawk carrier-based attack bomber.[11]

Defining the Limits

Now fitted with the new stabiliser and a high-speed picture manometer to record wing-pressure distribution over one span-wise station, NACA engineers readied the aircraft for a thorough investigation of the Skystreak's buffet boundary. To investigate the wing as a potential source of buffet, technicians installed a downwash vane for airflow analysis. NACA planned to record additional longitudinal-stability and trim data, and the strain gauges in the tail would measure the bending and twisting of the horizontal stabiliser and the twist of the elevators. In mid-October 1950, technicians completed final instrumentation and maintenance checks and the Skystreak again took to the air when John Griffith completed an instrument- and operation-check flight on 26 October. One of the modifications to the aircraft had been the addition of a nose boom. During the check flight, the boom performed satisfactorily and NACA engineers mounted an airspeed head having an angle-of-attack vane onto the boom for future flights. The Skystreak returned to the air on 29 November 1950 on a pilot familiarisation flight for NACA research pilot Scott Crossfield, who subsequently flew the first complete buffet, tail loads and longitudinal-stability investigation mission on 12 December 1950. During that month, NACA engineers added an investigation of dynamic longitudinal stability to the longitudinal-stability programme because the manometer had additional unused capabilities. The Skystreak flew this entire longitudinal-stability programme over the next year, one of the most extensive early transonic aerodynamic investigations.[12]

Early in the programme, elevator twists up to 2° occurred during pullups at Mach numbers around Mach 0.80. Additional data indicated undesirable elevator vibration. NACA engineers removed the elevators and X-rayed them for fatigue cracks, but no flaws appeared. Further investigation revealed that the vibrations increased in amplitude in direct relation to an increase in Mach number and occurred at all high-lift conditions over the entire Mach number range, even in a stall. NACA engineers contacted the Douglas Aircraft Company since Douglas had encountered similar difficulties during the contractor programme on the first Skystreak. Subsequent discussions with Douglas representatives revealed that during a dive to Mach 0.94, the first Skystreak had encountered vibrations so severe that Douglas placed a never-exceed limit of Mach 0.92 on the aeroplane – something that should have been reported to the agency. The company attempted to lessen the vibrations by shifting the location of the outboard elevator balance weights, but had not effectively evaluated the change, concluding the contractor programme without making additional flights. Douglas engineers conceded that the change had been more of a shot in the dark than a seriously considered fix. Ever-cautious, NACA engineers realised they had a potential safety-of-flight problem on their hands, for if the elevator surfaces had been long exposed to unexpected vibration, they might well fail under load. The vibration did not readily show on the elevator force and position indicator, and appeared unaffected by the elevator damper. High-Speed Flight Research Station engineers concluded that the vibration probably assumed undesirable proportions before showing up on the force trace. Subsequently, Douglas engineers calculated that the vibrations during the NACA flights caused stresses of approximately 70% of the maximum design stress and, not surprisingly, that such stresses could eventually lead to fatigue failure. To be safe, NACA engineers replaced the elevators used on the third Skystreak (these elevators had originally been on the first D-558-1 aircraft on over 100 flights, including the notorious dive to Mach 0.94) with the third Skystreak's original elevators, which had only been used for 22 NACA flights. NACA did not complete replacement of the elevators until late April 1951, and the first flight following replacement came on 2 May 1951.[13]

Towards the End

In September 1951, technicians removed the high-speed picture manometer from the aeroplane and in October, the Skystreak began a lateral-stability investigation programme in conjunction with the on-going buffet and tail-loads programme. Following flight 56 (9 November 1951), NACA grounded the aeroplane for extensive maintenance work due to severe fuel leaks and engine ignition problems until January 1952. Winter rains then kept the lakebed closed, precluding any research flights and HSFRS engineers decided to add strain gauges for vertical-tail-loads measurements. Design and installation of the strain-gauge instrumentation – which involved cutting two access doors in the base of the vertical fin – began immediately. Additionally, NACA technicians installed instrumentation for measuring rudder hinge-moments and transverse acceleration at the tail in preparation for the anticipated vertical-tail-loads programme. When the NACA D-558-1 resumed flying in June 1952, however, it embarked on a lateral-stability and aileron-control-effectiveness investigation. NACA felt the lateral-stability and control programme to be more important than the vertical tail-loads investigation and, consequently, postponed the vertical-tail-loads programme indefinitely. Further, NACA filed without analysis Skystreak flight data taken during the horizontal-tail-loads and buffeting programme because of a shortage of engineers. The agency needed all available engineers for the D-558-2 Skyrocket, Bell X-5 and the Douglas X-3 programmes, a sign that the Skystreak, however valuable it might be, was nevertheless no longer considered in the front-rank of NACA research aircraft. During the lateral-control programme, Scott Crossfield and a new NACA research pilot, Stanley P Butchart (who had piloted Grumman-General Motors TBM Avenger torpedo-bombers in the Pacific along with his best friend, a young ensign named George Bush, destined to be President of the United States) flew the jet aircraft in abrupt aileron rolls to obtain aileron effectiveness data over a Mach number range from 0.4 to the limiting Mach number at 10,000 feet, 25,000 feet and 35,000 feet. As indicated earlier, in other NACA and Douglas flights lateral stability decreased rapidly above Mach 0.85. Throughout the summer of 1952, the two pilots confirmed that aileron effectiveness sharply dropped at velocities above Mach 0.88 by making aileron rolls at speeds approaching Mach 0.9. Concurrently with the lateral-stability programme, the High Speed Flight Research Station continued the dynamic longitudinal stability investigation.[14]

NACA-Edwards engineers concluded the lateral-stability programme on the Skystreak with its sixty-third research flight, 12 August 1952. During a postflight inspection, maintenance personnel noticed a looseness in the horizontal stabiliser mountings. Further inspection revealed that the assembly required extensive repairs. During necessary repair work, NACA technicians discovered that the strain gauges installed for the vertical-tail-loads programme likewise needed extensive repair work. Project engineers decided, however, that the vertical-tail-loads programme would not return data valuable enough to justify the man-hours needed to repair the gauges and abandoned the planned vertical-tail-loads programme. Despite this, the Skystreak still had an active programme ahead of it. The dynamic stability programme had yet to be concluded, and the aircraft would then be used briefly as a pilot trainer for research pilots. After repairs to the horizontal stabiliser had been completed, the D-558-1 resumed flying in January 1953. During the dynamic-stability programme, NACA planned to record aircraft response after elevator and rudder pulses over a Mach range to o-go at 25,000 feet and 35,000 feet. Though agency engineers had acquired some dynamic-stability data during the lateral-stability programme, roughly ⅔ of the elevator pulse data remained to be gathered, as well as almost all of the rudder pulse data. Project engineers planned to record the data from these flights on IBM cards and send it to Langley Laboratory for computation on IBM machines. Stanley Butchart completed the first flight in the dynamic-stability programme on 29 January 1953 and made four more flights by mid-February. This rounded out the flight portion of the dynamic-stability investigation, though engineers believed that one or two additional flights might be necessary to provide fill-in data. On 27 March 1953, research pilot John B 'Jack' McKay checked out in the Skystreak, and on 1 and 2 April he completed two additional flights to provide dynamic-stability data. These last two flights marked the end of the aircraft's planned research programme, and NACA relegated it to a pilot training role.[15]

The Skystreak had not quite ended its research days, however, for the agency decided to conduct a brief investigation of the effects of wingtip fuel tanks up the plane's buffet characteristics, hoping that the tanks might act as 'endplates' on the wing, reducing the severity of buffet once it developed. During May and June, McKay completed an additional six flights, but data acquired from the tip tank buffet programme indicated that the tanks had no appreciable effect upon buffet up to a Mach number of 0.85. McKay completed his last flight in the aircraft on 3 June and one week later, Crossfield took the plane up for its final NACA mission, flight 78. In July 1953, after four-and-a-half years of active research flying, the agency retired the third Skystreak and removed its research instrumentation. NACA placed the Skystreak in dead storage in late 1953 like the first D-558-1, and in 1954 returned the aircraft to the Navy at Alameda Naval Air Station, on the shores of San Francisco Bay.[16]

In Sum . . .

Developed essentially as a turbojet-powered backup for the XS-1, the three D-558-1 Skystreaks completed a total of 225 flights. Unlike some of the other research aircraft, they had not required extensive changes in configuration as a result of flight testing. Douglas had installed a high-speed canopy and extended the aircraft's tail cone to improve engine cooling and prevent erosion of the skin around the tailpipe, but these changes did not demand any significant reworking of the airframe. The only other modifications had involved protection of control systems of the two remaining aircraft following the loss of Tick Lilly and the second Skystreak. Its loss produced an increased awareness of the hazards of research flying in NACA and industry personnel, and the changes made to the Skystreak were duplicated on other research aircraft as well, notably the Northrop X-4 and Bell

X-5. Generally speaking, pilots liked the flying qualities of the D-558-1, though they found the cockpit very cramped: the pilot could not turn his head and crash helmets had to be covered with chamois skin to prevent windshield scarring. The Skystreak had satisfactory flying characteristics until reaching about Mach 0.85 when aileron and elevator effectiveness decreased appreciably. In the low-speed spectrum, the Skystreak had good control characteristics as it approached a stall, but it encountered abrupt roll-off at the stall. Pilots therefore always executed the landing approach at about 210 mph, landing at no less than 143 mph, well above stall speed. Despite the high landing speed, pilots found its landing characteristics satisfactory.[17]

Some have denigrated the Skystreak as having been unnecessary in light of the success of the X-1 family and, for that matter, the rapid development of the slotted-throat wind tunnel. Undeniably, the Skystreak did not have the revolutionary impact on aviation that the X-1 did. Like the X-1 and the D-558-2, however, during 1948–50 the Skystreak provided virtually the only means of acquiring data on transonic flight conditions pending the development of improved ground-research techniques such as the slotted throat wind tunnel. Its construction enabled NACA and the industry to combine into one aircraft all the important and current developments of the aeronautical state of the art and then determine from flight testing whether or not the elements fitted together to produce a workable design. It freed the Bell X-1 rocket research aircraft to explore supersonic flight conditions during its short rocket flights without necessity to cruise at transonic speeds where its rocket potential would have been wasted. Further, it formed the departure point for an undeniably significant stablemate: the sweptwing D-558-2 which was, in many ways, the workhorse of NACA high-speed flight testing, bridging the era from the subsonic aeroplane almost to the era of the hypersonic X-15. Carrying 634 lbs of research instrumentation and flown by highly trained scientist-pilots, the D-558-1 represented an ideal first-generation simple transonic research aeroplane. Its value can be judged on the wide basis of research areas it explored: handling qualities, pressure distribution, buffeting, tail loads, static and dynamic longitudinal and lateral stability and control. Initiated as a research aeroplane, it ended its days as a pilot-familiarisation aircraft. Perhaps not as spectacular as the rocket research aircraft, the D-558 Phase One Skystreak nevertheless constituted a necessary step on the road to supersonic flight. Its creators were – and had every reason to be – proud of their accomplishments in bringing it from sketch to reality.

X-1 eventually reaching [illegible] of the flight qualities of the D-558 [illegible] though they found rocket [illegible] would not [illegible] both, and crash hazards [illegible] to [illegible] windshield [illegible]. The Skystreak had [illegible] until reaching about Mach 0.85 [illegible] and elevator effectiveness decreased [illegible]. In the low-speed [illegible] the Skystreak had good control characteristics as it approached a stall, but it encountered [illegible] roll at the stall. [illegible] the approach at about 200 mph [illegible] no less than 140 mph [illegible] stalled. Despite the high landing speed, [illegible] found its landing characteristics satisfactory.

Some have contrasted the Skystreak [illegible] in light of the success of the X-1 family and [illegible] the rapid development of the swept-wing [illegible] model. [illegible] the Skystreak did not have the revolutionary impact on aviation that the X-1 did [illegible] the D-558-2 [illegible]. [illegible] the Skystreak provided virtually the only means of acquiring [illegible] flight conditions [illegible] development of improved ground [illegible] wind tunnels [illegible] the slotted throat wind tunnel [illegible] and the industry [illegible] the [illegible] state of the art [illegible] the [illegible] rocket research aircraft [illegible] conditions during [illegible] at transonic speeds [illegible]. Further, [illegible] the D-558 [illegible] high speed flight testing [illegible] research instrumentation and flown [illegible] the D-558-1 [illegible] research airplane [illegible] handling qualities [illegible] loads, [illegible] lateral stability [illegible] acceptance [illegible] as the [illegible] research aircraft [illegible] the D-558 [illegible] superiority [illegible] really.

PART THREE
Through Mach 2

7
Nudging Towards Twice the Speed of Sound

When the first D-558-2 Skyrocket (BuNo. 37973) arrived at Muroc Dry Lake in early December 1947, the XS-1 had already shattered the myth of the 'sound barrier'. Much still remained, however, for aerodynamicists to learn about transonic and supersonic flight conditions, as evidenced by the detailed programmes the NACA conducted on their XS-1 and the second and third D-558-1s. As discussed previously, by the time the XS-1 began its initial probing towards the first supersonic flight in the fall of 1947, a new aircraft had already arrived at Muroc that could exceed the speed of sound in a slight dive: the prototype North American XP-86 Sabre which became the third American aeroplane to exceed the speed of sound, following the first and second XS-1s. While the Skyrocket filled the need for a research aeroplane to provide data from full-scale transonic flight testing on the behaviour of sweptwing aircraft, North American's bold gamble in developing the Sabre furnished the Air Force an operational fighter in 1949 having performance superior to any other fighter in the world. Not surprisingly, the NACA later acquired the fifth production F-86A, instrumented it for flight research, and shipped it to NACA's Ames Aeronautical Laboratory where NACA research pilots George Cooper and Rudolph Van Dyke flew it in an extensive programme involving dives to supersonic speeds. Curiously the F-86 and the Skyrocket complemented one another in sweptwing research though of the two, the D-558-2 had profoundly greater potential performance – as test pilots Bill Bridgeman, Marion Carl and Scott Crossfield would amply demonstrate. Designed to carry a minimum of 500 lbs of research instrumentation, the D-558-2 series as fully developed actually carried between 800 and 1,100 lbs of instruments. From the start of its research programme in February 1948, the Skyrocket constituted America's – and arguably the world's – most sophisticated flight-research tool for acquiring transonic sweptwing data.[1]

Douglas selected its chief pilot, John F Martin, as project pilot on the D-558-2. Martin had joined Douglas in 1940 after flying as a pilot for United Air Lines. Since then he had flown the initial prototype and engineering tests on a variety of Douglas aircraft, including the A-20 Havoc, A-26 Invader and the C-54, the latter the military version of the civil DC-4. The story, allegedly true, of how Martin became project pilot illustrates how the company's own test pilots viewed the Skyrocket project. At the time the Skyrocket neared completion, the 'sound barrier' still loomed as a dangerous unknown in the minds of most pilots: Yeager's epochal flight lay in the future. When it came time to select a project pilot for the aeroplane, Douglas El Segundo sent notice to the flight test office at Santa Monica for the company test pilots to submit bids for the flight programme. The pilots did not desire to fly the somewhat imposing plane and, after talking among themselves, decided to submit exceptionally high bids that almost certainly would not be accepted. Martin was the only pilot not in on the plot, for he was away delivering an aeroplane. Unaware of the conspiracy in Santa Monica, he submitted a reasonable bid, so the company designated him as project pilot.[2]

First Flights

The white Skyrocket waited out at Muroc during January 1948 while Douglas technicians completed installing its Westinghouse 24C (better known as the J34-WE-40) turbojet engine and research instrumentation. By the beginning of February, the plane was ready for flight and, on 4 February, Martin lifted the first Skyrocket off the lakebed at Muroc. The plane had a sluggish performance on just its turbojet engine (it only produced 3,000lb-thrust) and Douglas decided to improve takeoff performance by installing JATO solid-fuel booster equipment. These early flights also revealed a 'Dutch-roll'-type oscillation – a combined yawing and rolling motion as if the plane was hunting back and forth to find a comfortable position for itself – and the flush canopy guaranteed poor visibility for the pilot. Douglas remedied the two problems by increasing the height of the vertical fin by roughly 18 inches, raising it to an even 13 feet (coincidentally giving it an even more graceful line) and installing a raised cockpit similar to the D-558-1 Skystreak. After making 15 flights in the Skyrocket, Martin turned it over to Gene May. As it had with the Skystreak, Douglas retained the first Skyrocket for a contractor evaluation programme. Over the spring and summer of 1948, Douglas engineers at Muroc readied the second D-558-2 (BuNo. 37974) for flight. Like the first Skyrocket, this aeroplane lacked its RMI LR-8 rocket engine, but the NACA agreed to accept the aeroplane and utilise it with the provision that when the rocket engine became available, the agency could return the Skyrocket to Douglas for its installation. On 2 and 7 November, May completed two demonstration flights in the aeroplane and on 1 December, Douglas delivered it to the NACA for initiation of its flight-research programme. Equipped only with its J34, the Skyrocket was limited to about Mach 0.9, but since the NACA planned to use the jet-powered plane for general stability and control and air-loads research to Mach 0.85, this did not as yet constitute a serious limitation. Engine difficulties kept it grounded until May 1949, but NACA engineers took advantage of the time spent in the hangar to calibrate and install all instrumentation. On 24 May, research pilot Robert Champine completed the plane's first NACA flight, beginning its research programme with the second flight on 1 June 1949.[3]

Pitch-up: A Nasty Transonic Surprise

By early August, NACA had completed a total of six flights and Champine had reached Mach 0.87 in a dive, the fastest flight yet. During the plane's seventh NACA flight on 8 August 1949, Champine banked into a 4g turn at Mach 0.6. Suddenly, and without warning, the nose of the plane pitched upwards violently, attaining a positive acceleration of 6g. Shaken, Champine applied full down elevator and the Skyrocket responded rapidly. Not taking any chances, however, the NACA pilot landed immediately. Though model tests had indicated that sweptwing aeroplanes might experience longitudinal instability resulting in a 'pitch-up' phenomenon, this seventh NACA flight provided aerodynamicists with the first indications of the severity and seriousness of the problem, particularly those that had their horizontal tail located high on the vertical fin. The remainder of the plane's programme, prior to installation of the RMI engine, concentrated on the pitch-up problem, soon encountered with other high-speed swept-and-delta-wing aircraft. In September 1949, John Griffith checked out in the aeroplane and for the remainder of the year, Champine and Griffith flew the plane. On 1 November, Griffith entered a 4g turn at Mach 0.6 as Champine had done nearly three months previously. As on the earlier flight, the plane became longitudinally unstable and pitched up. Griffith attempted to fly beyond the point of instability and, as the plane's angle of attack increased, the Skyrocket commenced rolling and yawing, then spasmodically snap-rolled, turning completely over. Griffith recovered from the snap-roll and, after ascertaining that the plane and engine functioned normally, he continued the flight. During a stall approach with the flaps and landing gear extended and the wing slats closed at 14,000 feet, the Skyrocket remained stable down to 130 mph, but then pitched up. Again, Griffith attempted to fly beyond the point of instability and the bucking plane rolled into a spin. During the spin, Griffith found the plane would not recover with flaps and gear extended. After he retracted the gear and flaps and initiated spin recovery, the plane returned to normal flight at 7,000 feet. Altogether, it had been a dangerous and eventful flight, not one to be repeated too often. NACA continued the longitudinal-stability-investigation programme on the second Skyrocket until January 1950 when the High-Speed Flight Station returned the plane to Douglas for installation of its RMI rocket engine – and conversion to air-launch configuration. The pitch-up investigation programme, which would become one of the major aspects of NACA research on the Skyrocket (and other research aeroplanes as well), would continue, but with the third Skyrocket (BuNo 37975), an air-launched jet-and-rocket propelled model delivered to the agency a year later, in December 1950.[4]

Air Launching: The Obvious Progression

The decision to modify the Skyrocket for air launching, like the X-1, came about as a result of discussions between Douglas, the Navy and NACA. As early as the autumn of 1948, NACA had considered the possibility of modifying the D-558-2 aeroplanes for air launching. The X-1 had proven the feasibility of air launching, and it offered increased performance and a greater margin of safety. In 1948, neither NACA, Douglas nor the Navy knew the operational characteristics of the Skyrocket with both jet and rocket propulsion. The first flight of the aeroplane with its rocket engine was still in the future. At a conference on 21 October 1948, Navy, NACA and Douglas representatives decided to postpone a decision on modifying the D-558-2 aircraft for air launching until Douglas accumulated rocket experience. The first Skyrocket completed with rocket propulsion was the third aeroplane, BuNo. 37975. The rocket engine increased the plane's speed over 100 knots as compared to the jet-only Skyrockets, raising the maximum level-flight Mach number from approximately 0.82 at 20,000 feet to 0.99 at the same altitude. It could attain approximately Mach 1.08 at 40,000 feet in level flight with both turbojet and rocket propulsion. On 8 January 1949, Gene May completed the plane's first flight, making its first rocket flight over a month later, on 25 February. On 24 June 1949, the third D-558-2 exceeded the speed of sound for the first time, the first Skyrocket to do so.* As the plane went supersonic, May noted, 'the flight got glassy smooth, placid, quite the smoothest flying I had ever known.'[5]

* And the sixth aircraft type to do so, behind the XS-1 (October 1947), XP-86 (April 1948), D H 108 (September 1948), D-558-1 (September 1948), and La-176 (December 1948).

As soon as the Skyrocket began flying with its rocket engine, the potential advantages of air launching became obvious, for takeoff operations in the heavily loaded jet-and-rocket aeroplane were extremely hazardous. Laden with rocket fuel and jet fuel, the Skyrocket pilot began his takeoff roll using the jet engine, then added thrust from two of the four chambers on the rocket engine, and then, just before lift-off, fired four JATO bottles to kick the plane into the air. The takeoff run consumed *three miles* and imposed severe strain on the landing gear. Douglas recognised that if the landing gear should collapse, the plane would certainly be lost – along with the pilot – in a truly memorable explosive finale. Additionally, of course, using rocket fuel for takeoff decreased the plane's maximum attainable speed and altitude. NACA concluded that if it operated the D-558-2 with both jet and rocket power using standard ground takeoffs, the plane could attain approximately Mach 0.95 in level flight. Yet, on its jet engine alone, the plane could attain Mach 0.9 – clearly the 0.05 increase in performance did not warrant the risk of flying the jet- and rocket-powered plane on research missions. For its part, Douglas also recognised the safety and performance advantages accruing from air launching and studied modifying one Skyrocket by removing its turbojet engine and replacing the engine with increased fuel for the rocket engine. Thus modified, and air launching from a Boeing B-29, Heinemann's engineers estimated that the plane could attain airspeeds between Mach 1.46 and 1.6. NACA believed this modification highly desirable, for results from the sweptwing aeroplane could then be directly compared over the same speed range covered by the straight-wing X-1. Further, the supersonic behaviour of the D-558-2's NACA 63 series airfoil section could be compared with the supersonic behaviour of the unconventional biconvex airfoil then designed for the projected Bell X-2 sweptwing supersonic research aeroplane.

Hugh L Dryden, the NACA's Director of Research, firmly supported the Douglas plan, and proposed that Douglas modify the second Skyrocket to all-rocket air-launch configuration when NACA returned this plane to Douglas for installation of its RMI engine: clearly NACA thinking on both rockets and air launching had come a long way from a few short years before. The company could modify one of the two remaining Skyrockets to air-launch configuration as well, and would conduct a small demonstration programme on the aeroplanes before turning them over to the NACA for flight research. Dryden conveyed his recommendations to the Navy in a strongly worded letter on 1 September 1949. Nearly three months later, on 25 November, the Bureau of Aeronautics added Amendment 19 to the Douglas D-558 contract, providing for modification of the second and third Skyrockets to air-launch configuration. Additionally, the company would modify a Boeing B-29 Superfortress to act as launch aeroplane.[6]

After receiving the two Skyrockets back at El Segundo, Douglas disassembled the two planes for modification. The modifications to the third were not extensive, for the aeroplane retained both its turbojet and rocket powerplants. The company simply installed retractable mounts for the launch hooks. This aeroplane returned to Edwards Air Force Base for its contractor demonstration flights in September 1950. The modifications to the second D-558-2, the proposed all-rocket aeroplane were far more extensive. After receiving the plane back from the NACA in January 1950, company technicians removed its J34 turbojet and the gasoline storage tanks. In place of the jet engine and its fuel tanks, technicians installed a liquid oxygen tank and an alcohol-water tank. The plane had two tanks each for the liquid oxygen and alcohol-water propellants, a total capacity of 345 gallons of liquid oxygen and 378 gallons of alcohol-water. The company replaced the air intakes with smooth, flush fuselage panels, removed the jet exhaust outlet, and installed a Reaction Motors LR-8-RM-2 rocket engine rated at 6,000lb-thrust, together with its fuel-feed turbopump. Douglas completed modifications to the aeroplane by August 1950, then installed special NACA recording instrumentation. The all-rocket aeroplane returned to Edwards on 8 November 1950, by which time the modified third Skyrocket already had completed four successful air-launch flights. To launch the two Skyrockets, the Navy acquired a Boeing B-29A Superfortress from the Air Force and assigned it to Douglas. The Navy redesignated the bomber as a P2B-1S (BuNo. 84029) but to Douglas and NACA launch crews, the plane was known more informally as *Fertile Myrtle*.[7]

The company designated El Segundo test pilot George Jansen (who had enthusiastically volunteered for the position) as launch pilot. A former crop duster, Jansen served during the war as a B-24 Liberator bomber pilot with the 44th Bomb Group, the *Eight Balls*. During the first Ploesti raid, a daring low-level penetration by B-24's deep into Romania on 1 August 1943 in an attempt to deprive the Third Reich of its oil, Jansen put his low-level tactics to good use, enabling his bombardier to destroy a vital boiler house serving the Creditul Minier refinery five miles south of Ploesti. After the war he joined Douglas as a test pilot, and flew many of the initial performance investigations on the AD Skyraider and XA2D Skyshark, the latter a troublesome and dangerous experimental turboprop-powered attack aeroplane. Likewise, the Douglas pilot for the Skyrocket drop flights also had a multi-engine flying background. A tall, gregarious airman as much at home on Southern California's beaches as on the Edwards flightline, William B 'Bill' Bridgeman had trained as a Navy pilot, flying graceful Consolidated-Vultee PBY Catalina patrol seaplanes from Pearl Harbor and Australia for the first 18 months of the war. Frustrated by a lack of action, he developed stomach ulcers, and after three months treatment, requested assignment to a combat squadron. The Navy sent him to Bombing Squadron 109, flying Consolidated-Vultee PB4Y-1 Liberators, the Navy equivalent to the B-24. Commanded by the flamboyant Capt. Norman 'Buzz' Miller, VB-109 entered the war in December 1943 and quickly gained fame as *Miller's Reluctant Raiders*. The squadron's accomplishments belied its name, for it made anti-shipping strikes all over the Central Pacific, sinking over 180,000 tons of Japanese shipping in daring (and too often costly) low-level raids on heavily defended harbours and anchorages. By the end of the war, Bridgeman was a Lieutenant Commander, holder of two Distinguished Flying Crosses, four Air Medals, and a Purple Heart. For a short while, he remained in the Navy as a ferry pilot, then spent a year each with Hawaiian and Southwest Airlines. But such a life was too placid; he joined Douglas El Segundo in 1948 and flew acceptance tests – with the occasional in-flight emergency to spice up the flying – on AD Skyraider attack bombers and F3D Skyknight night fighters. In 1949, Bridgeman joined the Skyrocket programme, making a series of contractor investigations in the first Skyrocket. Having proven his mastery over the white jet, he was given the opportunity to master its air-launched sister as well.[8]

Out to Launch

Early in September, Douglas performed several captive flights of the third Skyrocket, the jet and rocket aeroplane, to check out its fuel-jettison system and whether the J34 could start at altitude. Then, on 8 September 1950, Bill Bridgeman completed the first Skyrocket drop flight. Jansen launched the third D-558-2 at 24,850 feet and 225 mph. Though the flight plan called for a series of level flight speed runs and coordinated turns, Bridgeman just flew down to a landing on the lakebed, for the airspeed system malfunctioned due to frozen moisture in the lines. Bridgeman completed three more jet-only flights by early October. On 17 November, he made the first air-launched rocket flight, exceeding Mach 1 in a shallow dive with both the turbojet and one chamber of the rocket engine firing. Douglas scheduled one more flight before turning the plane over to the NACA. On 27 November, Bridgeman dropped away from the P2B-1S at 30,000 feet, ignited two chambers and climbed on both turbojet and rocket power. He began a pushover into level flight at 40,000 feet and fired all four chambers, but negative g forces starved the rocket engine, shutting it down. With the

Skyrocket moving at approximately Mach 0.92, Bridgeman noted a buffeting not associated with normal Skyrocket transonic behaviour and he realised that the J-34 had compressor stall, a condition caused by air stagnation within the inlet and at the engine 'face'. He pulled the jet throttle to idle, but the engine flamed out. Soon a thin fog formed between the windshield panels, preventing Bridgeman from seeing out. The chase pilot, Major Chuck Yeager, realised the seriousness of Bridgeman's predicament and began talking him down for a 'blind' landing on the lakebed. At about 24,000 feet, Bridgeman succeeded in relegating the J-34 and, as cockpit pressurisation returned, the cockpit glass cleared. Bridgeman completed a normal landing, the flight a tribute to both men's airmanship. Douglas delivered the third D-558-2 to NACA on 15 December 1950.[9]

Things did not go so smoothly with the second Skyrocket, the all-rocket aeroplane. In late December 1950, Douglas attempted to air launch it on three separate occasions, but each time, after the P2B-1S had laboured to altitude, Douglas scrubbed the attempt, first because of radio failure in the Skyrocket, and then because of deteriorating weather conditions. On 26 January 1951, Douglas made another flight attempt, but after entering the Skyrocket's cockpit and priming the rocket system, Bridgeman noticed the fuel pressure dropping off slowly. He had less than a minute to go before the planned launch, but reluctantly decided to abort the mission. He radioed George Jansen, flying the P2B-1S, 'No drop. This is an abort'. He then began shutting down the various aircraft systems in preparation for the long trip back to Edwards. To his horror, Bridgeman heard Jansen intoning the 10-second countdown. He shouted, 'Don't drop me, George!', but Jansen was holding his thumb on the microphone transmission key and could not hear Bridgeman's protestations unless he took his thumb off the key. Frantically, Bridgeman prepared the plane for flight once again. In a snap judgement, he decided to attempt an engine light for, with the falling pressure, if he attempted jettison only one propellant tank might empty and the plane would then go completely out of control. Finally, Jansen launched the white rocket plane, and it fell away from the Superfortress like a large bomb as Bridgeman readied it for flight. The Skyrocket pilot tripped the first rocket switch, glad to feel the push of acceleration as it fired. He ignited the remaining chambers in rapid sequence, and began a shallow climb. Still upset, he radioed, 'George, I *told* you not to drop me.' Chase pilot Frank Everest, below Bridgeman in an F-86, responded with a laugh, replying, 'You got keen friends, Bridgeman.' Using stabiliser trim, Bridgeman nosed over into level flight at 40,000 feet and accelerated rapidly through Mach 1. He noted that the plane tended to nose up between Mach 0.9 and 1.0 and that the plane required a large increase in down elevator for trim. Bridgeman found that elevator effectiveness decreased markedly above Mach 1 and that the plane began a Dutch roll. While in a shallow dive, it attained Mach 1.28 at 38,890 feet The rocket then cut out from fuel starvation. Everest's F-86 joined up with the now-powerless Skyrocket and both aeroplanes descended to the lakebed, Bridgeman completing an uneventful landing. The realisation that the mistaken launch could have had more serious consequences dampened the satisfaction of a successful rocket flight. As a result, Douglas installed an electrical hookup from the research plane to the P2B-1S to indicate through a green light whether or not the Skyrocket was ready for flight. Additionally, Douglas shortened the pre-launch countdown sequence. The company then turned to completing the remaining contractor flights in the second aeroplane. During February and March 1951, Douglas made four more launch attempts, but had to abort each of them. On one attempt, Bridgeman could not jettison the Skyrocket's propellants and P2B-1S had to land on the lakebed with the rocket plane carrying a full load of fuel and oxidiser. On 5 April, Bridgeman completed another drop flight in the all-rocket aircraft. As the Skyrocket accelerated beyond Mach 1, it developed an abrupt rolling motion. This lateral oscillation became wilder at higher Mach numbers and Bridgeman shut off the engine because of its severity and the plane reached a peak Mach 1.36 at about 45,600 feet. Effectively, this flight constituted the last of the contractor's own proving flights. Douglas and the Navy were now on a quest for speed and altitude. [10]

Fastest and Highest

Despite evidence of the plane's decreasing lateral (e.g. roll) stability, Douglas now decided to revise the Skyrocket's flight plans to permit the plane to approach its maximum airspeed. This involved Bridgeman climbing to around 50,000 feet, then levelling off for a maximum speed run. On 18 May, Jansen dropped Bridgeman at 34,000 feet and the Skyrocket pilot climbed to approximately 55,000 feet and levelled off. During the speed run, the Skyrocket attained Mach 1.72 at 62,000 feet, approximately 1,130 mph, making it the fastest aeroplane in the world. Ostensibly the last flight before Douglas turned the second Skyrocket over to the NACA, it was actually the first of a series of maximum speed attempts by the Douglas company, for the company received enthusiastic permission from the Navy's Bureau of Aeronautics to make additional maximum speed flights. On 11 June 1951, Bill Bridgeman pushed the Mach number up to 1.79 at 64,000 feet, approximately 1,180 mph. On these two record flights, the lateral oscillation (the rolling motion) did prove completely controllable and, thus lulled into a feeling the trouble might not persist, Douglas aimed for even higher speeds. Normally, following the climb Bridgeman pushed over into level flight with about a .6 to .8g load factor. The Douglas test team felt that if he pushed over more rapidly, with a smaller load factor of about .25g, the Skyrocket would have increased acceleration. They did not know, however, that at this low load factor, the Skyrocket's lateral oscillation, previously controllable, would become wild. On 23 June, Bridgeman dropped away from the P2B-1S, ignited the four chambers of the LR-8 rocket engine and began his climb. Near 60,000 feet he started his .25g pushover. As the Mach number increased beyond 1.5, the Skyrocket started to roll violently, throwing Bridgeman from side to side in the small cockpit. The wings dipped as much as 75°, and the Skyrocket's roll rate approached 90° per second. The pilot attempted to control the roll with his ailerons, but the gyrations worsened: he had entered a classic 'PIO' (pilot-induced oscillation), where his own attempts at correction were getting out of phase with the motion and were in fact, contributing to it. In a matter of seconds the ride became so violent that Bridgeman felt it necessary to shut down the engine, though about 50 seconds of rocket time remained. But the loss of power had no appreciable effect upon the oscillation and, for a time, the rolling motions became even wilder. They prevented Bridgeman from turning around for the lakebed landing, and the Skyrocket, now in a high-speed dive, was getting further and further away from Edwards. In desperation, the pilot hauled back on the control wheel, making a 4g pull-out. The Skyrocket began to climb, and the uncontrollable gyrations decreased and then disappeared entirely as the speed fell off. Once again under control, the Skyrocket responded smoothly to Bridgeman's control movements as, shaken, he initiated a turn to the lakebed, joining the F-86 chase plane on the journey back.[11]

NACA postflight evaluation of the data indicated that the plane had attained Mach 1.85, approximately 1,220 mph, during its frenzied flight through the upper atmosphere and, chastened, Douglas decided to go back to a pushover load factor of about .8g in an attempt to control the Skyrocket's wild supersonic motions. During the remainder of June and into early July, the company added a liquid oxygen top-off system in the P2B-1S to replace liquid oxygen that boiled away during the Skyrocket's climb to launch altitudes, thus offering the potential of even greater performance. The next Skyrocket flight came on 7 August 1951 during which Bridgeman pushed over at .8g, then decreased the load factor to .6g. At the lower figure, the Skyrocket became left-wing heavy and corrective aileron brought no response; Bridgeman increased the load factor to .8g, whereupon lateral stability returned. Bridgeman went back to a .6g loading and held the Skyrocket at this loading until the rocket engine starved itself of fuel. On this flight, the Skyrocket attained Mach 1.88 at 67,000 feet, approximately 1,260 mph. Prudently, Douglas decided not to attempt flights faster than this (later, with greater knowledge and understanding, NACA would, with spectacular results). Instead, with Navy concurrence, Douglas decided to explore the all-rocket D-558-2's altitude capabilities, hoping to break the old *Explorer II* record that had eluded the X-1. The day after Bridgeman's Mach 1.88 flight, the Skyrocket test team held a meeting to plan a maximum altitude attempt. Charles Pettingall, chief aerodynamicist of the Douglas testing division, plotted the flight path Bridgeman should follow. On 15 August, George Jansen dropped the second Skyrocket from the P2B-1S. Bridgeman ignited all four chambers of the LR-8 engine in sequence and began climbing, attaining Mach 1.35. At an altitude of approximately 63,000 feet, the Skyrocket slowly rolled left. After using full opposite aileron, Bridgeman noted that the plane reluctantly returned to wings-level flight. The Skyrocket continued upwards. Finally the LR-8 engine exhausted the last of its propellants, and the supersonic climb slowed. Momentum forced the plane still higher and eventually the plane peaked at the top of its parabolic flight and started falling earthwards. NACA radar plotted the maximum altitude at 79,494 feet. From his vantage point higher than any man had previously flown, Bridgeman compared the earth to 'a vast relief map with papier-mâché mountains and mirrored lakes and seas'. The Skyrocket was now not only the world's fastest aircraft, but the world's highest as well. Later, in recognition of the talents that joined together to make the Skyrocket the record-breaker that it was, the Institute of Aeronautical Sciences presented designer Ed Heinemann with the Sylvanus Albert Reed award and Bill Bridgeman with the 1953 Octave Chanute award. Its time in the limelight ostensibly over, the all-rocket D-558-2 awaited its new owners, the NACA's High-Speed Flight Station.[12]

False Starts

Using the all-rocket D-558-2, Douglas had come close to exceeding Mach 2, twice the speed of sound, less than four years after Yeager and the X-1 first demolished the 'sound barrier' syndrome. Now, as the autumn of 1951 approached, attention shifted to two new aeroplanes on the Edwards scene: the X-1-3 and the X-1D. These two rocket-research aeroplanes represented the first of the advanced X-1s, turbopump-equipped aeroplanes having almost double the performance capabilities of the original X-1 aircraft. It appeared that the honour of being first through twice the speed of sound would fall to one of the two planes within the next few short weeks.

The X-1-3 (46-064) actually represented the fulfilment of the original XS-1 conception. Bell completed the first two XS-1 aeroplanes with high-pressure fuel systems because the company could not obtain a satisfactory turbopump to feed the propellants to the Reaction Motors rocket engine. The company planned to complete the third XS-1, the aeroplane later designated X-1-3, when a suitable turbopump became available for installation. Early in 1947, the Air Materiel Command requested that Bell discontinue all work on the aeroplane because of a lack of funds to cover the work. Then, in early 1948, the Air Force obtained the necessary funds to continue development, and Bell estimated the plane could be ready for flight at Muroc by the autumn of that year. However, Bell did not get the engine for the aeroplane, and the AMC once again suspended development because of lack of funds. Late in 1949, to fulfil a research contract with the Cornell Aeronautical Laboratory, the AMC authorised continuing development of the X-1-3. In April 1951, the X-1-3, known to its Bell test crew as *Queenie*, arrived at Edwards Air Force Base for its initial contractor test flights, immaculate in a glistening white paint scheme.[13]

Externally, there was little difference between the X-1-3 and its two sisters, the original X-1s. Internally, though, the X-1-3 differed markedly from the other aeroplanes. The X-1-3 used a steam-driven turbopump to suck propellants from the tanks into the engine. Hydrogen peroxide passed over a manganese dioxide catalyst, creating superheated steam that drove the pump turbine. The turbopump obviated the need for nitrogen pressurisation for fuel feed, together with the thick spherical liquid oxygen and alcohol tanks. Thus, the X-1-3's fuel tanks matched the external contours of the fuselage. It could carry 126 gallons of liquid oxygen and 205 gallons of alcohol more than the earlier X-1s. Translated into performance, this meant the X-1-3 had a powered endurance of 4.1 minutes. Bell estimated the plane's maximum performance as Mach 2.44 at 70,000 feet, equivalent to 1,612 mph, roughly four times the performance of the best propeller-driven fighters of the Second World War just six years before. The X-1D had a different lineage. In mid-November 1947, one month after Yeager's Mach 1 flight in the X-1, the Air Force authorised Bell to begin development of a series of advanced X-1 aeroplanes, issuing the company a contract to develop four of the aircraft on 2 April 1948, under project MX-984. Though similar in structure and performance characteristics, the planes would receive different alphabetical suffixes, i.e., X-1A, X-1B, X-1C and X-1D. The letters differed to indicate emphasis in each aeroplane's flight programme. The X-1C, for example, was to be a supersonic armament-test aeroplane. (The availability of the F-86 for transonic weapons testing caused the Air Force to drop procurement of the X-1C.) Bell continued development of the other three aeroplanes, the X-1A, X-1B and X-1D, which retained the same wing and tail planform as the first X-1, but had a completely different fuselage. They not only incorporated a low-pressure turbopump fuel system, but were longer than the original two X-1 aeroplanes, and had a fighter-like canopy. Bell engineer Richard Frost found that Bell would have to limit the length of the aeroplanes to permit the nose landing gear of a B-29 to retract and so the tail could clear the aft pressure bulkhead located behind the B-29's bomb-bay section. Within these restrictions, Bell could stretch the advanced X-1s slightly over 4½ feet. Rather than incorporate the side entrance hatch utilised in the earlier X-1s, Frost decided to use a hinged canopy that the pilot could enter from above, enabling the pilot to enter the rocket plane from the mothership without exposing himself to the slipstream. If he had to, he could jump back into the bomber's bomb bay quickly, without having to wriggle out of a side hatch and then up a ladder into the launch plane. Unlike the original X-1s, the advanced X-1s would incorporate a fighter-type control stick rather than the wheel on the earlier planes. Additionally, the advanced X-1s featured simplified construction over their predecessors. Routine maintenance had sometimes proven difficult on the first two X-1s because of their construction, and Bell designed the advanced X-1s so that test personnel could perform maintenance and inspection work more easily, using special separations and access panels. As completed, the advanced X-1s could carry 500 gallons of liquid oxygen and 570 gallons of alcohol, giving the plane a powered endurance of 4.65 minutes and a maximum Mach number of 2.47 at 70,000 feet, equivalent to 1,635 mph. The first advanced X-1 that the Bell company completed was the Bell X-1D (48-1386), which arrived at Edwards Air Force Base in July 1951.[14]

Close Calls

On 20 July 1951, strapped under its straining launch B-29, *Queenie* left Edwards on the long climb to launch altitude for its maiden glide flight. At 30,000 feet, 10 miles south-east of the lakebed, flight test centre commander General Albert Boyd, piloting the B-29, launched the X-1-3, which dropped away crisply. As the pristine white rocket plane settled toward its lakebed landing, Bell test pilot Joseph A Cannon found the control forces to be extremely light, with aileron control especially sensitive. While Colonel Frank 'Pete' Everest, flying photo chase in a Lockheed T-33 Shooting Star, looked on, Cannon made two stall checks, one in clean configuration, the other 'dirty' with flaps and landing gear extended. Alas, the landing gremlins that so often plagued the X-1 programme struck again, for upon landing on the lakebed, the nose wheel collapsed and Bell spent the rest of the summer repairing the damage. Attention turned to *Queenie*'s big brother, the X-1D. Four days later, the X-1D went aloft locked in the bomb bay of its launch aeroplane, a Boeing EB-50A Superfortress. After dropping from the B-50, Bell chief test pilot Jean 'Skip' Ziegler flew a brief nine-minute glide flight to check out the X-1D's low-speed handling qualities. Like *Queenie*, the X-1D broke its nose landing gear on touchdown. Now both rocket planes were down until late August. After Bell completed repairs, the company turned the X-1D over to the Air Force so that the service could attempt to exceed the speed and altitude marks of Mach 1.88 and 79,494 feet set by the Douglas Skyrocket earlier in the month.[15]

One week after Bill Bridgeman set his altitude record, the Air Force attempted the X-1D's first powered flight, an effort, as Everest recalled later, 'to see what it could do wide open'. On 22 August, the EB-50A climbed for altitude, piloted by Major Wilbur Sellers (who, sadly, would die less than 24 hours later in the crash of a Lockheed F-94B Starfire all-weather fighter) and Major Jack Ridley. Flying chase were General Albert Boyd in an F-86 and Colonel Gust Askounis, chief of the Experimental Flight Test Section at Edwards, in a T-33. At 7,000 feet, Pete Everest entered the cockpit of the X-1D from the bomb bay of the B-50. Though the X-1D did not use nitrogen gas to feed propellants into its rocket engine, it did use nitrogen stored at 4,800 psi to pressurise fuel tank regulators, flap actuators and extension of the landing gear. Upon checking the instrument panel, Everest noted that nitrogen source pressure had dropped from 4,800 psi to 1,500 psi. Something was seriously wrong. Disappointed, he climbed back into the B-50 and discussed the problem with Jack Ridley and Bell engineer Wendell Moore. They decided to abort the flight, and Everest re-entered the X-1D to jettison the propellants prior to returning to Edwards. He began pressurising the liquid oxygen tank, then stopped to see if the tank gauge was stuck. Satisfied that it was not, he reopened the pressurisation valve to bring the liquid oxygen tank pressure up to 46 psi, the required jettisoning pressure. As he did so, an explosion shook the X-1D's innards, nearly knocking the pilot down. Listeners in the Edwards tower, monitoring the flight, heard chase pilot Gust Askounis radio Everest, 'Hey Pete! Drop her, drop her, she's on fire!' In the B-50, Everest turned to see flames pouring from the top of the X-1D into the B-50's bomb bay. Here was where Dick Frost's canopy design paid off, for Everest jumped from the burning rocket plane into the bomber, knocking down Jack Ridley, who was about to pull the emergency drop handle to release the X-1D. Everest's knocking down of Ridley possibly saved the lives of all aboard the bomber, for the shackles holding the X-1D still contained their removable locking pins and, if Ridley had pulled the drop handle, the X-1D would have jammed in place. As it was, Ridley pulled the usual drop handle in the cockpit and the X-1D fell clear of the launch plane, still burning. Streaming flame and vapour, it arched downwards, impacting about two miles west of the south end of the lakebed – providentially one mile south of a trailer court – and exploded violently. In response to anxious queries by Askounis and Boyd, Sellers reported that Everest was safe, and the B-50 and its two chase planes slipped down to land on the concrete runway. Things could have been a great deal worse.[16]

Air Force and Bell personnel gathered together on an accident board investigation, a difficult task in light of the explosion when the stricken plane hit the desert. After studying what wreckage existed, the accident board concluded that X-1D had a leak in its fuel system that caused a mixture of alcohol and air to form. The board believed that an electrical spark from the rocket plane's radio transmitter or external power supply plug ignited the explosive mixture. Pending completion of the other advanced X-1s, one research aeroplane remained that could possibly break Mach 2: little *Queenie*, the X-1-3. On 9 November 1951, the X-1-3 went aloft tucked under the EB-50A – the same launch plane used for the ill-fated X-1D's flight attempt nearly three months previously – for a captive test of the rocket plane's jettison system. The X-1-3 carried a full load of alcohol and liquid oxygen. Early in the flight, test pilot Joseph Cannon jettisoned distilled water, simulating hydrogen peroxide, but loss of nitrogen source-pressure – as with the X-1D – forced the test crew to cancel the jettison test of the liquid oxygen and alcohol. Still carrying a full load of propellants, the EB-50A returned to Edwards. The launch plane taxied over to the propellant loading area to take on nitrogen gas to purge the propellants still in the X-1-3's tanks. After acquiring source pressure, the test crew had a tractor tow the EB-50A, with *Queenie* still attached, to the east end of the concrete ramp. The ground crew cleared the area behind the Superfortress and, as a precaution, fire trucks stood by for the jettisoning procedure. The test crew gave Cannon an all-clear signal and the pilot, sitting inside the X-1-3, began tank pressurisation. When the liquid oxygen tank pressure reached 42 psi, Cannon heard a muffled explosion from deep within the rocket plane. At once he suspected the worst, and began to exit rapidly from the X-1-3's side hatch while the plane emitted a deep hissing sound and liquid oxygen vapour swathed it like fog. As Cannon left the cockpit, a violent secondary explosion knocked him to his hands and knees. He tried to get up, but succeeding blasts pinned him to the ground, exposing him to supercold liquid oxygen. Various Bell personnel, including William H Means, the X-1-3 project foreman, and Walter Myers, were near the X-1-3's cockpit when the first explosion occurred, and they saw Cannon attempting to crawl away from the burning plane. Heroically risking their own lives, Means and Myers pulled Cannon to safety, carried him to a Bell pick-up truck, and drove to the base hospital. Though badly burned, he recovered from his injuries, resuming a long and productive career with Bell. However, the explosions and subsequent fire destroyed both *Queenie* and its launch aeroplane.[17]

The X-1-3 accident investigation board officially attributed the explosion to a failure in the high-pressure nitrogen system. The nitrogen storage system, known as a tube bundle assembly, utilised 410 stainless steel, a type of steel having an extremely low impact rating at low temperatures, so low in fact, that steel manufacturers did not recommend use of this type steel in such an environment. The nitrogen tubes were next to the supercold liquid oxygen tank, and thus prone to cold-soaking that would lower their temperature greatly. The accident board theorised that the cold liquid oxygen lowered the temperature of the surrounding structure, so that when the test pilot pressurised the system, the nitrogen tubing shattered, tearing the plane's propellant tanks to pieces and triggering the secondary explosion that destroyed both aeroplanes and seriously injured Joe Cannon. (For example, investigators located pieces of the tubing as much as 250 feet from the wreckage.) Following the X-1-3 accident, NACA operations head Joseph R Vensel and NACA research pilot Scott Crossfield conducted tests dropping a 1lb weight through two inches onto a bottle of chilled nitrogen: the bottle exploded like a bomb.[18]

Thus it was that the year 1951, which had held bright promise of seeing the first Mach 2 flight, passed into history leaving behind shattered hopes and the wreckage of three aeroplanes. Mach 2 remained an elusive goal for the next two years.

8
Mach 2 – and Beyond

Slightly over a week after the X-1D dropped to destruction, Douglas delivered both the first and second Skyrockets to the NACA High-Speed Flight Research Station at Edwards, where they joined the third, delivered almost nine months earlier. Douglas had not modified the first to air-launch configuration and when NACA received the plane, it had completed a total of 122 contractor test flights. Flown primarily to check the results of wind tunnel studies, the first Skyrocket's flight testing had generally confirmed wind tunnel performance estimates, though the plane exhibited less drag above Mach 0.85 than tunnel tests had predicted. The NACA placed it in storage, having no desire to operate the plane in the ground-launched mode. Then, in October 1951, the Navy Bureau of Aeronautics amended the Douglas D-558 contract to provide for converting it to all-rocket air-launch configuration, for NACA hoped that eventually the plane might be utilised for testing external stores – bomb shapes and drop tanks – at supersonic speeds. The NACA gave all three Skyrockets special call signs; the first was 143, the second 144 and the third 145. The planned NACA flight programmes on the three aeroplanes differed according to the capabilities of the individual aircraft.[1]

NACA's Skyrocket Programme

After receiving the third Skyrocket in December 1950, NACA readied it for flight and, on 22 and 27 December, Scott Crossfield completed two familiarisation flights, dropping away from the launch P2B-1S and flying down to the lakebed on turbojet power only. During January and February of 1951, High-Speed Flight Research Station technicians modified the plane to record loads upon the wing slats via pressure-distribution and strain-gauge measurements. On 26 March 1951, Crossfield launched from the Superfortress on the first slat-loads investigation, but the J34 engine flamed out and he lost considerable altitude before getting a re-light, preventing the Skyrocket from acquiring data at altitudes above 25,000 feet. On 17 May 1951, NACA attempted their first jet-and-rocket flight in the aeroplane, but it was virtually identical to Bill Bridgeman's last contractor flight in the plane. During the climb, the J34 experienced flame instability and Crossfield shut it down, whereupon the cabin depressurised and ice formed on the windshield. Chase pilots Fitzhugh Fulton and John Konrad began talking Crossfield down for a 'blind' landing, but at lower altitudes the ice began to melt and he brushed away the frozen slush. Unable to get an airstart on the J34, Crossfield completed a dead-stick landing on the lakebed. NACA technicians pulled the turbojet engine for inspection and replacement. In July, the plane returned to the air, and NACA pilot Walter P Jones checked out. In August, Crossfield flew it to Mach 1.14, and General Albert Boyd completed a familiarisation flight to Mach 1.05. After one more check flight by Jones, NACA prepared the third Skyrocket for an extensive pitch-up research programme.

As will be recalled, the Skyrocket had first encountered pitch-up in August 1949 during NACA flights by Robert A Champine and John Griffith in the second D-558-2, before modification of that aeroplane to all-rocket air-launch configuration. NACA recognised the seriousness of the pitch-up problem and desired to make a complete investigation. In conventional aircraft, pitch-up could produce a limiting and dangerous restriction on aircraft performance. For example, a pilot might encounter pitch-up at low speeds at takeoff and landing, where he would have insufficient altitude to effect a recovery before the plane plunged into the ground. A fighter pilot might execute a tight wind-up turn to place his aeroplane on the tail of an enemy plane and his fighter might pitch-up, tumbling out of control, possibly exceeding its load limitations or putting the pilot in a disadvantageous position with his opponents. In September 1951, Walter Jones and Scott Crossfield began a pitch-up investigation programme on the third Skyrocket that lasted for nearly the next two years and involved flying the plane with various wing slat and wing fence configurations in both steady and manoeuvring flight at transonic speeds, as will be discussed more fully later in this chapter.[2]

As always, however, it was the all-rocket research aeroplanes that captured the imagination rather than the exacting data-gathering missions of their slower compatriots. The spectacular speed and altitude marks set by the rocket-research aircraft, figures far in excess of those obtainable in conventional high-performance aeroplanes, sometimes obscured the fact that these aircraft too were highly instrumented test vehicles designed to acquire aerodynamic data of use to aircraft designers, and not simply expensive experiments in and of their own right. After Douglas delivered the all-rocket D-558-2 to NACA on 31 August 1952, NACA High-Speed Flight Research Station engineers decided to perform exploratory flights aimed at defining the plane's operational limitations. On 28 September 1951, Scott Crossfield completed its first NACA flight. By the end of the year he made four more familiarisation flights in the plane. Drenching annual rains then closed the lakebed until the following summer. NACA technicians utilised the time spent on the ground by installing research instrumentation, including a 36-channel oscillograph. When the all-rocket D-558-2 resumed its flight programme in June 1952, Crossfield piloted the aircraft on a series of research flights at moderate supersonic speeds to approximately Mach 1.5. By the end of October 1952, NACA had completed 13 flights in the plane, primarily to acquire data on longitudinal stability and control, wing and tail loads and evaluation of lift, drag and buffeting characteristics at supersonic speeds. (During turns at supersonic speeds, for example, the Skyrocket's notorious pitch-up problem reappeared, particularly in high-lift conditions around Mach 1.2.) Crossfield found that during these and other manoeuvres at low supersonic speeds, the Skyrocket exhibited a general decrease in longitudinal stability, resulting in pronounced pitching motions. In October 1951, early rains flooded the lakebed and grounded the Skyrocket through the winter. Agency engineers decided that on future Skyrocket flights the plane would explore lateral stability and control at supersonic speeds, particularly the severe rolling motions at high Mach numbers that plagued the Douglas contractor flights in 1951.[3]

Hint of Promise, Shock of Tragedy

When the new year opened, the Skyrocket was no longer the only rocket research aeroplane flying at Edwards. Bell's test team arrived at Edwards on 7 January 1953, with the X-1A (48-1384), the second of the advanced turbopump-equipped X-1s to be completed. Originally, the Air Force planned for the Cornell Aeronautical Laboratory to conduct a stability investigation on the X-1A, but, to prevent delays in acquiring the aeroplane, the Air Force cancelled the Cornell stability programme and offered it to the NACA pending successful completion of a Bell contractor programme on the aeroplane. The NACA quickly agreed to the Air Force proposal. Bell assigned company test pilot Jean 'Skip' Ziegler to the X-1A project. Ziegler, Bell's chief test pilot, had flown contractor flights on the two Bell X-5 variable-wing-sweep aircraft, the X-2, and the X-1D. On 24

January 1953, he completed a captive flight with a full load of propellants, following this on 14 and 20 February with two uneventful glide flights. The X-1A completed its first powered flight on 21 February 1953. Ziegler dropped away from the B-29 launch ship and fired up the XLR-11 engine, but after igniting three chambers, a fire-warning light glared in the cockpit. Ziegler did not know the alarm was false, so as a precaution he shut off the engine, jettisoned the remaining propellants and glided down to a landing. He completed three more powered flights in the X-1A by the end of April, none of them faster than Mach 0.94, noting an annoying elevator buzz – a mild form of control surface aerodynamic flutter that could in more extreme cases lead to the elevator separating from the aeroplane – and Bell halted further flight tests in order to instrument the plane to examine the buzz in detail. Ziegler returned to Buffalo to participate in fuel system functional checks on another Bell rocket plane, the second X-2.[4]

Early in 1953, the Bell company completed the second X-2 (46-675) a potential Mach 3 sweptwing supersonic research aeroplane. Previously, the X-2 had completed some low-speed glide flights at Edwards Air Force Base but its engine, a complex Curtiss-Wright design, delayed it literally for years. Now at last, the programme seemed on the verge of success. In the late afternoon of 12 May, Bell flew a captive flight of the second X-2 locked under a B-50 launch plane, over Lake Ontario to test liquid oxygen topping-off procedures. As the B-50 cruised back and forth over the lake at 30,000 feet with a chase Republic Thunderbolt off its left wing, Skip Ziegler prepared to check the topping-off system. At 6.05pm, the X-2 suddenly exploded in a huge ball of red flame, slamming the B-50 100 feet vertically and throwing the Thunderbolt into a left bank, a big section of the X-2's wing narrowly missing the chase plane. On the ground, Bell engineers in the company radio station heard a voice radio, 'We lost the beast!' and the chase pilot call, 'Okay B-50? Got her under control B-50?' Inside the B-50, Bell pilots William Leyshon and D W Howe struggled to retain control of the mangled bomber. The blast had blown Skip Ziegler out of the bomb bay, probably killing him instantly, and what remained of the X-2 dribbled out of the sky into Lake Ontario. A flash fire swept through the bomb bay, slightly injuring crew chief Robert Walters and one aircrewman, Frank Wolko, bailed out. Either through shock or hypoxia, he failed to open his parachute and was lost. Shrapnel from the explosion shredded the inboard sections of the landing flaps and tore into the engine nacelles though, by a miracle, it did not puncture the Superfortress's vulnerable fuel tanks. In a superb display of airmanship, Leyshon and Howe brought the crippled launch plane back to an emergency landing at Niagara Falls Airport. That the B-50 survived at all was a testimony to the rugged structure Boeing engineers designed and to the capabilities of its crewmen. The explosion had damaged the bomber so extensively that it never flew again. An accident board, formed to trace the cause of the fatal explosion, concluded that an electrical malfunction triggered an explosion of alcohol and oxygen vapours in the propellant tank section of the plane, and that the remaining X-1 and X-2 aeroplanes should undergo thorough inspection before the Air Force returned them to flight status. As a precaution against further accidents, Bell replaced the nitrogen-tube bundle assemblies in the X-1A, X-1B and surviving first X-2 with simple spherical nitrogen containers. The X-1A left Edwards for the Bell plant in June 1953 and did not return to the test centre until mid-October 1953. It was a fatal delay for the hopes of those who thought the advanced X-1 might be the first aeroplane to exceed Mach 2.[5]

Skyrocket Triumphant

While the X-1A underwent modifications in Buffalo, the all-rocket D-558-2 soldiered on. Early in April 1953, the NACA High-Speed Flight Research Station began working up to higher Mach numbers as part of its planned lateral-stability investigation programme. Crossfield found that the plane remained fully controllable at moderate angles of attack, but at low angles of attack and low load factors – near or at 0g – lateral stability decreased markedly. On 5 August 1953, Scott Crossfield reached Mach 1.878, the fastest Skyrocket flight since the Douglas demonstration programme. Once again, the D-558-2 was nudging Mach 2. In mid-summer the Bureau of Aeronautics requested that Marine Colonel Marion Carl, who had set the world airspeed record in the Skystreak in August 1947, should fly the all-rocket D-558-2 to high altitudes for pressure suit and physiological research – hoping, not coincidentally – to set both a new altitude record and then use the opportunity to beat the Air Force through Mach 2 as well. Carl arrived at Edwards in July 1953 and made two familiarisation flights in the jet and rocket D-558-2 before trying his hand with the all-rocket aeroplane. The Navy hoped to exceed 80,000 feet on these flights. Carl aborted the first flight attempt because of oxygen-equipment difficulties with the pressure suit, and subsequent flights on 14 and 18 August failed to attain the desired altitudes. But on 21 August 1953 Carl dropped away from the launch P2B-1S Superfortress, tripped the rocket switches controlling the LR-8 engine and nosed upwards. The Skyrocket kept climbing, beyond Bill Bridgeman's mark of 79,494 feet. The engine consumed the last of its propellants and shut down, but still the rocket plane moved onwards, finally peaking at 83,235 feet – a new unofficial world altitude record. The Navy now requested an additional two flights for attainment of the plane's maximum Mach number. Neither attempt, however, succeeded. On the first flight on 31 August 1953, Carl pushed over at a low angle of attack and the rocket engine cut out. Travelling at Mach 1.5, the Skyrocket began a violent rolling motion reminiscent of Bridgeman's earlier experiences and he wisely backed off. On the second attempt, the Skyrocket only reached Mach 1.728.[6]

After Carl completed the Navy flights in the aeroplane, NACA technicians installed nozzle extensions on the LR-8 engine's rocket chambers. The extensions prevented the expanding rocket gases of the exhaust from impinging upon the rudder at supersonic speeds, but they also had the added beneficial effect of boosting the engine thrust by 6.5% at Mach 1.7 at 70,000 feet. Crossfield completed the D-558-2's first nozzle extension flight on 17 September, streaking to Mach 1.85 at 74,000 feet. On 25 September, he made another high-Mach flight. Pushing over at a low angle of attack with a 0g load factor, Crossfield noted that the plane developed a severe rolling motion and rolled over on its back at about Mach 1.7. He attempted to recover using the ailerons, but the ailerons did not have sufficient effectiveness so, boldly, he used the stabiliser trim to execute a split-S recovery manoeuvre like some form of supersonic dive bomber. In October, he flew a further four lateral-stability investigations in the plane, reaching Mach 1.96 on 14 October. So it was that six years to the day since Chuck Yeager's first Mach 1 flight, a piloted aeroplane first came close to Mach 2. While the NACA had not formulated any firm studies of the Skyrocket's behaviour at speeds near Mach 2, beyond tentative data taken during the Douglas contractor programme, agency engineers found that changing the Skyrocket's angle of attack at high Mach numbers from around –2° to +2° improved the aeroplane's lateral stability and the characteristic rolling motions became controllable up to Mach 1.96. Beyond this, NACA engineers did not know what might take place. All indications were, however, that the Skyrocket could operate satisfactorily to Mach 2.[7]

As the winter of 1953 approached, therefore, the band of NACA engineers and research pilots at Edwards conceived flying the all-rocket D-558-2 through Mach 2, the first flight at twice the speed of sound. It was something Ed Heinemann and his group of engineers at Douglas El Segundo would have thought hardly possible back in 1946 – they had designed it to go only half as fast. It was also something completely out of character for the normally staid and reserved NACA. Now the agency had a 'dark horse' chance to achieve what had eluded the major services, the Navy and Air Force. Somewhat surprisingly then, early in November 1953, the High-Speed Flight Research Station requested and secured the approval of Dr

Hugh Dryden, NACA Director, to attempt the world's first Mach 2 flight. Once more interest sharpened at Edwards, for if the Skyrocket failed to attain its desired goal, the Air Force's X-1A, with its listed Mach 2+ performance, had returned to the base following the calamitous events of the summer.

NACA had reduced the complexities of Skyrocket operations to routine. Agency pilots and crewmen operated the P2B-1S launch aircraft, and agency technicians readied both the research plane and mothership for flight. Agency engineers, working with the pilot, planned the flights in the rocket plane, and operated the telemetry-receiving and radar tracking equipment during the actual flight itself. They reduced the oscillograph film and instrument records to meaningful data in usable form. In preparing the Skyrocket for the world's first Mach 2 manned flight, NACA technicians drew on all their past experience with the plane. NACA D-558-2 project engineer Herman O Ankenbruck devised a special flight plan for the attempt whereby Crossfield would climb to approximately 72,000 feet, push over gradually, hopefully reaching Mach 2 in a shallow dive. The flight plan followed the shape of a parabola, similar in general method to that followed by Douglas during the contractor programme on the D-558-2 in 1951.

To squeeze the maximum performance possible from the Skyrocket, the NACA flight team learned to 'cold soak' the plane by loading liquid oxygen in the plane about five hours prior to the scheduled launch, so that even more liquid oxygen could be added just before launch in order to increase the burn time of the engine, thereby boosting the plane's performance. Likewise, the team chilled the alcohol so that more could be carried. In preparation for the Mach 2 attempt, the ground crew covered all panel cracks with masking tape, and gave the plane a heavy coat of wax. They replaced the standard stainless steel jettison pipes with lightweight aluminium tubes bent to protrude in the blast area of the rocket exhaust. During the climb after drop, these tubes (needed only in the event of an abort) would burn away, further lightening the Skyrocket for its record attempt.

Early in the morning of 20 November 1953, NACA prepared the Skyrocket and the P2B-1S launch plane for flight. During filling of the plane's hydrogen peroxide tanks, a spillage occurred showering John Moise, launch panel operator of the NACA P2B-1S, with the volatile liquid. A quick-thinking mechanic sprayed Moise with water, diluting the peroxide, then rushed him to a dispensary. Aside from bleaching his skin white, the peroxide did not seriously injure Moise and the discolouration disappeared within a few days. Moise's accident did not delay the Mach 2 flight attempt. In mid-morning, NACA pilot Stanley Butchart guided the P2B-1S from the runway at Edwards and climbed into the sky. Crossfield entered the Skyrocket from the bomb bay of the Superfortress and completed preparations for flight in the rocket plane. After climbing for over an hour, the launch plane reached 32,000 feet. Butchart positioned the rocket plane for launch and, finally, the Skyrocket dropped clear of the bomber. As it fell away, Crossfield ignited the LR-8 rocket engine and began his climb, carefully watching the flight instruments so that the plane would not stray from its pre-selected flight path, wasting energy and fuel. The Skyrocket accelerated up into the stratosphere, streaming a broad white contrail. At 72,000 feet, Crossfield initiated his pushover, and the Skyrocket arced over into a shallow dive. During the dive the plane picked up speed rapidly, edging closer to Mach 2 as the engine continued to fire, the cold soaking having added vital seconds of engine burn time. At 62,000 feet, the D-558-2 attained Mach 2.005, 1,291 mph: A Scott Crossfield had become the first man to travel at twice the speed of sound. After 207 seconds of burn time, the LR-8 engine starved itself of fuel and shut down. The Skyrocket decelerated and Crossfield began his recovery. During aileron rolls at Mach 1.95, he felt that the aileron effectiveness was as good as that near Mach 1. Gliding downwards, he set up his approach to the lakebed, extended the landing gear and flaps and set down in a smooth dead-stick landing. For his 'important contributions in aeronautical flight research, especially at transonic and supersonic speeds up to Mach 2', the Institute of Aeronautical Sciences awarded Scott Crossfield the 1954 Lawrence B Sperry Award. The D-558-2 never again approached Mach 2 during the remainder of its flight research programme, for the flight attempt had required extensive and unusual pre-flight preparations that could not be repeated for every high-Mach flight-research flight. NACA soon planned to receive two aeroplanes with Mach 2.5 performance, the X-1A and X-1B. With the Mach 2 flight behind them, NACA engineers at the High-Speed Flight Research Station transferred their attention to the Air Force, then preparing the X-1A for flight.

Chuck Yeager's Tempestuous Ride

Following Jean Ziegler's death in the second X-2, the Air Force Air Research and Development Command moved rapidly to keep the advanced X-1 and the lagging X-2 programmes near schedule. Veteran rocket pilot Colonel Frank K Everest requested and received assignment as project pilot on the X-2 and the Air Force Flight Test Center at Edwards assigned none other than Major Chuck Yeager to complete the remaining contractor flights, commonly termed 'Phase I' flights, in the X-1A and Lieutenant

Colonel Jack Ridley to serve as test planner. After the X-1A arrived back at Edwards AFB from Bell's Buffalo plant on 16 October 1953, Air Force technicians installed nitrogen storage tanks to replace the controversial 'tube bundle' storage system removed by Bell. Then, NACA engineers from the Edwards High-Speed Flight Research Station installed research instrumentation in the plane. On 21 November 1953, the day after Crossfield's Mach 2 flight, Yeager went aloft on the X-1A's first powered flight since modification, reaching Mach 1.15. On his second flight, he boldly extended this to Mach 1.5. On the latter flight he found the stick forces for aileron control so high that he subsequently recommended installation of a control wheel for better pilot advantage. The Air Force now decided to initiate the plane's high-speed research programme. On 8 December 1953, Yeager reached Mach 1.9 at 60,000 feet during a slight climb. Four days later on 12 December 1953, Major Harold Russell took off from Edwards with the B-29 and X-1A. During the climb to altitude, Ridley and Major Arthur 'Kit' Murray joined up in two F-86 Sabre chase aeroplanes. At 30,500 feet, Yeager dropped free from the B-29, fired three chambers of the XLR-11 engine and accelerated away from the bomber as NACA radar tracked him from the ground and Bell company technicians listened in from a mobile radio van.[8]

Climbing rapidly, Yeager fired the remaining rocket chamber at 45,000 feet. Under full power, the X-1A climbed past its planned test altitude of 70,000 feet, for Yeager was flying into a blinding sun that caused him to overshoot the planned climb angle of 45° and fly a 55° climb profile instead. He had started to lower the X-1A's nose as he passed through 60,000 feet, and he finally reached level flight at approximately 76,000 feet. The X-1A accelerated rapidly through Mach 2 and beyond, easily exceeding the Skyrocket's maximum speed and pressing on through Mach 2.3. During pre-flight briefings, Bell engineers had warned Yeager that the X-1A might lose stability and go out of control at speeds above Mach 2.3. The rocket aeroplane violently proved the correctness of the Bell engineers' hunches. Ten seconds after levelling out on its high-speed run, the X-1A began a slow roll and yaw to the left. Yeager applied aileron and then rudder to correct the motion and the X-1A rolled violently to the right. Attempting to correct this, Yeager found the plane snapped into a left roll again. Sensing that the plane was going rapidly out of control, he shut off the rocket engine. Still moving at about Mach 2.44, the X-1A tumbled completely out of control, throwing Yeager around in the cockpit. The X-1A's motions resembled an oscillatory spin, with frequent reversals in roll direction.

On the ground, the Bell radio truck monitored the urgent calls of Ridley and Murray as they tried to raise Yeager by radio. As the plane fell it decelerated, finally winding up in a subsonic inverted spin at 34,000 feet. In the cockpit, Yeager was fighting for his life as the rocket plane lost altitude rapidly, his struggle complicated because the face-plate of his helmet was fogging up with moisture. Yeager managed to turn on a switch that heated the helmet's face-plate, clearing the fog in time to see that he was in a fully developed spin over the Sierra Nevada mountains. At 29,000 feet, Yeager – groggy, battered and semi-conscious – recovered to a normal upright spin, then recovered from that into level flight at 25,000 feet. He then called the chase: he was over Tehachapi at 25,000 feet, uncertain he could limp back to Edwards. Incredulous, Murray and Ridley asked him for his position again, then swung their Sabres off for Tehachapi, 40 miles from the dry lake. Still dazed – for during the 50,000 feet drop, he had cracked the hard inner shell of the canopy with his helmet – Yeager jettisoned the remaining propellants, dropped his landing gear and flaps and descended to the lakebed, with Kit Murray riding off his wing to call out the altitude on landing.

As the X-1A touched down in a plume of dust, Yeager summed up his feelings about the flight. 'You know,' he radioed, 'if I'd a [ejection] seat, you wouldn't still see me in this thing.' The young pilot got out of the rocket plane, having completed one of the wildest, roughest yet shortest flights in aviation history, having survived, as he wrote later, 'on sheer instinct and pure luck'. On the flight, NACA radar tracking records indicated the plane attained a peak speed of 1,612 mph at 74,200 feet. Yeager later received the Harmon Trophy for the flight, which was his last in the X-1 family of aeroplanes. Later analysis by NACA led to the conclusion that decreasing directional stability and damping in roll at high Mach numbers led to a phenomenon of 'coupled' motions, combined rolling, pitching and yawing motions first predicted by NACA scientist William H Phillips in 1948. The plane's difficulties led to immediate recognition that aeroplanes flying beyond Mach 2 would require much larger vertical and horizontal stabiliser surfaces to retain adequate stability at high Mach numbers, often together with ventral fins on an aircraft's underside as well. (The trend towards two vertical fins, and sometimes ventral fins, too, – as on the Grumman F-14A Tomcat, McDonnell-Douglas F-15 Eagle, or Mikoyan MiG-29 – represents the logical outcome of such thought.)

The Drive to Altitude

Yeager's Mach 2.44 flight represented the high-water mark of the entire X-1 programme. No X-1 aeroplane ever equalled or exceeded this speed mark. Following Yeager's wild flight, the Air Force decided to limit the aeroplane to speeds no greater than Mach 2 and initiate a high-altitude research programme in an attempt to attain altitudes in excess of 90,000 feet. In February 1954, the Air Force secured permission from the NACA to conduct limited tests with the plane before turning it over to the NACA, and assigned Major Arthur 'Kit' Murray to make the altitude flights. Murray had a reputation as a skilled and cautious pilot. During a delivery flight of a North American B-45 Tornado jet bomber, the plane caught fire in mid-air. Rather than abandon the stricken plane over Los Angeles, Murray stayed with the plane until he could make an emergency landing at Los Angeles International. On another occasion, during a night accelerated service test flight in a Lockheed F-94 interceptor, Murray struck an unlighted radio range tower, shearing off the top 10 feet. He remained with the jet and made a wheels-up landing on the lakebed with severe wing damage.

During the spring and summer of 1954, the test centre's B-29 mothership hauled the X-1A and Murray up to altitude for a number of flight attempts. Out of 14 attempts, only four flights were really successful. One was Murray's checkout flight. The rest were altitude tries. On 28 May 1954, Murray exceeded Marion Carl's Skyrocket record by shooting to 87,094 feet. Slightly over a week later, on 4 June, Murray reached 89,750 feet, but on this flight he experienced the same violent tumbling Yeager had felt more than six months previously. Slightly over 80,000 feet and Mach 1.97, the X-1A began a roll to the left. Murray corrected with his ailerons and rudder, and the plane rolled to the right. He shut down the rocket engine, but the plane continued rolling as it climbed. Finally, as with Yeager, it went completely out of control, but at a higher altitude and lower airspeed than the earlier flight. After tumbling over 20,000 feet, Murray finally recovered control at 65,000 feet before continuing down to a landing. More than two months later, on 26 August, he cautiously extended the X-1A's altitude record to 90,440 feet. The next month, with the altitude record secure, the Air Force turned over the X-1A to the NACA. The engineers at the NACA facility – now operating out of new buildings of their own rather than leased Air Force structures, and renamed the NACA High-Speed Flight Station – returned the plane to Bell for installation of an ejection seat as a precaution against the X-1A's poor stability and control characteristics at high Mach numbers. In retrospect, given the experiences of Yeager and Murray, it is hard to believe that the X-1A could have been designed without an ejection seat, much less that it could have then been flown without one.[9]

By the autumn of 1954, both the D-558-2 and the advanced X-1 series had passed the peak of their performance potential. The Skyrocket never exceeded Carl's altitude figure or Crossfield's maximum speed. The advanced X-1s never flew beyond Yeager's record Mach 2.44 or Murray's 90,440 feet. The image of the two aircraft types extending the borders of speed and altitude died away, therefore, for all their subsequent flights fell well within the earlier marks mapped out by Carl, Crossfield, Yeager and Murray. As the speed and altitude marks set by the Skyrocket fell to the X-1A, so did the latter's fall to another research aeroplane, the surviving all-rocket X-2 . The X-1A did not bequeath its laurels as quickly as had the Skyrocket, however, for the X-2 did not break the earlier marks until the summer of 1956, two and a half years after Yeager reached Mach 2.44 and slightly more than two years after Murray approached the purple-black void of space. Then, in the hands of Iven Kincheloe and Milburn 'Mel' Apt, it reached an altitude of 126,200 feet and Mach 3.2. The exclusivity of the supersonic club, however, previously restricted to highly exotic research aeroplanes, was clearly beginning to wane. On 27 April 1956, a preproduction prototype of the world's first double-sonic fighter, the Lockheed F-104A Starfighter, had reached Mach 2.2. The first prototype British supersonic fighter, the English Electric P.1B Lightning, would exceed Mach 2 slightly over two years later on 25 November 1958 (piloted, appropriately enough, by Roland Beamont).

Though the Skyrocket and the X-1s are best known for their pioneering flights into the unknown, the records they set in 1953 and 1954 did not terminate their usefulness. None of the research aeroplanes built were fabricated for setting records. Rather, they were designed as research tools. Though they set some spectacular records in the course of their research, their main function remained unchanged: the acquisition via flight instrumentation of data on a variety of areas. While the record-breaking flights served to bring the Skyrocket and X-1 programmes into sharp focus, they often tended to diminish, in the public eye, the very great benefits accruing from dozens of 'routine' flights at now unspectacular speed and altitude figures. The new glamour ship awaited at Edwards was the X-2 (and, already in hazy planning beyond it, the North American X-15). The X-1s and Skyrockets, gradually faded from the limelight. Out of mind did not equate to out of sight, though, as the Skyrockets and X-1s kept going well into the latter half of the decade.

NACA's Skyrockets: The Twilight Years

Of the two, the Skyrocket series flew most extensively. The jet-and-rocket D-558-2 did not have the performance potential of its all-rocket sibling; accordingly, engineers at the NACA High-Speed Flight Station utilised the jet-and rocket aeroplane primarily for transonic research. By the autumn of

1954, NACA had concluded the pitch-up investigation programme started on this aeroplane in 1951. Engineers had tested the plane with its wing slats locked open in various positions, with inner and outer wing fences, even with a 'sawtooth' leading edge extension. All these attempts to find possible cures for the pitch-up problem were at best partially successful. Only with wing slats locked fully open did an improvement in the plane's longitudinal stability take place: fully open slats eliminated pitch-ups except around Mach 0.8–0.85. With the slats half open, and with or without wing fences, no improvement occurred. Wind tunnel tests of a Skyrocket model indicated that a chord extension over the outer 32% of the wing panels might eliminate pitch-up, but flight tests of the extensions on the third D-558-2 disproved the wind-tunnel test results. (In fact, Crossfield felt the pitch-up even more objectionable with the extensions than without.) The best solution – one subsequently adopted by most high-performance aeroplanes – but beyond hope of application to the Skyrocket family – was to lower the horizontal tail location from above the wing-wake to below. Aircraft that had high horizontal tail locations (for example the McDonnell F-101 Voodoo) always exhibited the worst pitch-up tendencies. On the other hand, aircraft with a lower tail location – for example the North American F-100 Super Sabre, the Republic F-105 Thunderchief or, in Great Britain, the English Electric Lightning F.1 – flew far more satisfactorily. NACA's Research Aeroplane Projects Panel actually recommended changing the Skyrocket's horizontal tail location for research purposes, but such a modification would have been so complex and costly as to not warrant its undertaking.[10]

Pitch-up lessons learned, NACA restored the D-558-2 to its original configuration and embarked on another research programme – on the transonic and supersonic behaviour of external stores. The need for supersonic external stores testing was a clear indication that, at least as far as the military services were concerned, the time had come to tailor the design of aircraft and the weapons that they carried for the rigours of the supersonic environment. It was a measure, really, of how far the transonic and supersonic revolution had already progressed, for it made little sense to develop high-performance aeroplanes if the shapes they carried were themselves so awkwardly designed that they compromised the carrying plane's aerodynamic performance. In the summer of 1954, agency technicians installed external-stores pylons on the third D-558-2 and began flying it to investigate the effects of external stores upon the aeroplane's transonic behaviour. The tests started using 120-inch 1,000lb bomb shapes, but the engineers felt that these were too small and later substituted Douglas-designed 180-inch, 150-gallon external fuel tanks. Scott

Crossfield began the stores investigation, but when he left the agency to join North American Aviation on the X-15 programme, NACA assigned research pilots John B 'Jack' McKay and Stanley Butchart to the programme. After over a year of research flying, McKay and Butchart completed the stores-investigation programme in early December 1955, and NACA scheduled only two more flights in the plane before its retirement, to acquire data on its behaviour in 'clean' configuration for performance comparisons with the transonic stores data already gathered. Due to a flooded lakebed and maintenance and repair work on the P2B-1S mothership, the High-Speed Flight Station did not complete these flights until August 1956. The third D-558-2 made its last research flight on 28 August 1956, piloted by Jack McKay.[11]

The retirement of the third D-558-2 did not bring an end to the agency's interest in stores research, however. NACA engineers sought to investigate the effects of external stores upon aeroplane behaviour at moderate supersonic speeds, but could not modify the second D-558-2, the all-rocket aeroplane, for stores testing because of the need for this plane for purely aerodynamic research. However, provision had been made in the Skyrocket contract for modification of the first Skyrocket, the original ground-launched jet aeroplane, to air-launch configuration. In 1954, NACA delivered the plane to Douglas for all-rocket air-launch modification, with provisions for external-stores pylons. It arrived back at Edwards on 15 November 1955. Operational and instrumentation work kept it grounded until September 1956. On 17 September 1956, John B McKay completed this plane's first air-launch flight, flying it in 'clean' configuration without either the pylons or external stores. NACA then set about preparing the plane for a stores investigation programme beginning in January 1957, but a wet lakebed forced cancellation of planned flights, and shifting priorities caused NACA to cancel the entire Skyrocket flight programme. In March 1957, the first Skyrocket returned to the limbo of retirement from which it had so briefly emerged. Following Scott Crossfield's epic Mach 2 flight in November 1953, NACA continued stability and control research in the now-historic second D-558-2. Engineers obtained data on lateral and longitudinal stability and control, wing and tail loads, pressure distribution on the wing and horizontal tail and lift and drag. Air Force Lieutenant Colonel Frank K Everest flew the Skyrocket on 5 May 1955, attaining Mach 1.46 at 68,000 feet, on a training flight in preparation for the upcoming X-2 programme. In June 1955, the all-rocket D-558-2 concluded its pressure-distribution research, and NACA removed the plane's manometer installation. Scott Crossfield left the Skyrocket programme in September 1955 to join North American in designing the X-15; as a result, Jack

McKay carried the NACA Skyrocket programme onward. NACA attempted a brief structural-heating programme on the plane in November 1955, but concluded it in January 1956 when test engineers discovered the Skyrocket could not remain at high Mach numbers long enough for the temperatures of the plane's structure to stabilise, permitting accurate measurements.[12]

On 22 March 1956, the second Skyrocket left the Edwards runway tucked under the P2B-1S launch aeroplane for its 67th flight since modification to all-rocket air-launch configuration. What no one suspected was that the plane would undergo the most serious in-flight emergency to occur during the entire high-speed flight-research programme, with the exception of the inflight rocket-plane explosions. Aside from the rocket-research pilot, the most difficult job was that of launch-plane commander. He had to have the rocket plane ready for flight, the launch plane at proper altitude and be over the correct geographic position in case the rocket pilot had a 'no light' condition, all completed simultaneously by launch. Unlike in normal heavy-aircraft operations, the left-seat pilot served as mission commander, while the right-seat pilot acted as pilot of the launch plane. It was impossible for the left-seat pilot both to fly the launch plane and to control the mission simultaneously. During launch flights, the B-29 used for the X-1s and the P2B-1S used for the Skyrockets did not experience control problems, but they did have potentially serious engine problems. Engine cooling was a traditional problem for the B-29, going back to its days as a strategic bomber used in the war against Japan. Heavily loaded, the Superfortress climbed very slowly, almost too slowly to ensure adequate cooling of its large radial engines, which could become quite hot around their rear banks of cylinders in the back of the engine cowling. As a result, the launch planes occasionally lost engines on the climb to altitude. On the March 22 flight attempt, Stanley Butchart rode the left seat as mission commander, and research pilot Neil A Armstrong, a young Korean War veteran, rode the right seat as P2B-1S pilot. A few minutes before the scheduled drop, the Skyrocket experienced a systems malfunction, scrubbing the launch. Then the Superfortress' number four engine, the outer engine on the right wing, 'ran away', the propeller rotating faster and faster. As previously agreed, should an emergency occur to the launch ship, Butchart and Armstrong immediately released McKay in the Skyrocket. As the rocket plane fell earthwards, McKay jettisoned propellants and glided down to a safe landing on the lakebed. No sooner had the launch crew dropped the D-558-2, than the propeller from the ailing engine tore loose from its shaft, sawing across the underside of the wing and fuselage and cutting into the bomb bay before scything off into space. Had the propeller cut into the Skyrocket before it dropped away, the rocket plane would undoubtedly have exploded, dooming both aeroplanes and perhaps their crews. As it was, the propeller cut all the controls on the left side of the P2B-1S so that Butchart's control column flopped back-and-forth uselessly. But Armstrong still retained control and brought the P2B-1S down for a safe emergency landing on the dry lake, in a display of the same kind of cool airmanship that would enable him to execute an unplanned but critically important last-minute course-correction while landing the lunar module *Eagle* on the Sea of Tranquillity 13 years later. Repairs to the launch plane took four months.[13]

In August 1956, the second Skyrocket returned to the air, completing its loads and stability programme by November. NACA then instrumented the plane to measure overall noise levels at supersonic speeds. After two noise investigations in December by Jack McKay, NACA engineers removed the research instrumentation from the plane and prepared it for a brief pilot-familiarisation programme for NACA and military pilots. NACA planned familiarisation flights in all the D-558-2 aeroplanes. In March 1957, however, NACA cancelled the planned familiarisation programme and retired all three planes. After almost nine years of research flying, the D-558-2s joined the two surviving D-558-1s in retirement.

The End of a Mystery

Though not as extensive as the Skyrocket programme, the X-1 programme after 1954 lasted through December 1958, until the dawn of the hypersonic X-15. The Bell X-1B, 48-1385, arrived at Edwards on 20 June 1954. NACA planned to use this aeroplane for aerodynamic heating research, following a brief Air Force programme for pilot familiarisation. The Air Force had little need for the plane since it offered no performance advantages over the X-1A; further, because of the marked instability encountered by Yeager and Murray in the X-1A, the service had no desire to explore the X-1B's maximum performance limitations. Instead, mission planners mapped out a brief programme of pilot-familiarisation flights in the plane in preparation for the upcoming X-2 programme, then just over a year away from its first powered flight. In September and October 1954, the X-1B completed two glide flights. On 8 October, Kit Murray completed the plane's first powered flight. Between then and the end of November, the plane made six more flights for pilot familiarisation.[14]

Late in November, after making a familiarisation flight in the X-1B, Lieutenant Colonel Frank K Everest secured permission to attempt a maximum speed run in the X-1B. Everest felt that the high-Mach experience would serve him well in the upcoming X-2 programme, since he was the designated X-2 project pilot. Everest flew the mission on 2 December 1954, attaining approximately Mach 2.3 at 65,000 feet. During the flight, at speeds above Mach 2, the X-1B began rolling and yawing, the wings dipping as much as 70°. After Everest reduced power, the motions slowly ceased as the X-1B decelerated. The next day, the Air Force delivered the plane to the NACA High-Speed Flight Station. NACA engineers planned to use the X-1B for heating research and the sister X-1A for high-altitude high-Mach research; they shipped the X-1B to NACA's Langley Laboratory on 14 December for instrument installation. It did not arrive back at Edwards until 1 August 1955.

As will be recalled, following Yeager's and Murray's harrowing flights, the Bell X-1A had been returned to Bell for installation of an ejection seat. It did not return to Edwards until 23 February 1955, its bare aluminium look traded in for a glossy white paint scheme, save for the skin around the liquid oxygen tank. NACA engineers began preparations for its flight-research programme, but instrument installation and systems inspection kept the plane grounded until the summer. On 20 July 1955, research pilot Joe Walker made a familiarisation flight in the aeroplane, encountering severe aileron buzz from Mach 0.90 to 0.92, echoing Skip Ziegler's experience two years before, but he continued the flight to Mach 1.45 at 45,000 feet. Later in the month he attempted a second flight, but operational difficulties forced an abort before launch. On 8 August 1955, Walker went aloft in the B-29 preparatory to another X-1A flight attempt. At 8,000 feet, he entered the cockpit of the rocket plane, closed the canopy and readied it for flight. At 31,000 feet, one minute prior to launch, the X-1A's liquid oxygen tank exploded. A white cloud of liquid oxygen vapour erupted from the lower centre section of the rocket plane and hurtling debris from the explosion shattered the canopy on Kit Murray's chase F-86. Fortunately, the explosion was of low order, nothing like the catastrophic blast with Skip Ziegler's X-2 over Lake Ontario. Walker scrambled from the X-1A's cockpit into the bomb bay of the B-29 launch plane. Then B-29 pilot Stanley Butchart, co-pilot Jack McKay, Joe Walker and Kit Murray, together with the rest of the B-29 drop crew and NACA chase pilot Neil Armstrong in an F-51 Mustang tried to figure how to save the crippled X-1A.

The explosion had caused the X-1A's landing gear to extend, and the rocket plane appeared to have dropped a few inches. Further, the plane still retained a full load of fuel and volatile hydrogen peroxide that would have to be jettisoned if Butchart and McKay hoped to make a safe landing. NACA X-1A crew chief Richard Payne entered the B-29 bomb bay and examined the cockpit of the X-1A, noting that the landing gear handle was still in the retracted position. He attempted to jettison the fuel and peroxide using the remaining nitrogen supply, but Kit Murray observed only a small amount of fuel before the flow stopped. If Butchart wanted to land with the X-1A, he would have to do it with a dragging landing gear and a full load of propellants in the rocket plane. On the ground, NACA operations chief Joseph R Vensel, ever a realist, radioed Butchart with the only possible answer: 'Butch, you might as well drop it. Pick a good place'. Butchart swung the bomber over the Edwards bombing range and jettisoned the research plane. Out of balance from the explosion, the X-1A entered a tail-down flat spin and exploded when it hit the desert, starting a brush fire. NACA and the Air Force immediately established a joint accident board to investigate the cause of this incident, the fourth of a long series of mysterious explosions that had claimed the X-1-3, X-1D, X-2 #2 and now the X-1A.

Unlike the three earlier victims, the X-1A did not have a tube-bundle nitrogen storage system. Bell had replaced the tube-bundle system on the X-1A and X-1B with three cylindrical nitrogen storage tanks. These tanks survived the in-flight explosion and subsequent ground impact and thus could not be the cause of the explosion. NACA engineers assembled the wreckage of the X-1A in the High-Speed Flight Station and placed the X-1B next to it to facilitate examination of the wreckage. During removal of the X-1B's liquid oxygen tank-access doors, technicians discovered an oily substance inside the tank and the liquid oxygen supply lines to the rocket engine. Analysis revealed the liquid to be tricresyl phosphate, a substance used to impregnate leather. The treated leather, called Ulmer leather, found application on the X-1s as gasket material in their liquid oxygen tanks. The accident board consulted a commercial air-products company, which advised that Ulmer leather should not be used in proximity to liquid oxygen, since at low temperatures it readily detonated under comparatively low impact. The accident board then placed pieces of Ulmer leather and frozen drops of tricresyl phosphate on an anvil and dropped a 5lb steel bar from 10 feet above the anvil. In each of 30 tests, the leather and frozen tricresyl phosphate exploded when hit by the bar. The board theorised that under the jolt of pressurisation, the gaskets exploded, fatally crippling the X-1A. The accident board noted the great similarity between the loss of the previous three research aeroplanes and the X-1A. In each case, the explosion occurred near the liquid oxygen tanks during pressurisation. The accident report suggested Ulmer leather as the possible cause of the previous accidents since all of the destroyed planes used Ulmer leather gaskets. The board recommended in its accident report that the Ulmer leather gaskets be removed from the remaining research aeroplanes, and NACA and the Air Force quickly complied.

There was then the matter of the flight itself and the conduct of a crew who valiantly tried, no matter how frustrated, to save the X-1A for another day. NACA awarded B-29 crewmen Charles W Littleton and John W Moise the NACA Distinguished Service Medal for 'outstanding bravery beyond the call of duty'. Stanley Butchart, Joseph Walker and Richard Payne received the NACA Exceptional Service Medal; and Jack McKay, Rex L Cook, Richard A DeMore and Merle C Woods received letters of commendation. Jerome Hunsaker, former NACA chairman, sent Major Arthur Murray a letter of commendation for aiding the B-29 crew despite severe damage to his own chase plane. While the discovery was a relief, it could not compensate for the death of two men, the injury of a third and the destruction of six aeroplanes, four of which were irreplaceable research aeroplanes and two of which were much-modified motherships.

Closing an Era: The X-1B and X-1E

NACA still had two advanced X-1s left, the X-1B and the X-1E. The X-1B was a sister ship to the X-1A. Following the X-1A accident, technicians removed its Ulmer leather gaskets and its aerodynamic-heating programme did not get under way until the autumn of 1956. In August, Jack McKay completed two familiarisation flights marred by minor malfunctions, including the recurrence of the X-1's traditional nosewheel-collapse-on-landing problem. McKay completed the heating programme in January 1957 after a Mach 1.94 flight. Preliminary data showed a maximum heating rate of about 3° F per second. Maximum recorded temperature was 185°F. on the forward point of the X-1B's nose. Calculated temperatures agreed closely with those recorded in flight and were felt to be representative of heating conditions that could be expected on future fighter and interceptor aircraft.[15]

Due to the loss of the X-1A, NACA decided to employ the X-1B for high-altitude research in areas of low dynamic pressure where ordinary aerodynamic control surfaces lost effectiveness, preventing the pilot from controlling the motion of the aeroplane. This had been shown most dramatically on Air Force Captain Iven Kincheloe's 126,200 feet altitude flight in the surviving Bell X-2 on 7 September 1956. On the flight, the X-2 described a ballistic arc, during which Kincheloe could not have altered its path in the slightest by use of his ailerons, elevator or rudder controls. Only when the X-2 descended to lower, denser levels, could the control surfaces once more generate changes in airflow to manoeuvre the aeroplane. The obvious solution for control at high altitudes where ordinary aerodynamic controls proved ineffective was in the use of reaction control jets, small thruster rockets utilised to govern the attitude of the plane. The NACA

High Speed Flight Station decided to install a prototype reaction-control system on the X-1B to gain information and experience that might be of great value in preparation for the upcoming X-15 programme, for the X-15 had been designed with reaction controls in addition to aerodynamic ones.

Pending installation of the reaction controls, Jack McKay completed a brief series of stability and control investigations to Mach 1.65. On 8 August 1957, he made his last flight in the plane. Neil Armstrong checked out on 15 August attaining Mach 1.32 at 45,000 feet. In preparation for the reaction-control programme, Armstrong 'flew' a ground simulator, the 'Iron Cross'. A beam-type structure in the shape of an 'X', it matched the dimensions and inertial characteristics of the X-1B aircraft. One 'fuselage' leg consisted of a 'cockpit' for the pilot at one end and pitch and yaw reaction controls at the other end. The 'wing' bar housed reaction controls for rolling (lateral) motion in the right 'wing tip'. The simulator used nitrogen gas for motion and rested on a short tower that limited permitted motion about all three axes: pitch, roll and yaw.

Armstrong began the X-1B's reaction control programme in November 1957. In support of the X-1B programme he flew a NACA F-100A Super Sabre on zoom simulation flights to duplicate X-1B trajectories and positioning over the lakebed. On 23 January 1958, he completed the X-1B's 17th NACA flight. Winter rains then closed the lakebed to further operations and NACA installed a modified engine and ventral fins, the latter to improve directional control above Mach 1.6. During a pre-flight inspection in May 1958, NACA technicians then discovered four fatigue cracks in the bottom of the X-1B's liquid oxygen tank. They welded the cracks shut, but X-ray inspection revealed flaws in the welds. Even if the the welds had been perfect, NACA safety engineers still felt that safety factors for tank pressure were not high enough to permit use of the tank with welded cracks. So, in June 1958, NACA retired the X-1B. Engineers transferred the reaction-control programme from the X-1B to a NACA Lockheed F-104A Starfighter, and this plane later served as a pilot trainer for the X-15 programme.

Though it had not made any spectacular contributions to high-speed flight research, the X-1B did make two important contributions to the sum of knowledge acquired through the research-aircraft programme. It returned the first detailed data on structural heating at supersonic speeds, and it was the first aeroplane to employ hydrogen peroxide reaction controls for use during semi-ballistic flight in regions of low dynamic pressure. While events prevented the X-1B from using its reaction controls in high-altitude research, the development of the controls for the aeroplane and the limited flight testing with them provided valuable experience that aided the initiation of the X-15 programme in 1959, an important predecessor itself to subsequent hypersonic vehicles.

The remaining X-1 aircraft, the X-1E, was informally known among NACA personnel as *Little Joe*.[16] Perhaps a better appellation could have been *The Phoenix*. After grounding the second X-1 in late 1951 because of fears its nitrogen pressurisation spheres might fail catastrophically, NACA engineers concluded the plane should be rebuilt with a low-pressure turbopump fuel-feed system. In April 1952, NACA engineers at Edwards began design of the new fuel system using ideas gleaned from the turbopump installations on the D-558-2 and the ill-fated X-1D and X-1-3. At the same time, the Air Force and NACA reached a joint agreement whereby the Air Force would procure a new extremely thin – 4% thickness:chord ratio – wing with an equally radical aspect ratio of only four. With the low-pressure fuel system and the new thin wing in contrast to the old 'thick' wing, the modified second X-1 would be able to fly around Mach 2.5. To develop the special wing, the NACA and Air Force turned to the Stanley Aviation Corporation, founded by Robert M Stanley, the former Bell Vice-President for Engineering so closely associated with the original XS-1 programme. Among its personnel, the firm counted Richard H Frost, former X-1 project engineer at Muroc. The Stanley engineers designed the wing with a span of 22.79 feet, a root chord of 7.62 feet, a tip chord of 3.81 feet and used a modified NACA 64A-004 symmetrical airfoil section. The construction of the wing posed difficult problems for the Stanley technicians. Maximum thickness of the wing at the root was only 3⅜ inches, and the Stanley engineers had to guard against aeroelastic effects – the tendency of the wing to twist – at high Mach numbers. They achieved maximum torsional stiffness by using multiple spars of rectangular cross section with tapered, milled wing-skins bolted to the spars and tips. The wing contained over 200 orifices for pressure distribution studies, as well as 343 gauges baked onto the wing surface for measurement of structural loads and aerodynamic heating.

As well as changing the wing configuration and propellant feed system, NACA modified the cockpit section of the plane. Technicians installed a hooded enclosure closely resembling that of the D-558-2 Skyrocket and installed an ejection seat scrounged from the second Northrop X-4 semi-tailless research aeroplane after NACA retired it. The resulting plane was a composite aircraft: NACA designed and fabricated the rocket system, and reworked and rebuilt the cockpit; Stanley furnished the wing; Bell, the original builders of the aeroplane, only claimed responsibility for those components that Bell had furnished, namely the tail group, forward fuselage shell and landing gear. Everything else was new. The plane owed much to the Skyrocket rocket system – in fact, the modified research aircraft used a RMI LR-8 engine originally assembled for the Navy D-558-2. Rather than any indigenous Bell product, the modified plane's development closely paralleled that of the composite General Motors XP-75 fighter, the 'Fisher Eagle' of the Second World War, though, blessedly, with more success.

Recognising the distinct character of the aeroplane, NACA formally designated it as the X-1E in March 1954. NACA completed engineering work and assembly of the plane in mid-1955. Following instrumentation and final checks on the plane's systems, the High-Speed Flight Station readied it for its flight programme. The first attempt came on 3 December 1955, but loss of source pressure forced an abort before launch. On 12 December 1955, nine years after its original powered flight, the X-1E dropped from its B-29 mothership – the same one, 45-21800, that launched it nine years earlier on its first glide flight – piloted by NACA research pilot Joe Walker. Three days later, on 15 December 1955, Walker completed the X-1E's first powered flight. After further general handling-qualities and familiarisation flights in April and May 1956, Walker completed the X-1E's first faster-than-sound flight on 7 June 1956, accelerating in the little rocket aeroplane to Mach 1.55, approximately 1,020 mph. Two months later, on 31 August 1956, Joe Walker reached Mach 2. Following a Mach 2 flight on 15 May 1957, the X-1E suffered severe damage during a landing accident. Walker's helmet face plate distorted his vision upon landing, and he landed hard; the nose gear snapped off and the X-1E also received damage to its main gear, fuselage skin and ventral fairing. Engineers at the High-Speed Flight Station decided to replace the nitrogen cockpit-pressurisation system with an air system so that the pilot could remove his face piece to improve his vision for landing. While technicians repaired and modified the aeroplane, engineers published preliminary data from its aerodynamic heating investigations made over the previous year. In August 1957, the aeroplane returned to flight status. On 8 October 1957, Walker streaked to Mach 2.24, approximately 1,480 mph. Because Langley Laboratory wind tunnel studies had shown the X-1E to have very low directional stability at speeds above Mach 2, NACA engineers at the High-Speed Flight Station devised a ventral fin configuration for the plane to improve its directional stability at high Mach numbers. Technicians installed two ventral fins on the aeroplane in December 1957 and Walker made the first flight in the plane with the ventrals added in May 1958. He found that the ventral fins markedly improved stability, but decreased the plane's manoeuvrability over the entire speed range.

The loss of the surviving Bell X-2 in September 1956, following Captain Milburn G Apt's ill-fated flight to Mach 3.2 – the first Mach 3 manned flight ever made – left the X-1B and the X-1E the only two high-performance rocket-research aeroplanes still flying, pending the delivery of the North American X-15. Following its Air Force programme, the X-2 was to have joined the NACA stable as an aerodynamic heating research aircraft. Incredible as it may seem, some in NACA actually entertained *repairing* Apt's X-2, which, pilotless, had spun into the desert without suffering extreme damage. Cooler heads – and good taste – prevailed. To fill the gap caused by its loss, High-Speed Flight Research Station engineers Hubert M Drake and Donald R Bellman suggested in November 1957 that the performance capabilities of the X-1E be boosted from its design speed of Mach 2.7 to about Mach 3 by boosting engine chamber pressure in the RMI LR-8 engine from 250 psi to 300 psi. They would also replace the alcohol fuel with a more powerful fuel called Hidyne or U-deta, consisting of a mixture of 60% unsymmetrical dimethylhydrazine and 40% diethylene triamine. Following another landing accident to the X-1E on 10 June 1958, when it again lost its nose gear, NACA decided to undertake the U-deta programme and install a new stabiliser bell crank to increase stabiliser travel, thus improving longitudinal manoeuvrability that the ventral fins had decreased. In September 1958, the plane flew again, the last two flights by Joe Walker. John McKay checked out in the rocket plane on 19 September 1958, and flew it on the remainder of its research programme. In October, he made two flights in the plane at increased engine pressure. On 6 November 1958, McKay launched from the Boeing B-29 Superfortress on the first check flight using U-deta as a propellant. Fourteen minutes after launch the X-1E touched down on the smooth surface of the dry lake, having completed its 26th flight.

Though no one knew it, it was the last research flight ever made by the X-1 series of aeroplanes. In November 1958, High-Speed Flight Station engineers grounded the plane for replacement of its ejection system, for the ejection seat, a holdover from the X-4 programme, was not compatible with a pilot wearing a high-altitude pressure suit. In December 1958, an X-ray inspection revealed cracks in the plane's alcohol tank. Serious enough separately, the two deficiencies spelled the end of the X-1E's research programme. The X-1E joined *Glamorous Glennis* and the X-1B in retirement in April 1959. After more than 12 years of research flying, the X-1s disappeared from the skies.

Epilogue

Today the surviving X-1 and D-558 aeroplanes are on exhibit across the United States. Chuck Yeager's *Glamorous Glennis* holds a special place of honour in the Milestones of Flight Gallery at the Smithsonian Institution's National Air and Space Museum. Near it, angling upwards, is Scott Crossfield's Mach 2-busting Skyrocket. The X-1B resides in the Air Force Museum, several hundred yards from where Ezra Kotcher undertook his 'Mach 0.999' study at Wright-Patterson AFB, Ohio. The X-1E, *nee* the second XS-1, is mounted in characteristic attitude on a concrete pedestal in front of the former NACA High-Speed Flight Station, now the NASA Dryden Flight Research Center at Edwards AFB, a fitting reminder of the X-1s and D-558s that flew from the exotic dry lake. The Navy displays the first Douglas D-558-1 Skystreak at the Naval Aviation Museum, Pensacola, Florida having repainted this record-setter in its original scarlet. The only other surviving Skystreak, the third D-558-1, NACA 142, is on exhibit at the Marine Corps Air-Ground Museum, Quantico, Virginia. The first Skyrocket is in storage at the Planes of Fame Museum, Ontario, California. The workhorse third D-558-2 is fittingly perched, pointing upwards, in front of the Antelope Valley College, Lancaster, California.

The X-1 and D-558 aircraft could serve as models for research aircraft procurement. Generally speaking, leading aerodynamicists and engineers like Ezra Kotcher and John Stack recognised the need for a transonic and supersonic research aeroplane at least five years prior to government and industry efforts to procure such a plane. When the NACA, AAF and the Navy issued authorisation for companies to develop the research aeroplanes, the goals stipulated were modest, despite their radical mission.

Essentially, the companies had to design the aircraft to attain or approach Mach 1, the speed of sound. The design should not be hindered by existing requirements for design or by military requirements. Though the contractor had a free hand in the development of an airframe, the plane should be as simple and conventional as possible, so that its results from research could be applied in the design of new operational-type aircraft. This approach led to an almost phenomenally rapid development and delivery schedule. Within one year from Robert Woods' original discussion with Ezra Kotcher, Bell had the first XS-1 ready for its initial glide flights. Within two years from initial discussions with the Navy, Douglas had the first D-558-1 flying at Muroc.

Both designs reflected good engineering and careful development. Neither the XS-1 nor the D-558-1 and D-558-2 required extensive modifications or corrections following flight testing. They all met their contract specifications. The only difficulty in the two programmes centred around the designing of a reliable turbopump installation to feed propellants to the rocket engines. Bell wisely decided to adopt an interim high-pressure system that could still permit the aeroplane to attain the requisite 800 mph at 35,000 feet. Since Skyrocket development started at a later date, Douglas could afford the luxury of waiting for the turbopump to come through. Both teams kept the designs flexible enough to permit adoption of the basic airframe for additional capabilities. In the X-1s, this meant lengthening the basic design to incorporate additional fuel giving the plane Mach 2.5 performance. In the D-558-2, this meant that Douglas could modify the plane into an all-rocket air-launched aircraft. The discovery that the modified or advanced aeroplanes encountered severe stability and control problems around Mach 2 did not reflect on the design teams, but rather indicated new problems in supersonic flight, potentially affecting all high-performance aircraft, that scientists and engineers would have to solve or alleviate before practical Mach 2 aircraft became a reality. It must be remembered that these aircraft, the D-558s and the X-1s, were originally designed for speeds around Mach 1. This was the design goal, despite predictions by some that they might go even faster. Finally, engineers not only designed the aircraft to fulfil their research mission, but at the same time did not sacrifice the plane's flying qualities. Pilots of both the X-1 and D-558 series continually stated that the planes had very satisfactory flying qualities both at subsonic and transonic velocities, though the X-1 undoubtedly flew better, a tribute to the wisdom of the Kotcher-Bell partnership.

Of the original two designs, the XS-1s were the most radical, since they incorporated a rocket power plant. This itself implied the planes might undergo a prolonged flight programme before the rocket engine proved itself a reliable system. However, the Reaction Motors engineers developed a highly satisfactory engine that did not require extensive development testing and that did not disrupt the X-1 or D-558 development programmes. Drawing on their previous experience with liquid oxygen-and-alcohol-fuelled engines, the small firm built a reliable power plant that, more than a decade later, propelled the North American X-15 on its initial contractor flights. Though frozen valves and the like occasionally plagued X-1 and D-558-2 flights, the difficulties did not stem from any inherent flaw or fault in the design of the engine. This simple approach taken in the design of the X-1s and D-558-2s, together with the resulting success of the two programmes, contrasts sharply with other research aircraft programmes undertaken at later dates and plagued by problems in airframe and power plant design.

At the time of their initial formulation, both the X-1 and D-558 aircraft were conceived for research in the vicinity of Mach 1. It would seem a duplication of effort to develop both planes. However, the D-558 represented closely a service-type aeroplane, while the XS-1 was more radical in that it required air launching, special fuels and the like. In effect, the D-558 was insurance against the possible failure of the XS-1 to emerge as a successful design. When the two types began flying, however, the D-558, with its high transonic speeds, freed the XS-1s to concentrate on supersonic flight research. This meant that the XS-1s could obtain data on supersonic flight that much faster, without having to fly an extensive transonic flight programme as well. Further, the development of the sweptwing D-558-2 opened up a whole new field of research, that of sweptwing behaviour at transonic and supersonic speeds. Had the D-558 been dropped in favour of the X-1, derivation of data on supersonic sweptwing behaviour would have depended on the ill-starred X-2 which did not make its first research flights until late 1955, far too late to assist in the exploitation of the sweptwing for supersonic flight.

Aside from direct technical contributions, the most important contribution of the X-1 was in eliminating the myth of a 'sound barrier' from the minds of most aeronautical engineers. It changed thinking from subsonic to supersonic, and forced engineers and pilots to look upon supersonic flight as just as real a possibility as subsonic flight. The fact that the nation possessed an aeroplane that could actually fly through the 'barrier' caused the barrier to melt away. It gave aircraft designers and manufacturers confidence in developing transonic and supersonic aeroplanes. Less than a year after Yeager exceeded Mach 1, Air Force pilots regularly dived faster than the speed of sound in production F-86 Sabres. On 25 May 1953,

North American test pilot George 'Wheaties' Welch completed the first flight of the prototype North American F-100A Super Sabre, the first Western supersonic fighter. Its design dated to 1949 and with its thin wing and low-placed all-moving horizontal tail, it reflected lessons already learned with both the D-558 and X-1 aircraft. Before the X-1s retired in 1958, production F-100s and Mach 2 Lockheed F-104s regularly flew chase for the rocket planes.

Technically, the X-1 and D-558 series made valuable contributions to the state of aeronautical science. Primarily, the development of the XS-1 enabled NACA and Air Force engineers to give focus to the whole problem of transonic and supersonic flight. It gave them a real vehicle to work on and enabled them to bring together in one design the best aerodynamic data and ideas concerning transonic and supersonic aerodynamics that they had at the time. The X-1 came at a time when ground-based research methods could not generate the necessary data to permit designers to configure aeroplanes for safe transonic and supersonic flight. Aerodynamicists sought to reach a desired technology level permitting them to design aeroplanes for flight at Mach 1. Not content to wait for ground-research methods to reach the-state-of-the-art whereby reliable high-speed aerodynamic data could be acquired, they elected to develop a research aircraft that could jump above the state-of-the-art until ground-research methods could catch up to the same level of technological sophistication.

One can only speculate what the course of aeronautical science might have been had the X-1 and D-558 family not been produced and had the United States, for example, made the same decision as Great Britain: namely to forego piloted supersonic testing in favour of reliance on models and ground research methods. The likelihood is that American aviation might well have followed a basically similar path. Certainly delay and lack of focus are two probable results, not pleasant to contemplate given the pace of aeronautical development and the growing postwar nuclear threat that accompanied the occasionally-hot Cold War. To this end, it is worth noting the conclusions of Sir Roy Fedden, one of Great Britain's most distinguished aero-engine designers and a long-time student of American aviation policy and development. Writing in 1957–in exasperation at what he perceived to be the timidity and lack of resolve afflicting the British aeronautical industry at that time–he rightly stated:[1]

> No single act set back Britain's aircraft development quite so drastically, however, as the Government's decision in 1947 not to allow manned supersonic investigations . . . The Miles supersonic aircraft project was abandoned, and we turned to an abortive programme of

> supersonic investigation through models powered by a rocket motor and launched in the air from a Mosquito. This was followed by another equally fruitless plan to launch models of possible supersonic shapes from the ground . . . Our high-speed subsonic types succeeded in diving past the speed of sound, but the ban on manned supersonic investigation continued for four years. Its effect, however, was a greater loss than that. *This unfortunate decision cost us at least ten years in aeronautical progress. It put us at least a generation behind the United States in fighters, and more than that in the basic knowledge necessary for the production of Mach 2 machines.* [Emphasis added]

In time, ground-research methods did catch up. In part, this was in response to an increased need for accurate aerodynamic data in the design of the X-1 and D-558. Wind tunnel technology underwent an evolutionary process aimed at developing transonic-research methods. First came the development of the 'splitter plate' method, then the emergence of the sting model support system, used for testing the XS-1 and D-558. In 1945, Coleman du Pont Donaldson invented the annular transonic tunnel for airfoil testing to Mach 1.1. The next year, Ray Wright and Vernon Ward at NACA's Langley Laboratory arrived at the basic theory supporting the 'slotted throat' wind tunnel. Testing began on 8 January 1947, and during December 1949 NACA achieved the first continuous transonic flow using the slotted throat technique in the Langley 8-foot high-speed tunnel. The mysterious gap from Mach 0.75 to Mach 1.3 in wind tunnel testing was no more. John Stack and his associates at Langley received the Robert J Collier trophy for 1951 for the development of the slotted throat wind tunnel. The place of the X-1 and D-558, then, can be set on a graph of time versus the level of technological capability:

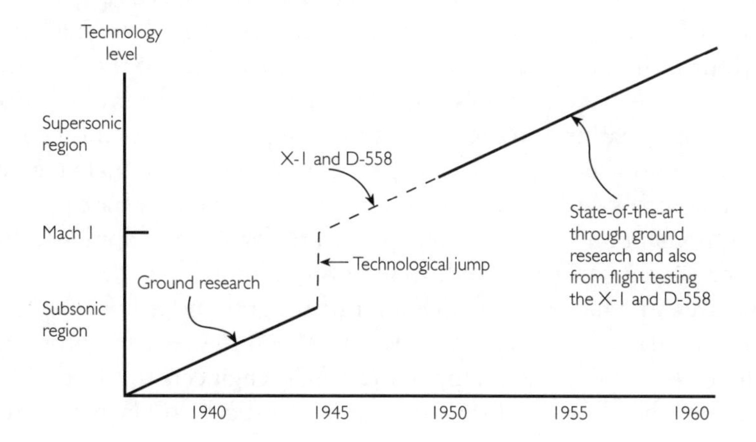

During the flight testing of the X-1 and D-558 research aeroplanes, engineers gained insight into new problems affecting aircraft design. One was the marked loss of elevator effectiveness at Mach 1 and above. X-1 pilots found the stabiliser to be 20 times more effective than the elevator. This led directly to the development of the all-moving or 'flying' tail for supersonic aircraft that appeared first on the North American F-86D Sabre interceptor, and then on the first Western supersonic jet fighter, the F-100A Super Sabre. Provisions for an adjustable stabiliser together with the elevators on the regular F-86 fighter contributed greatly to its excellent transonic behaviour in contrast to the MiG-15; the MiG lacked an adjustable horizontal stabiliser and, unlike the Sabre, was limited to no more than Mach 0.92, a serious combat restriction. Transonic buffet investigations with the X-1 and D-558-1 confirmed the desirability of using thin wing sections on high-speed aeroplanes. Using a thin wing, for example, the Lockheed F-94C Starfire interceptor could exceed Mach 1 in a dive, something earlier F-94 aircraft could not. The pitch-up phenomena, first encountered and studied by the D-558-2 Skyrocket, plagued subsequent high-performance sweptwing aircraft, particularly the McDonnell F-101 Voodoo fighter-interceptor. The only cure appeared in changing the tail configuration so that the horizontal stabiliser emerged very low on the side of the fuselage. This became a standard design feature on such aircraft as the North American F-100, Grumman F11F-1 Tiger and Vought F8U-1 Crusader. At one point, as mentioned, NACA gave serious consideration to adding a low horizontal tail to the Skyrocket.

Once the advanced X-1s and the all-rocket D-558-2 began explorations at speeds above Mach 1.5, other problems appeared. The most serious were the loss of directional stability and the phenomenon of coupled motions, such as combined pitching, rolling and yawing motions. This led to greater research on tail effectiveness at supersonic speeds and the introduction of analogue studies using flight-test results to predict aircraft behaviour at higher speeds. The advanced capabilities of the aircraft for speeds at Mach 2 and altitudes above 80,000 feet led to the investigation of aerodynamic heating and the development of prototype reaction controls for control of the aircraft in a near-space environment. The X-15 employed reaction controls during its flight-test programme and they have been standard equipment on virtually every spacecraft, foreign and domestic, flown since the Mercury, Gemini and Apollo programmes.

One area in which the research aircraft programme contributed to the future was that of research organisation. When the research programme began in 1946, with the detailing of the NACA engineers to Muroc, Walter C Williams, head of the little group, reported back to Hartley A Soule,

head of the Stability Research Division at Langley Laboratory. As the research activities of the X-1 and D-558 at Muroc expanded, so did the need for tighter organisation, with the activities of the NACA Muroc unit tied into the activities of the other NACA research centres. Accordingly, NACA designated Soule as Research Aeroplane Projects Leader in August 1948. A month later, he recommended formation of a Research Aeroplane Projects Panel with representatives from the various NACA laboratories. NACA formed the panel and appointed Soule as chairman. Later, in 1954, when initiating development of the X-15, NACA, the Air Force, and the Navy formed a special three-man X-15 steering committee better to administer the development programme on the aeroplane. Of particular importance, however, was the tight organisational relationship between NACA, the military services and private industry in the development and testing of the research aeroplanes. In the meantime, at Edwards, the personnel of the NACA High-Speed Flight Station and the Air Force Flight Test Center developed organisation and mission planning to a high level of competence, from engineering planning on the aeroplanes, through administration, flight planning, flight testing, ground tracking, interpretation of data and maintenance. This found its fullest expression in the special 'High Range' tracking network established specifically for the X-15, a flight testing corridor running from Utah through Nevada and on into California. It was a long way from the original two SCR-584 gun radars that tracked the XS-1 and D-558 back in 1946 and 1947.

All of this backlog of programme management paid off in the manned spacecraft programme, a strange byproduct of an aerodynamic programme beginning with the transonic D-558 and supersonic XS-1. When the time came for NASA to form a team for Project Mercury, the first American manned spacecraft venture, the agency called on those individuals most closely associated with the research aircraft programme. Robert R Gilruth, who developed the wing-flow method of research and who favoured the use of a thin wing for the X-1, became Director of the Space Task Group, with responsibility for the development of Project Mercury. Walter C Williams, Chief of the NACA High-Speed Flight Station, later the NASA Flight Research Center, became Operations Director of Project Mercury and Associate Director of the Manned Spacecraft Center. Hartley A Soule, former Research Aeroplane Projects Leader, became project director for establishing a world-wide tracking network for Project Mercury. Gerald M Truszynski, Chief of the Instrumentation Division at the NACA High-Speed Flight Station, became Deputy Associate Administrator for Tracking and Data Acquisition for the Mercury programme. It seems fitting, then, that the first man to set foot on the moon was a former NACA research pilot who had flown the X-1B, X-5, and X-15 and who had served as launch pilot on the NACA B-29 and P2B-1S, Neil Armstrong.

Writing in the early 17th century, Sir Francis Bacon stated,

> By far the greatest obstacle to the progress of science and to the undertaking of new tasks and provinces therein, is found in this–that men despair and think things impossible.

The men who designed, maintained and flew the X-1s and D-558s did not despair and think their task impossible. Through them, aviation science crossed the invisible threshold to flight faster than sound.

Appendix One: Technical Specifications for the X-1s and D-558s

BELL XS-1 (X-1)

Engine: One Reaction Motors Inc. XLR-11-RM-3 (Model A6000C4) four-chamber rocket engine rated at 6,000lbs static thrust.

Propellants: Liquid oxygen (311 gallons) and diluted ethyl alcohol (293 gallons).

Weights:
Launch configuration with 8% wing: 12,250 lbs.
Launch configuration with 10% wing: 12,000 lbs.
Landing configuration with 8% wing: 7,000lbs.
Landing configuration with 10% wing: 6,750 lbs.

Height: 10.85 ft.

Length: 30.90 ft.

Wing: Straight with a NACA 65-108 (XS-1-1) or NACA 65-110 (XS-1-2) airfoil section.
Span: 28 ft.
Area: 130 sq. ft. (including fuselage section)
Root Chord: 74.2 in.
Tip Chord: 37.1 in.
Flap Area: 11.6 sq. ft.
Aileron Area (each aileron): 3.15 sq. ft

Horizontal Tail: Area: 26.0 sq. ft. Elevator Area: 5.2 sq. ft.

Vertical Tail: Area: 25.6 sq. ft. Rudder Area: 5.2 sq. ft.

Approximate Maximum Performance: Mach 1.45 at 35,000 ft (960 mph).

Specification is applicable to the XS-1-1 (46-062) and XS-1-2 (46-063).

BELL X-1-3

Engine:	One Reaction Motors Inc. XLR-11-RM-5 (Model E6000C4) four-chamber rocket engine rated at 6,000lbs static thrust.
Propellants:	Liquid oxygen (437 gallons) and diluted ethyl alcohol (498 gallons). Hydrogen peroxide (for turbine pump) (31 gallons).
Weights:	Launch configuration: 14,751lbs. Landing configuration: 6,847 lbs.
Height:	10.85 ft.
Length:	30.90 ft.
Wing:	Straight with a NACA 65-108 airfoil section. Span: 28 ft. Area: 130 sq. ft. (including fuselage section) Root Chord: 74.2 in. Tip Chord: 37.1 in. Flap Area: 11.6 sq. ft. Aileron Area (each aileron): 3.15 sq. ft.
Horizontal Tail:	Area: 26.0 sq. ft. Elevator Area: 5.2 sq. ft.
Vertical Tail:	Area: 25.6 sq. ft. Rudder Area: 5.2 sq. ft.
Approximate Maximum Performance:	Mach 2.44 at 70,000 ft. (1,612 mph).

Specification is applicable to the X-1-3 (46-064)

BELL X-1A

Engine:	One Reaction Motors Inc. XLR-11-RM-5 (Model E6000C4) four-chamber rocket engine rated at 6,000lbs static thrust.
Propellants:	Liquid oxygen (500 gallons) and diluted ethyl alcohol (570 gallons). Hydrogen peroxide (for turbine pump) (37 gallons).
Weights:	Launch configuration: 16,487lbs. Landing configuration: 7,266 lbs.
Height:	10.70 ft.
Length:	35.55 ft.
Wing:	Straight with a NACA 65-108 airfoil section. Span: 28 ft. Area: 130 sq. ft. (including fuselage section). Root Chord: 74.2 in. Tip Chord: 37.1 in.

Flap Area: 11.46 sq. ft.
Aileron Area (each aileron): 3.21 sq. ft.
Horizontal Tail: Area: 26.0 sq. ft. Elevator Area: 5.2 sq. ft.
Vertical Tail: Area: .25.6 sq. ft. Rudder Area: 5.2 sq. ft.
Approximate Maximum Performance: Mach 2.47 at 70,000 ft (1,635 mph).

Specification is applicable to the X-1A (48-1384), X-1B (48-1385), and X-1D (48-1386).

DOUGLAS D-558-1 SKYSTREAK

Engine: One Allison J35-A-11 (developed by General Electric as the TG-180) turbojet rated at 5,000lbs static thrust.
Propellants: Aviation fuel (kerosene) (230 gallons).
Weights: Takeoff configuration: 10,105lbs. Landing configuration: 7,711 lbs.
Height: 12 ft 1.7 in.
Length: 35 ft 8.5 in.
Wing: Straight with a NACA 65-110 airfoil section.
Span: 25 ft.
Area: 150.7 sq. ft.
Approximate Maximum Performance: Mach 0.832 at sea level (632 mph).

Specification is applicable to the D-558-1 #1, #2 and #3 (BuNos. 37970, 37971, and 37972).

DOUGLAS D-558-2 SKYROCKET

(Turbojet Ground-Launched Version)

Engine: One Westinghouse J34-40 turbojet engine rated at 3,000lbs static thrust.
Propellants: Aviation fuel (gasoline not kerosene) (260 gallons).
Weights: Takeoff configuration: 10,572lbs. Landing configuration: 7,914lbs.
Height: 12 ft 8 in.
Length: 42 ft.
Wing: Swept (35° at 30% chord) with a NACA 63-010 airfoil section at the root and a NACA 63-012 airfoil section at the tip.
Span: 25 ft.
Area: 175 sq. fit.

	Root Chord: 108.51 in.
	Tip Chord: 61.18 in.
	Flap Area: 12.58 sq. ft.
	Aileron Area: 9.8 sq. ft.
Horizontal Tail:	Area: 39.9 sq. ft. Elevator Area: 9.4 sq. ft.
Vertical Tail:	Area: 36.6 sq. ft. Rudder Area: 6.15 sq. ft.
Approximate Maximum Performance:	Mach 0.825 at 20,000 ft (585 mph).

Specification is applicable to the ground-launched, turbojet-powered D-558-2 #1 (BuNo. 37973) prior to installation of its RMI rocket engine. It is also applicable to the ground-launched D-558-2 #2 (BuAer 37974) prior to its modification in 1950 to all-rocket air-launched configuration.

DOUGLAS D-558-2 SKYROCKET

(All-Rocket Air-Launched Version)

Engine:	One Reaction Motors Inc. XLR-8-RM-6 (Model A6000C4) four-chamber rocket engine rated at 6,000lbs static thrust.
Propellants:	Liquid oxygen (345 gallons) and diluted ethyl alcohol (378 gallons).
Weights:	Launch configuration: 15,787 lbs. Landing configuration: 9,421 lbs.
Height:	12 ft 8 in.
Length:	42 ft.
Wing:	Swept (35°at 30% chord) with a NACA 63-010 airfoil section at the root and a NACA 63-012 airfoil section at the tip.
	Span: 25 ft.
	Area: 175 sq. ft.
	Root Chord: 108.51 in.
	Tip Chord: 61.18 in.
	Flap Area: 12.58 sq. ft.
	Aileron Area: 9.8 sq. ft.
Horizontal Tail:	Area: 39.9 sq. ft. Elevator Area: 9.4 sq. ft.
Vertical Tail:	Area: 36.6 sq. ft. Rudder Area: 6.15 sq. ft.
Approximate Maximum Performance:	Mach 2.0 at 62,000 ft (1,290 mph)

Specification is applicable to the air-launched D-558-2 #2 (BuNo. 37974), but also applies to the D-558-2 #1 (BuNo. 37973) as modified in 1955.

DOUGLAS D-558-2 SKYROCKET

(Jet-and-Rocket Air-Launched Version)

Engines: One Reaction Motors Inc. XLR-8-RM-5 (Model E6000C4) four-chamber rocket engine rated at 6,000lbs static thrust. One Westinghouse J34-40 turbojet engine rated at 3,000lbs static thrust.

Propellants: Rocket Engine: Liquid oxygen (170 gallons) and diluted ethyl alcohol (192 gallons). Turbojet Engine: Aviation fuel (gasoline not kerosene) (260 gallons).

Weights: Launch configuration: 15,266lbs. Landing configuration: 9,500lbs.

Height: 12 ft 8 in.

Length: 42 ft.

Wing: Swept (35°at 30% chord) with a NACA 63-010 airfoil section at the root and a NACA 63-012 airfoil section at the tip.
Span: 25 ft.
Area: 175 sq. ft.
Root Chord: 108.51 in.
Tip Chord: 61.18 in.
Flap Area: 12.58 sq. ft.
Aileron Area: 9.8 sq. ft.

Horizontal Tail: Area: 39.9 sq. ft. Elevator Area: 9.4 sq. ft.

Vertical Tail: Area: 36.6 sq. ft. Rudder Area: 6.15 sq. ft.

Approximate Maximum Performance: Mach 1.08 at 40,000 ft (720 mph).

Specification is applicable to the air-launched, jet-and-rocket-powered D-558-2 #3 (BuNo. 37975), but may be considered applicable to the ground-launched D-558-2 #1 (BuNo. 37973) prior to its modification in 1955 to all-rocket air-launched configuration.

Appendix Two: Flight Chronologies of the Research Aeroplanes

I

XS-1 #1 (X-1-1, 46-062) FLIGHTS

A. BELL CONTRACTOR FLIGHTS

At Pinecastle AAF, Florida (Plane equipped with 10% wing, 8% tail)

Jan. 25 1946 Bell flight 1, Jack Woolams, pilot. Familiarisation.
Feb. 5 1946 Bell flight 2, Woolams.
Feb. 5 1946 Bell flight 3, Woolams.
Feb. 8 1946 Bell flight 4, Woolams. Gear retracted, left wing damaged.
Feb. 19 1946 Bell flight 5, Woolams. Nosewheel retracted on landing runout. Landing-gear door damaged.
Feb. 25 1946 Bell flight 6, Woolams. Static directional stability investigation.
Feb. 25 1946 Bell flight 7, Woolams. Longitudinal and directional stability investigation.
Feb. 26 1946 Bell flight 8, Woolams. Dynamic stability check.
Feb. 26 1946 Bell flight 9, Woolams. Rate of roll investigation.
March 6 1946 Bell flight 10, Woolams. Static longitudinal stability investigation.

At Muroc Dry Lake, California (Plane equipped with 8% wing, 6% tail)

April 10 1947 Bell flight 11, Chalmers Goodlin, pilot. Glide flight and stall check.

April 11 1947	Bell flight 12, Goodlin. Nosewheel damaged. First powered flight of the XS-1 #1 aircraft.
April 29 1947	Bell flight 13, Goodlin. Handling qualities check.
April 30 1947	Bell flight 14, Goodlin. Same as flight 13.
May 5 1947	Bell flight 15, Goodlin. Same as flight 13.
May 15 1947	Bell flight 16, Goodlin. Buffet-boundary investigation. Aileron-damper malfunction.
May 19 1947	Bell flight 17, Goodlin. Buffet boundary investigation.
May 21 1947	Bell flight 18, Goodlin. Same as flight 17.
June 5 1947	Bell flight 19, Goodlin. Demonstration flight for Aviation Writers Association.

B. AIR FORCE FLIGHTS

Aug. 6 1947	AF glide flight 1, Capt. Charles E. Yeager. For pilot familiarisation.
Aug. 7 1947	AF glide flight 2, Yeager. Same as flight 1.
Aug. 8 1947	AF glide flight 3, Yeager. Same as flight 1.
Aug. 29 1947	AF powered flight 1, Yeager. Mach 0.85.
Sept. 4 1947	AF flight 2, Yeager. Approx. Mach 0.89. Telemeter failure required a repeat of this flight.
Sept. 8 1947	AF flight 3, Yeager. Repeat of flight 2.
Sept. 10 1947	AF flight 4, Yeager. Mach 0.91. For stability and control investigation.
Sept. 12 1947	AF flight 5, Yeager. Mach 0.92. Check of elevator and stabiliser effectiveness. Also buffet investigation.
Oct. 5 1947	AF flight 6, Yeager. Same as flight 5.
Oct. 8 1947	AF flight 7, Yeager. Airspeed calibration flight. Plane attained Mach 0.925.
Oct. 10 1947	AF flight 8, Yeager. Stability and control investigation. Plane attains Mach 0.997.
Oct. 14 1947	AF flight 9, Yeager. World's first supersonic flight by a manned aircraft. XS-1 #1 attained Mach 1.06 at 43,000 ft., approximately 700 mph.
Oct. 27 1947	AF flight 10, Yeager. Electrical power failure. No rocket.
Oct. 28 1947	AF flight 11, Yeager. Telemeter failure.
Oct. 29 1947	AF flight 12, Yeager. Repeat of flight 11.
Oct. 31 1947	AF flight 13, Yeager.
Nov. 3 1947	AF flight 14, Yeager.
Nov. 4 1947	AF flight 15, Yeager.

Nov. 6 1947	AF flight 16, Yeager. Mach 1.35 at 48,600 ft.
Jan. 16 1948	AF flight 17, Yeager. Airspeed calibration. Mach 0.9.
Jan. 22 1948	AF flight 18, Yeager. Pressure distribution survey. Mach 1.2.
Jan. 30 1948	AF flight 19, Yeager. Same as flight 18. Mach 1.1.
Feb. 24 1948	AF flight 20, Captain James T Fitzgerald, Jr pilot. Engine fire after launch forced jettisoning of propellants; completed as a glide flight.
Mar. 11 1948	AF flight 21, Yeager. Attained Mach 1.25 in dive.
Mar. 26 1948	AF flight 22, Yeager. Attained Mach 1.45 at 40,130 ft. (approx. 957 mph) during dive. Fastest flight ever made in original XS-1 aeroplanes.
Mar. 31 1948	AF flight 23, Yeager. Engine shutdown after launch. Propellants jettisoned, completed as glide flight.
Apr. 6 1948	AF flight 24, Fitzgerald. Pilot check flight. Mach 1.1 during 4-chamber run at 41,000 ft.
Apr. 7 1948	AF flight 25, Major Gustav E Lundquist. Glide flight only. AF flight 26, Fitzgerald. Familiarisation flight.
Apr. 9 1948	AF flight 27, Lundquist. Powered pilot-check flight.
Apr. 16 1948	AF flight 28, Lundquist. Pressure distribution survey. Only chambers 2 and 4 ignited.
Apr. 26 1948	AF flight 29, Fitzgerald. Aborted because of inconsistent rocket operation. Reached Mach 0.9.
Apr. 29 1948	AF flight 30, Lundquist. Pressure distribution survey Attained Mach 1.18.
May 4 1948	AF flight 31, Fitzgerald. Same as flight 30. Mach 1.15
May 21 1948	AF flight 32, Lundquist. Stability and control, and buffeting investigation, Mach 0.92.
May 25 1948	AF flight 33, Fitzgerald. Buffet investigation, Wing and tail loads. Mach 1.08.
May 26 1948	AF flight 34, Yeager. Same as flight 33. Mach 1.05.
June 3 1948	AF flight 35, Lundquist. Left main gear door opened in flight. Nosewheel collapsed on landing.
Dec. 1 1948	AF flight 36, Yeager. Handling qualities and wing and tail loads at Mach 1.
Dec. 13 1948	AF flight 37, Yeager. Same as flight 36.
Dec. 23 1948	AF flight 38, Yeager. Wing and tail loads during supersonic flight at high altitudes. Mach 1.09 at 60,000 ft.
Jan 5 1949	AF flight 39, Yeager. Rocket takeoff from the ground. Mach 1.03 at 23,000 ft. in 80 seconds.
Mar. 11 1949	AF flight 40, Captain Jack Ridley, pilot. Familiarisation

	flight. Mach 1.23 at 35, 000 ft. Small engine fire due to loose igniter.
Mar. 16 1949	AF flight 41, Colonel Albert Boyd, pilot. Familiarisation flight. Mach 1.04. Inflight engine fire and shutdown.
Mar. 21 1949	AF flight 42, Major Frank Everest, pilot. Familiarisation flight. Mach 1.22 at 40,000 ft.
Mar. 25 1949	AF flight 43, Everest. Check of pressure suit for altitude operation. Mach 1.24 at 48,000 ft. Rocket fire and automatic engine shutdown.
Apr. 14 1949	AF flight 44, Ridley. Accelerated stall check at transonic speeds. Mach 1.1 at 40,000 ft.
Apr. 19 1949	AF flight 45, Everest. Altitude attempt. Only two chambers fired.
Apr. 29 1949	AF flight 46, Ridley. Stability and control investigation to high Mach numbers. Mach 1.43, 51,700 ft.
May 2 1949	AF flight 47, Everest. Engine chamber explodes, jamming rudder. Everest lands safely.
July 25 1949	AF flight 48, Everest. Altitude attempt. Everest attains 66,846 ft.
Aug. 8 1949	AF flight 49, Everest. Altitude attempt. Everest attains 71, 902 ft.
Aug. 25 1949	AF flight 50, Everest. First use of partial pressure suit to save life of pilot during flight at high altitude. X-1 #1 lost cockpit pressurisation about 69,000 ft. Everest made safe emergency descent.
Oct. 6 1949	AF flight 51, Lieutenant Colonel Patrick Fleming, pilot. Pilot familiarisation; attained Mach 1.2.
Oct. 26 1949	AF flight 52, Major Richard L Johnson, pilot. Pilot familiarisation.
Nov. 29 1949	AF flight 53, Everest. High-altitude wing-and-tail-loads investigation.
Dec. 2 1949	AF flight 54, Everest. Same as flight 53.
Feb. 21 1950	AF flight 55, Everest. Wing-and-tail-loads investigation.
April 25 1950	AF flight 56, Yeager. Lateral stability and control investigation.
April (?)1950	AF flight 57, Ridley. Buffeting, wing and tail loads.
May 8 1950	AF flight 58, Ridley. Same as flight 57.
May 12 1950	AF flight 59, Yeager. Last flight of the X-1. Flight made for camera footage for the motion picture *Jet Pilot*. Aircraft subsequently retired and presented to the Smithsonian Institution.

II

XS-1 #2 (X-1-2, 46-063) FLIGHTS

A. BELL CONTRACTOR FLIGHTS

Oct. 11 1946	Bell flight 1, Chalmers Goodlin pilot. Glide flight for pilot familiarisation.
Oct. 14 1946	Bell flight 2, Goodlin. Glide flight.
Oct. 17 1946	Bell flight 3, Goodlin. Glide flight. Stall check.
Dec. 2 1946	Bell flight 4, Goodlin. Glide flight for check of fuel jettison system.
Dec. 9 1946	Bell flight 5, Goodlin, First XS-1 powered flight. Mach 0.79 at 35,000 ft.; minor engine fire.
Dec. 20 1946	Bell flight 6, Goodlin. Familiarisation powered flight.
Jan. 8 1947	Bell flight 7, Goodlin. Buffet boundary investigation. Mach 0.80 at 3,5,000 ft.
Jan. 17 1947	Bell flight 8, Goodlin. Same as flight 7. Full power climb. Plane reached Mach 0.82.
Jan. 22 1947	Bell flight 9, Goodlin. Same as flight 8. Telemetry failure.
Jan. 23 1947	Bell flight 10, Goodlin. Same as flight 8.
Jan. 30 1947	Bell flight 11, Goodlin. Accelerated stalls. Partial power due to faulty engine igniters. Mach 0.75.
Jan. 31 1947	Bell flight 12, Goodlin. Same as flight 7. Mach 0.7.
Feb. 5 1947	Bell flight 13, Goodlin. Machmeter calibration.
Feb. 7 1947	Bell flight 14, Goodlin. Same as flight 7.
Feb. 19 1947	Bell flight 15, Goodlin. Accelerated stalls.
Feb. 21 1947	Bell flight 16, Goodlin. Flight aborted after drop because of low engine-chamber pressure.
May 22 1947	Bell flight 17, Alvin M Johnston, pilot. Pilot familiarisation flight. Mach 0.72. 8g pullout.
May 29 1947	Bell flight 18, Goodlin. Airspeed calibration flight to Mach 0.72. End of Bell contractor programme.

B. NACA FLIGHTS

Sept. 25 1947	NACA acceptance flight. Captain Charles E Yeager, pilot. Number 4 chamber burned out.
Oct. 21 1947	NACA glide-familiarisation flight for NACA pilot Herbert H. Hoover. Stall check. Nosewheel collapsed on landing.
Dec. 16 1947	NACA powered flight 1, Hoover. For familiarisation. Mach 0.84. No telemetry record.

Dec. 17 1947	NACA flight 2, Hoover. Same as flight 1. Mach 0.8.
Jan. 6 1948	NACA flight 3, Hoover, Turns and pull-ups to buffet. Mach 0.74
Jan. 8 1948	NACA flight 4, Hoover. Turns and pull-ups to buffet. Mach 0.83.
Jan. 9 1948	NACA flight 5, Howard C. Lilly, pilot. For pilot familiarisation.
Jan. 15 1948	NACA flight 6, Lilly. Turns and pull-ups to buffet. Sideslips. Mach 0.76.
Jan. 21 1948	NACA flight 7, Hoover. Stabiliser effectiveness investigation. Mach 0.82 at 29,000 ft.
Jan. 23 1948	NACA flight 8, Hoover. Attempted high-speed run aborted at Mach 0.83 due to drop in chamber pressure.
Jan. 27 1948	NACA flight 9, Hoover. High-speed run to Mach 0.925 at 38,000 ft. Chambers 2 and 3 failed to fire.
Mar. 4 1948	NACA flight 10, Hoover. High-speed run to Mach 0.943 at 40,000 ft.
Mar. 10 1948	NACA flight 11, Hoover. First NACA supersonic flight. Mach 1.065. First civilian supersonic flight. Nosewheel failed to extend for landing. Minor damage.
Mar. 22 1948	NACA flight 12, Hoover. Stability and loads investigation. Mach 1.12.
Mar. 30 1948	NACA flight 13, Hoover. Same as flight 12. Mach 0.90.
Mar. 31 1948	NACA flight 14, Lilly. Same as flight 12. Plane attained Mach 1.1.
Apr. 5 1948	NACA flight 15, Lilly. Engine failed to ignite. Propellants jettisoned, completed as glide flight.
Apr. 9 1948	NACA flight 16, Lilly. Same as flight 12. Mach 0.89.
Apr. 16 1948	NACA flight 17, Lilly. Same as flight 12. Mach 0.94, 41,000 ft. Plane's nosewheel collapsed on landing. Moderate damage.
Nov. 1 1948	NACA flight 18, Hoover. Stability and control. Mach 0.9. Number 4 chamber failed to fire.
Nov. 15 1948	NACA flight 19, Hoover. Same as flight 18. Also pressure distribution survey. Mach 0.98.
Nov. 23 1948	NACA flight 20, Robert A Champine, pilot. For pilot familiarisation. Check on handling qualities and pressure distribution. Mach 0.70.
Nov. 29 1948	NACA flight 21, Champine. Check on handling qualities and pressure distribution. Mach 0.88.
Nov. 30 1948	NACA flight 22, Champine. Same as flight 21.

Dec. 2 1948 NACA flight 23, Champine. Same as flight 21. Plane exceeded Mach 1 briefly.

May 6 1948 NACA flight 24, Champine. Check on aeroplane instrumentation. Mach 0.88 at 40,000 ft.

May 13 1948 NACA flight 25, Champine. Spanwise pressure distribution, stability and control. Mach 0.91.

May 27 1948 NACA flight 26, Champine. Same as flight 25. Mach 0.91.

June 16 1949 NACA flight 27, Champine. Same as flight 25, Rolls and pull-ups around Mach 0.91.

June 23 1949 NACA flight 28, Champine. Same as flight 25, Rolls, pull-ups, check of stabiliser effectiveness.

July 11 1949 NACA flight 29, Champine. Same as flight 25. Rolls, pull-ups, check of stabiliser effectiveness. Mach 0.91. Number 2 chamber failed to fire.

July 19 1949 NACA flight 30, Champine. Same as flight 29. Rolls, pull-ups, check of stabiliser effectiveness. Mach 0.91. Number 2 chamber failed to fire.

July 27 1949 NACA flight 31, Champine. Same as flight 25. Rolls, pull-ups, check of stabiliser effectiveness.

Aug. 4 1949 NACA flight 32, Champine. Same as flight 25. Sideslips, rolls, check-of stabiliser effectiveness.

Sept. 23 1949 NACA flight 33, John H. Griffith, pilot. For pilot familiarisation. Mach 0.9.

Nov. 30 1949 NACA flight 34, Griffith. Same as flight 33. Mach 0.93.

May 12 1950 NACA flight 35, Griffith. Same as flight 25. Pull-ups and rolls. Mach 0.95.

May 17 1950 NACA flight 36, Griffith. Same as flight 25. Push-downs and pull-ups. Mach 1.13 at 42,000 ft.

May 26 1950 NACA flight 37, Griffith. Same as flight 25. Push-downs, pull-ups, rolls. Mach 1.20, 45,000 ft. Nosewheel collapsed on landing.

Aug. 9 1950 NACA flight 38, Griffith. For pressure distribution and stability and control data. Check of stabiliser effectiveness. Mach 0.98.

Aug. 11 1950 NACA flight 39, Griffith. Same as flight 38. Mach 0.92.

Sept. 21 1950 NACA flight 40, Griffith. Same as flight 38. Also drag investigation. Pull-ups. Mach 0.90.

April 6 1951 NACA flight 42, Captain Charles E Yeager, pilot. Flight for RKO film *Jet Pilot*. Slight engine fire but no damage.

April 20 1951 NACA flight 43, A. Scott Crossfield, pilot. For pilot familiarisation. Reached Mach 1.07.

April 27 1951 NACA flight 44, Crossfield. Plane and instrument check. Mach 1.12, 45,000 ft.

May 15 1951 NACA flight 45, Crossfield. Wing loads and aileron effectiveness. Aileron rolls at Mach 0.90.

July 12 1951 NACA flight 46, Crossfield. Same as flight 45. Aileron rolls at Mach 1.07, 45,000 ft.

July 20 1951 NACA flight 47, Crossfield. Same as flight 45. Abrupt rudder fixed aileron rolls left and right, from Mach 0.70 to Mach 0.88; max. Mach 1.12, 45,000 ft.

July 31 1951 NACA flight 48, Crossfield. Same as flight 45. Mach 0.89, 45,000 ft.

Aug. 3 1951 NACA flight 49, Crossfield. Same as flight 45. Mach 0.90, 45,000 ft.

Aug. 8 1951 NACA flight 50, Crossfield. Same as flight 45. Elevator and stabiliser pull-ups, Mach 0.90, 45,000 ft.

Aug. 10 1951 NACA flight 51, Crossfield. Same as flight 45. Elevator and stabiliser pull-ups, clean stalls, Mach 0.90, 45,000 ft..

Aug. 27 1951 NACA flight 52, Joseph A Walker, pilot. For pilot familiarisation. Reached Mach 1.16 at 44,000 ft. during four-chamber run.

Sept. 5 1951 NACA flight 53, Crossfield. Fuselage pressure distribution survey. Number 1 chamber failed to fire. Stabiliser pull-ups at Mach 1.07.

Oct. 23 1951 NACA flight 54, Walker. Vortex-generator investigation. Engine cut out after two ignition attempts; propellants jettisoned and flight completed as a glide flight. Flap actuator failed, so landing made flaps-up. Plane subsequently grounded because of possibility of fatigue failure of nitrogen spheres. Later rebuilt as the Mach 2+ X-1E.

III

X-1 #3 (X-1-3, 46-064) FLIGHTS

July 20 1951 Bell flight 1, Joseph Cannon, pilot. Glide flight for familiarisation. Nosewheel collapse on landing.

Nov. 9 1951 Bell flight 2, Cannon. Captive flight with B-50 for propellant Jettison test. X-1-3 destroyed in postflight explosion and fire on the ground. B-50 launch plane also lost and Cannon injured.

IV
X-1A (48-1384) FLIGHTS

A. BELL CONTRACTOR FLIGHTS

Feb. 14 1953 Bell flight 1, Jean 'Skip' Ziegler, pilot. For familiarisation. Fuel jettison test. Glide flight only.

Feb. 20 1953 Bell flight 2, Ziegler. Planned as powered flight, but completed as glide flight following propellant-system difficulties.

Feb. 21 1953 Bell flight 3, Ziegler. First powered flight. False fire warning.

Mar. 26 1953 Bell flight 4, Ziegler. Plane demonstrated successful four-chamber engine operation.

April l0 1953 Bell flight 5, Ziegler. Pilot noted low-frequency elevator buzz at Mach 0.93, and did not proceed above this speed pending buzz investigation.

April 25 1953 Bell flight 6, Ziegler. Buzz again noted at Mach 0.93. Turbopump overspeeding caused pilot to terminate power and jettison remaining fuel.

B. AIR FORCE FLIGHTS

After USAF took over remaining Bell programme on X-1A and initiated their own flight programme

Nov. 21 1953 Flight 7, Yeager, pilot. First Air Force flight. Reached Mach 1.15 on this flight, made for familiarisation purposes.

Dec. 2 1953 Flight 8, Yeager. Mach 1.5.

Dec. 8 1953 Flight 9, Yeager. First high-Mach flight attempt by X-1A. Mach 1.9 attained at 60,000 ft. during a slight climb.

Dec. 12 1953 Flight 10, Yeager. Plane attained Mach 2.44, but encountered violent instability above Mach 2.3. Tumbled 50,000 ft., ended up in subsonic inverted spin. Yeager recovered to upright spin, then into normal flight at 25,000 ft.

Fourteen Air Force flight attempts for high altitudes were made in the spring and summer of 1954. Of these, only four flights were successful. The rest were aborted for various malfunctions, including ruptured canopy seal, failure of gear doors to close fully, turbine overspeed and faulty ignition operation. Of the four successful flights, one was Major Arthur Murray's checkout flight. The rest were successful high-altitude tries by Murray. The successful altitude flights were:

May 28 1954	Flight 16, Murray. X-1A attained 87,094 ft., an unofficial world altitude record for manned aircraft.
June 4 1954	Flight 17, Murray, X-1A reached 89,750 ft. Encountered same instability Yeager had, but at Mach 1.97. Murray recovered after tumbling 20,000 ft. down to 66,000 ft.
Aug. 26 1954	Flight 24, Murray. X-1A attained 90,440 ft. Air Force then turned over the X-1A to NACA.

C. NACA FLIGHTS

July 20 1955	NACA flight 1, Joseph A Walker, pilot. For familiarisation purposes. Walker attained Mach 1.45 at 45,000 ft. Noted severe aileron buzz at Mach 0.90 to 0.92.
Aug. 8 1955	Planned as NACA flight 2. Shortly before launch, X-1A suffered a low-order explosion later traced to detonation of Ulmer leather gaskets. Walker exited into B-29 bomb bay. Extent of damage prohibited landing with the crippled X-1A, and the NACA B-29 launch crew jettisoned it into the desert. It exploded and burned on impact.

V

X-1B (48-1385) FLIGHTS

A. AIR FORCE FLIGHTS

Sept. 24 1954	X-1B Air Force flight 1, Lieutenant Colonel Jack Ridley pilot. Glide flight, due to turbopump overspeeding.
Oct. 6 1954	X-1B Air Force flight 2, Ridley. Glide flight, aborted power flight due to evidence of high lox tank pressure.
Oct. 8 1954	X-1B Air Force flight 3, Major Arthur 'Kit' Murray pilot. First powered flight.
Oct. 13 1954	X-1B Air Force flight 4, Major Robert Stephens pilot.
Oct. 19 1954	X-1B Air Force flight 5, Major Stuart R Childs pilot.
Oct. 26 1954	X-1B Air Force flight 6, Colonel Horace B Hanes pilot.
Nov. 4 1954	X-1B Air Force flight 7, Captain Richard B Harer pilot.
Nov. 26 1954	X-1B Air Force flight 8, Brigadier General J Stanley

	Holtoner pilot. (Commander, Air Force Flight Test Center).
Nov. 30 1954	X-1B Air Force flight 9, Lieutenant Colonel Frank K Everest pilot.
Dec. 2 1954	X-1B Air Force flight 10, Everest. Mach 2.3 (approx. 1,520 mph) at 65,000 ft.

B. NACA FLIGHTS

John B McKay pilot on flights 1–13. Neil A Armstrong pilot on flights 14-17.

Aug. 14 1956	X-1B NACA flight 1. Pilot check; nose landing gear failed on landing, minor damage.
Aug. 29 1956	X-1B NACA flight 2. Cabin-pressure regulator malfunction causes inner canopy to crack; only lowspeed, low-altitude manoeuvres made.
Sept. 7 1956	X-1B NACA flight 3. Speed run to 56,000 ft. and Mach 1.8. Limited heating data gathered.
Sept. 18 1956	X-1B NACA flight 4. Glide flight, due to erratic engine start.
Sept. 28 1956	X-1B NACA flight 5. Three chamber engine run to 60,000 ft. to obtain heating data.
Jan. 3 1957	X-1B NACA flight 6. Mach 1.94 aerodynamic heating investigation (end of heating programme).
May 22 1957	X-1B NACA flight 7. Control pulses at Mach 1.45 at 60,000 ft. Flight made for instrumentation check.
June 7 1957	X-1B NACA flight 8. Supersonic manoeuvres to Mach 1.5 at 60,000 ft. to determine the dynamic and static stability and control characteristics.
June 24 1957	X-1B NACA flight 9. Supersonic manoeuvres to Mach 1.5 at 60,000 ft. to determine the dynamic and static stability and control characteristics.
July 11 1957	X-1B NACA flight 10. Aborted after launch, due to indication of open landing-gear door. Propellants jettisoned, completed as a glide flight.
July 19 1957	X-1B NACA flight 11. Mach 1.65 at 60,000 ft. Control pulses, sideslips, and a 2g wind-up turn.
July 29 1957	X-1B NACA flight 12. Enlarged wing tips installed to simulate wing tips to be used with reaction controls. Mach 1.55 at 60,000 ft.
Aug. 8 1957	X-1B NACA flight 13. Stability and control investigation.

	Mach 1.5 at 60,000 ft., accelerated manoeuvres, control pulses, and pull-ups.
Aug. 15 1957	X-1B NACA flight 14. Pilot check for Armstrong. Nose landing gear failed on landing, minor damage.
Nov. 27 1957	X-1B NACA flight 15. First reaction-control flight.
Jan. 16 1958	X-1B NACA flight 16. Low altitude, low Mach reaction-control investigation.
Jan. 23 1958	X-1B NACA flight 17, Reaction-control investigation. Mach 1.5 at 55,000 ft. Last NACA flight.

VI

X-ID (48-1386) FLIGHTS

A. BELL CONTRACTOR FLIGHTS

July 24 1951	Bell flight 1, Jean 'Skip' Ziegler pilot. Glide flight for familiarisation. Nose landing gear broken on landing. Following repairs, plane turned over to the Air Force.

B. AIR FORCE FLIGHTS

Aug. 22 1951	AF flight 1, Lieutenant Colonel Frank K Everest plot. Launch aborted, but X-1D suffered a low-order explosion during pressurisation for fuel jettison. Plane jettisoned from B-50. X-1D exploded on impact with desert. Everest managed to get into B-50 bomb bay before drop. B-50 not damaged, no personal injuries.

VII

X-1E (46-063) FLIGHTS

Joseph Walker pilot for flights 1-21. John McKay pilot for flights 22-26.

Dec. 3 1955	Captive flight.
Dec. 12 1955	X-1E NACA flight 1. Glide flight for pilot checkout and low speed evaluation.
Dec. 15 1955	X-1E NACA flight 2. First powered flight. Engine running at excessive pressure, four overspeeds of turbopump and two automatic shutdowns. Power terminated by pilot.
Apr. 3 1956	X-1E NACA flight 3. Mach 0.85 at 30,000 ft. Damping characteristics good; number one chamber failed to fire.

April 30 1956 X-1E NACA flight 4. Turbopump did not start; no engine operation.

May 11 1956 X-1E NACA flight 5. Wind-up turns to $C_{L_{max}}$ from Mach 0.69 to 0.84; also control pulses.

June 7 1956 X-1E NACA flight 6, Mach 1.55 at 45,000 ft. (approx. 1,020 mph). Longitudinal and lateral trim changes in transonic region found annoying to pilot.

June 18 1956 X-1E NACA flight 7. Mach 1.74 at 60,000 ft. (approx. 1,150 mph). Damaged on landing.

July 26 1956 X-1E NACA flight 8. Subsonic because chambers 3 and 4 would not fire.

Aug. 31 1956 X-1E NACA flight 9. Mach 2.0 at 6000 ft. (approx. 1,340 mph). Sideslips, pulses, rolls.

Sept. 10 1956 X-1E NACA flight 10. Mach 2.1 at 62,000 ft. (approx. 1,385 mph). Stabiliser, rudder, and aileron pulses.

Sept. 20 1956 X-1E NACA flight 11. Brief engine power only; flight aborted due to unspecified engine malfunction.

Oct. 3 1956 X-1E NACA flight 12. Only 60 seconds of rocket operation; intermittent pump operation. Flight aborted, and turbopump and engine replaced.

Nov. 20 1956 X-1E NACA flight 13. No engine operation due to ignition failure and lack of manifold pressure.

Apr. 25 1957 X-1E NACA flight 14. Mach 1.71 at 67,000 ft. (approx. 1,130 mph). Aileron and rudder pulses.

May 15 1957 X-1E NACA flight 15. Mach 2.0 at 73,000 ft. (approx. 1,325 mph). Aileron pulses and rolls, sideslips, and wind-up turns. Plane severely damaged upon landing.

Sept. 19 1957 X-1E NACA flight 16. Planned Mach number not attained due to loss of power during pushover from climb.

Oct. 8 1957 X-1E NACA flight 17. Mach 2.24 (approx. 1,480 mph).

May 14 1958 X-1E NACA flight 18. First flight with ventral fins; longitudinal and lateral stability and control manoeuvres. Engine airstart made at approx. 70,000 ft.

June l0 1958 X-1E NACA flight 19. Flight aborted after only one chamber of engine fired. Plane damaged on landing.

Sept. 10 1958 X-1E NACA flight 20. Stability and control investigation with ventral fins,

Sept. 17 l958 X-1E NACA flight 21. Stability and control with ventral fins and a new stabiliser bell crank permitting greater stabiliser travel.

Sept. 19 1958 X-1E NACA flight 22. Checkout flight for John McKay.

Sept. 30 1958 X-1E NACA flight 23. Checkout flight for McKay, also check of low-speed stability and control.
Oct. 16 1958 X-1E NASA flight 24. First flight with elevated chamber pressure; cut short because overcast obscured pilot's view of lakebed.
Oct. 28 1958 X-1E NASA flight 25. Elevated chamber pressure; good stability and control data gathered.
Nov. 6 1958 X-1E NASA flight 26. Elevated chamber pressure; low-altitude and low-Mach investigation of U-Deta fuel. Last NASA flight.

VIII

D-558-1 # 1 (BuNo. 37970) FLIGHTS

This aircraft completed 101 flights during its Douglas contractor programme. Highlights of the contractor programme were:
April 14 1947 Douglas flight 1, Eugene F 'Gene' May pilot. For familiarisation. Partial power loss forced immediate landing after takeoff.
July 17 1947 Douglas flight 14, May. Beginning of performance investigations. Attained Mach 0.81.
Aug. 20 1947 Douglas flight 25, Commander Turner F Caldwell, Jr, USN, pilot. Set new world airspeed record of 640.663 mph.
Sept. 29 1948 Douglas flight (?), May. Plane exceeded Mach 1 during a 35° dive, only time a Skystreak attained Mach 1.

Douglas delivered the D-558-1 #1 to the NACA on 21 April 1949. NACA never flew it, relegating it to spares support for the D-558-1 #3.

IX

D-558-1 #2 (BuNo. 37971) FLIGHTS

A. DOUGLAS CONTRACTOR FLIGHTS

Prior to the NACA flight programme on this aircraft, Douglas, Navy, and Marine test pilots flew 27 flights. Highlights included:
Aug. 15 1947 Gene May, pilot. First flight.
Aug. 20 1947 Major Marion Carl, USMC. Unsuccessful attempt to exceed world airspeed record.
Aug. 25 1947 Major Marion Carl, USMC. Set new world airspeed

record of 650.796 mph, breaking the record set on 20 August by Commander Turner Caldwell in the D-558-1 #1.

Aug. 28 1947 Captain Frederick M Trapnell, USN. Bureau of Aeronautics evaluation flight; Trapnell concluded D-558-1 was 'a very sound airplane for research purposes . . .'

B. NACA FLIGHTS

Howard C 'Tick' Lilly, pilot

Nov. 25 1947 NACA flight 1. Pilot familiarisation; instrumentation malfunction.

Nov. 26 1947 NACA flight 2. Landing gear would not lock up.

Feb. 16 1948 NACA flight 3. Attempted airspeed calibration; instrumentation malfunction.

Mar. 31 1948 NACA flight 4. Landing gear door would not lock.

Mar. 31 1948 NACA flight 5. Landing gear door would not lock.

Apr. 1 1948 NACA flight 6. Landing gear door would not lock.

Apr. 7 1948 NACA flight 7. Landing gear door would not lock.

Apr. 8 1948 NACA flight 8. Attempted airspeed calibration; radar beacon failure,

Apr. 8 1948 NACA flight 9. Airspeed calibration, 30,000 ft.

Apr. 9 1948 NACA flight l0. Airspeed calibration, 30,000 ft.

Apr. 12 1948 NACA flight 11. Airspeed calibration, tower fly-by.

Apr. 12 1948 NACA flight 12. Airspeed calibration, 30,000 ft.

Apr. 14 1948 NACA flight 13. Smoke in cockpit after takeoff necessitated landing. Smoke due to burning 400-cycle inverter in nose compartment; inverter replaced.

Apr. 20 1948 NACA flight 14. Sideslips at 10,000 ft. from Mach 0.50 through 0.85, for static directional stability.

Apr. 23 1948 NACA flight 15. Sideslips at 30,000 ft. from Mach 0.50 through 0.85, for static directional stability.

Apr. 28 1948 NACA flight 16. Right landing gear would not retract.

Apr. 29 1948 NACA flight 17. Two speed runs; Mach 0.70 at 41,000 ft-, Mach 0.88 at 36,000 ft. Left and right rudder kicks at 10,000 ft.

May 3 1948 NACA flight 18. Landing gear would not retract.

May 3 1948 NACA flight 19. Crash after takeoff due to compressor disintegration; Lilly killed.

X

D-558-1 #3 (BuNo. 37972) FLIGHTS

Four flights made in early 1948 by Douglas pilots and Howard Lilly

Apr. 22 1949 NACA flight 1, Robert A. Champine pilot, for pilot familiarisation purposes.
Apr. 28 1949 NACA flight 2, Champine. Pilot check; dive to Mach 0.87.
Aug. 12 1949 NACA flight 3, Champine. Handling qualities (rudder kicks, aileron rolls, sideslips); dive to Mach 0.9.
Aug. 18 1949 NACA flight 4, Champine. Handling qualities; dive to Mach 0.875.
Aug. 19 1949 NACA flight 5, John H. Griffith, pilot check, handling qualities; trim run to Mach 0.84.
Aug. 23 1949 NACA flight 6, Griffith. Airspeed calibration using tower passes.
Aug. 24 1949 NACA flight 7, Champine. Handling qualities; dive to Mach 0.87.
Aug. 31 1949 NACA flight 8, Champine. Aileron effectiveness investigation; no records taken.
Sept. 28 1949 NACA flight 9, Griffith. Aileron effectiveness investigations; 16 rolls made, 4 above Mach 0.87.
Oct. 30 1949 NACA flight 10, Griffith. Beginning of pressure-distribution survey.
Nov. 21 1949 NACA flight 11, Griffith. Pressure-distribution investigation.
Nov. 23 1949 NACA flight 12, Champine. Pressure-distribution investigation.
Jan. 26 1950 NACA flight 13, Champine. Check of airspeed system.
Feb. 15 1950 NACA flight 14, Champine. Aborted due to engine malfunction.
Apr. 5 1950 NACA flight 15, Griffith. Pressure-distribution investigation. Mach 0.95 attained.
April 11 1950 NACA flight 16, Griffith. Pressure-distribution investigation. Mach 0.98 attained.
May 3 1950 NACA flight 17, Griffith. Vortex generator investigation as part of pressure-distribution investigation. Mach 0.97 attained.
May 5 1950 NACA flight 17A, Griffith. Vortex generator investigation as part of pressure-distribution investigations.
May 11 1950 NACA flight 18, Griffith. Vortex generator investigation. Mach 0.87 attained.

May 18 1950	NACA flight 19, Griffith. Vortex generator distributor investigation. Mach 0.98 attained.
May 31 1950	NACA flight 20, Griffith. Vortex generator distribution investigation.
June 8 1950	NACA flight 21, Griffith. Vortex generator distribution investigation.
June 13 1950	NACA flight 22, Griffith. Vortex generator distribution investigation. Mach 0.98–1.0. Conclusion of pressure-distribution investigation.
Oct. 26 1950	NACA flight 23, Griffith. Instrument and operational check flight in preparation for the buffeting tail loads and longitudinal stability investigation
Nov. 29 1950	NACA flight 24, A Scott Crossfield, pilot check. Beginning of buffeting, tail loads, and longitudinal stability programme.
Dec. 12 1950	NACA flight 25, Crossfield. Buffeting, tail loads, longitudinal stability investigation.
Dec. 18 1950	NACA flight 26, Crossfield. Buffeting, tail loads, dynamic longitudinal stability added to longitudinal stability programme.
Dec. 20 1950	NACA flight 27, Crossfield. Longitudinal stability programme.
Dec. 26 1950	NACA flight 28, Crossfield. Longitudinal stability programme.
Jan. 5 1951	NACA flight 29, Crossfield. Longitudinal stability programme.
Jan. 23 1951	NACA flight 30, Crossfield. Longitudinal stability programme. (Aborted due to fuel leak.)
Jan. 25 1951	NACA flight 31, Crossfield. Longitudinal stability programme
Feb. 8 1951	NACA flight 32, Crossfield. Airspeed calibration, five tower passes.
Feb. 13 1951	NACA flight 33, Walter P Jones. Pilot check, but some buffeting, tail loads, and longitudinal stability data taken.
Feb. 20 1951	NACA flight 34, Jones. Aborted after Jones suffered anoxia due to a faulty O_2 regulator.
May 2 1951	NACA flight 35, Jones. Buffeting, tail loads, longitudinal stability investigation.
June 1 1951	NACA flight 36, Crossfield. Buffeting, tail loads, longitudinal stability investigation. Mach 0.84

June 13 1951 NACA flight 37, Crossfield. Buffeting, tail loads, longitudinal stability investigation. Mach 0.86.
June 21 1951 NACA flight 38, Crossfield. Buffeting, tail loads, longitudinal stability investigation. Mach 0.835.
June 28 1951 NACA flight 39, Jones. Buffeting, tail loads, longitudinal stability investigation. 0.85.
June 29 1951 NACA flight 40, Joseph A Walker. Pilot check. Mach 0.82
July 5 1951 NACA flight 41, Walker. Buffeting, tail loads, longitudinal stability
July 17 1951 NACA flight 42, Walker. Buffeting, tail loads, longitudinal stability. (Cut short, made without tip tanks).
July 20 1951 NACA flight 43, Walker. Buffeting, tail loads, longitudinal stability.
July 26 1951 NACA flight 44, Walker. Buffeting, tail loads, longitudinal stability. Mach 0.83; cut short due to bad cloud formation.
July 30 1951 NACA flight 45, Walker. Buffeting, tail loads, longitudinal stability. Mach 0.85.
Aug. 2 1951 NACA flight 46, Walker. Buffeting, tail loads, longitudinal stability. Mach 0.84.
Aug. 7 1951 NACA flight 47, Jones. Buffeting, tail loads, longitudinal stability. Mach 0.86.
Aug. 10 1951 NACA flight 48, Walker. Flight cut short due to a fuel leak.
Aug. 20 1951 NACA flight 49, Walker. Buffeting, tail loads, longitudinal stability. Mach 0.875.
Aug. 22 1951 NACA flight 50, Walker. Flight cut short due to hydraulic line breaking.
Aug. 30 1951 NACA flight 51, Walker. Instrument malfunction. Mach 0.86.
Sept. 6 1951 NACA flight 52, Walker. Buffeting, tail loads, longitudinal stability. Mach 0.86.
Sept. 14 1951 NACA flight 53, Walker. Buffeting, tail loads, longitudinal stability. Mach 0.84.
Oct. 18 1951 NACA flight 54, Walker. Buffeting, tail loads, longitudinal stability. Beginning of lateral stability investigation. Mach 0.86.
Oct. 19 1951 NACA flight 55, Stanley P Butchart. Pilot check.
Nov. 9 1951 NACA flight 56, Butchart. Pilot check.
June 27 1952 NACA flight 57, Crossfield. Beginning of lateral stability and control (aileron effectiveness) investigation.

July 2 1952	NACA flight 58, Crossfield. Lateral stability and control. Mach 0.85.
July 17 1952	NACA flight 59, Butchart. Lateral stability and control. Also beginning of a simultaneous dynamic longitudinal stability investigation.
July 22, 1952	NACA flight 60, Butchart. Lateral stability and control. Simultaneous dynamic longitudinal stability investigation.
July 31 1952	NACA flight 61, Butchart. Lateral stability and control. Simultaneous dynamic longitudinal stability investigation.
Aug. 6 1952	NACA flight 62, Butchart. Lateral stability and control. Simultaneous dynamic longitudinal stability investigation.
Aug. 12 1952	NACA flight 63, Butchart. Lateral stability and control. Completion of lateral stability (aileron effectiveness) programme.
Jan. 29 1953	NACA flight 64, Butchart. Dynamic stability investigation.
Feb. 6 1953	NACA flight 65, Butchart. Dynamic stability investigation.
Feb. 11 1953	NACA flight 66, Butchart. Dynamic stability investigation.
Feb. 17 1953	NACA flight 67, Butchart. Dynamic stability investigation.
Feb. 20 1953	NACA flight 68, Butchart. Dynamic stability investigation. Conclusion of dynamic stability flights.
Mar. 27 1953	NACA flight 69, John B McKay. Pilot check.
Apr. 1 1953	NACA flight 70, McKay. Flight for dynamic stability fill-in data.
Apr. 2 1953	NACA flight 71, McKay. Flight for dynamic stability fill-in data.
May 7 1953	NACA flight 72, McKay. Beginning of investigation of tip tanks upon the Skystreak's buffet characteristics. Aborted due to leak in tip tank.
May 12 1953	NACA flight 73, McKay. Tip tank/buffet investigation. No records taken.
May 13 1953	NACA flight 74, McKay. Tip tank/buffet investigation.
May 20 1953	NACA flight 75, McKay. Tip tank/buffet investigation.
June 2 1953	NACA flight 76, McKay. Tip tank/buffet investigation Also low-speed stability-and-control-in coordinated-turns investigation.
June 3 1953	NACA flight 77, McKay, Tip tank/buffet investigation. Also low-speed stability-and-control-in coordinated-turns investigation.
June 10 1953	NACA flight 78, Crossfield. Tip tank/buffet investigation. Also low-speed stability-and-control-in coordinated-turns investigation. Last research flight flown by the Skystreak.

XI

D-558-2 #1 (BuNo. 37973) FLIGHTS

This aircraft completed 122 flights during its Douglas contractor programme. The first flight was on 4 February 1948 by John F Martin. After initial flight testing, and addition of its rocket engine, Douglas commenced the performance investigation programme in the aeroplane on. 25 October 1949. Douglas delivered the plane to NACA on 31 August 1951. NACA sent the plane to Douglas in 1954 for all-rocket air-launch modification and for external stores tests at supersonic speeds. The plane returned to Edwards on 15 November 1955. NACA research pilot John McKay completed a familiarisation flight in the plane on 17 September 1956, but NACA subsequently cancelled the remaining planned programme on the aeroplane.

XII

D-558-2 #2 (BuNo. 37974) FLIGHTS

A. NACA JET-POWERED FLIGHTS

May 24 1949	NACA flight 1, Robert A Champine. Pilot and instrument check, general handling qualities. Mach 0.74.
June 1 1949	NACA flight 2, Champine. Longitudinal and lateral stability and control, wing bending and twist. Mach 0.85.
June 13 1949	NACA flight 3, Champine. Longitudinal and lateral stability and control, wing and tail loads.
July 21 1949	NACA flight 4, Champine. Unsuccessful airspeed calibration due to airspeed/altitude recorder failure.
July 27 1949	NACA flight 5, Champine. Successful airspeed calibration, using tower passes.
Aug. 3 1949	NACA flight 6, Champine. Lateral control investigation.
Aug. 8 1949	NACA flight 7, Champine. Longitudinal stability and control; inadvertent pitch-up to 6g during a 4g turn at Mach 0.60.
Aug. 24 1949	NACA flight 8, Champine. Longitudinal stability and lateral control investigation during manoeuvring flight. M = 0.855.
Aug. 30 1949	NACA flight 9, Champine. Aborted after takeoff due to fluctuations in engine RPM and oil pressure.
Sept. 12 1949	NACA flight 10, Griffith. Longitudinal and lateral stability and control. Only partial completion of mission, for one JATO bottle failed to drop.

Sept. 13 1949 NACA flight 11, Griffith. Longitudinal and lateral stability and control. High engine temperatures.
Oct. 10 1949 NACA flight 12, Champine. Longitudinal and lateral stability and control, stall characteristics.
Oct. 14, 1949 NACA flight 13, Griffith. Same as flight 12.
Nov. 1 1949 NACA flight 14, Griffith. Same as flight 12. Inadvertent pitch-up and snap-roll, later pitch-up followed by stall/spin.
Nov. 21 1949 NACA flight 15, Champine. Lateral stability and control, and directional stability investigation (Aileron rolls), Mach 0.855.
Nov. 22 1949 NACA flight 16, Griffith. Same as flight 15.
Nov. 23 1949 NACA flight 17, Griffith. Same as flight 15.
Dec. 7 1949 NACA flight 18, Champine. Same as flight 15,
Dec. 30 1949 NACA flight 19, Griffith. Stall investigation with tufts.
Jan. 6 1950 NACA flight 20, Griffith. Same as flight 19.
Jan. 6 1950 NACA flight 21, Griffith. Same as flight 19.

B. DOUGLAS AIR-LAUNCH ROCKET FLIGHTS

William B 'Bill' Bridgeman, pilot.
Nov. 8 1950 D-558-2 #2, (BuNo. 37974) arrives at Edwards from Douglas via B-29 (P2B-1S) launch aircraft.
Jan. 26 1951 Douglas flight 1. Air launch at 32,000 ft., climb to 41,000 ft., level run to Mach 1.28. Dutch-roll oscillation, loss of elevator effectiveness noted.
Apr. 5 1951 Douglas flight 2. Drop at 34,000 ft., maximum Mach of 1.36 at 46,500 ft. Severe lateral oscillation forces Bridgeman to shut off engine prematurely. Rudder lock subsequently installed to control rapid rudder oscillation.
May 18 1951 Douglas flight 3. Launch at 34,000 ft., maximum Mach of 1.7 at 62,000 ft. Loss of rocket power occurred. Rudder locked at all speeds above Mach 1.
June 11 1951 Douglas flight 4. Mach 1.79 at 64,000 ft. Low lateral stability, also a lightly damped longitudinal oscillation noted after burnout.
June 23 1951 Douglas flight 5. Mach 1.85 at 63,000 ft. Violent lateral oscillation necessitates engine shutdown. Wing rolling ± 80°/sec. (1.5 radians per second.)
Aug. 7 1951 Douglas flight 6. Mach 1.88 at 66,000 ft. Dynamic lateral instability not as severe on this flight, for Bridgeman did

	not push over to as low an angle of attack as on previous flights.
Aug. 15 1951	Douglas flight 7. Altitude flight to 79,494 ft. Unofficial world altitude record. Last contractor flight.

C. NACA AIR-LAUNCH ROCKET FLIGHTS

Aug. 31 1951	Plane delivered to the NACA High-Speed Flight Research Station, designated NACA 144.
Sept. 28 1951	NACA flight 1, A Scott Crossfield. Pilot check, Mach 1.2, rough engine operation.
Oct. 12 1951	NACA flight 2, Crossfield. Stick impulses and rudder kicks, Mach 1.28.
Nov. 13 1951	NACA flight 3, Crossfield. Mach 1.11. Longitudinal and lateral stability and control, loads data and aileron effectiveness.
Nov. 16 1951	NACA flight 4, Crossfield. Same as flight 3. Maximum Mach 1.65 at 60,000 ft.
June 13 1952	NACA flight 5, Crossfield. Lateral stability and control, vertical tail loads. Mach 1.36.
June 18 1952	NACA flight 6, Crossfield. Stability and control, loads in low supersonic flight. Mach 1.05.
June 26 1952	NACA flight 7, Crossfield. Same as flight 6, Mach 1.35.
July 10 1952	NACA flight 8, Crossfield. Longitudinal stability and tail loads. Mach 1.68 at 55,000 ft.
July 15 1952	NACA flight 9, Crossfield. Longitudinal stability and tail loads. Mach 1.05, engine malfunction caused low Mach.
July 23 1952	NACA flight 10, Crossfield. High lift investigation at maximum Mach. Mach 1.53.
Aug. 13 1952	NACA flight 11, Crossfield, Aborted after launch due to lox prime valve remaining open.
Oct. 10 1952	NACA flight 12, Crossfield. Longitudinal stability at supersonic speeds to Mach 1.65. Pitch-up noted in turns.
Oct. 23 1952	NACA flight 13, Crossfield. Same as flight 12, Mach 1.10.
Mar. 26 1953	NACA flight 14, Crossfield. Same as flight 12.
Apr. 2 1953	NACA flight 15, Crossfield. Lateral stability and handling qualities investigation. Beginning of series of flights to evaluate lateral stability at various angles of attack above Mach 1.
Apr. 3 1953	NACA flight 16, Crossfield. Lateral stability investigation.
Apr. 21 1953	NACA flight 17, Crossfield. Lateral stability investigation.

June 9 1953 NACA flight 18, Crossfield. Lateral stability investigation.
June 18 1953 NACA flight 19, Crossfield. Aborted after drop; engine running rough, so was shut down.
Aug. 5 1953 NACA flight 20, Crossfield. Lateral stability investigation. Mach 1.878.
Aug. 14 1953 NACA flight 21, Lieutenant Colonel Marion Carl, USMC. Unsuccessful altitude attempt.
Aug. 18 1953 NACA flight 22, Carl. Unsuccessful altitude attempt.
Aug. 21 1953 NACA flight 23, Carl. Successful altitude flight to 83,235 ft.
Aug. 31 1953 NACA flight 24, Carl. Maximum Mach flight attempt. Mach 1.5. Violent lateral motions.
Sept. 2 1953 NACA flight 25, Carl. Maximum Mach flight attempt, to Mach 1.728 at 46,000 ft.
Sept. 17 1953 NACA flight 26, Crossfield. 1st flight with nozzle extensions. Mach 1.85 at 74,000 ft.
Sept. 25 1953 NACA flight 27, Crossfield. Lateral stability investigation. Mach 1.8 at 55,000ft. Severe lateral instability.
Oct. 7 1953 NACA flight 28, Crossfield. Lateral stability investigation.
Oct. 9 1953 NACA flight 29, Crossfield. To obtain data on the effect of rocket-nozzle extensions upon rudder hinge moments.
Oct. 14 1953 NACA flight 30, Crossfield. Lateral stability investigation. Attained Mach 1.96.
Oct. 29 1953 NACA flight 31, Crossfield. Lateral stability investigation. No. 2 chamber failed to ignite and engine shut down prematurely. Subsonic flight only.
Nov. 4 1953 NACA flight 32, Crossfield. Aerodynamic loads and longitudinal control research flight.
Nov. 6 1953 NACA flight 33, Crossfield. Lateral and longitudinal stability and control, loads research.
Nov. 20 1953 NACA flight 34, Crossfield. First Mach 2.0 flight; attains Mach 2.005 in slight dive at 62,000 ft.
Dec. 11 1953 NACA flight 35, Crossfield. Aborted due to fire warning light. Engine shut down due to frozen valve.
Dec. 23 1953 NACA flight 36, Crossfield. For rudder-hinge-moment data with rocket-nozzle extensions.
July 9 1953 NACA flight 37, Crossfield. Dynamic lateral stability investigation.
July 14 1953 NACA flight 38, Crossfield. Same as flight 37, also structural loads investigation, and wing pressure-distribution survey.

July 21 1954	NACA flight 39, Crossfield. Same as flight 38.
July 26 1954	NACA flight 40, Crossfield. Static and dynamic stability and control, loads, and pressure distribution. Mach 1.7 at 60,000 ft.
Aug. 6 1954	NACA flight 41, Crossfield. Same as flight 40.
Aug. 13 1954	NACA flight 42, Crossfield. Same as flight 40. Pitch-up encountered in turn at Mach 1.08, plane pitched to 5.8g with heavy buffeting.
Aug. 20 1954	NACA flight 43, Crossfield. Same as flight 40.
Sept. 17 1954	NACA flight 44, Crossfield. Same as flight 40.
Sept. 22 1954	NACA flight 45, Crossfield. Same as flight 40.
Oct. 4 1954	NACA flight 46, Crossfield. Dynamic lateral stability data to Mach 1.5.
Oct. 27 1954	NACA flight 47, Crossfield. Same as flight 46. Engine shut down due to pump overspeed during climb.
Mar. 18 1955	NACA flight 48, Crossfield. For pressure distribution and buffeting data at transonic speeds.
Apr. 29 1955	NACA flight 49, Joseph A. Walker. For pilot familiarisation.
May 5 1955	NACA flight 50, Lieutenant Colonel Frank K Everest, Jr., USAF. For pilot familiarisation in preparation for the X-2 programme. Mach 1.46 at 68,000 ft.
May 6 1955	NACA flight 51, Walker. For lateral stability and control data at low supersonic speeds.
May 12 1955	NACA flight 52, Crossfield. For wing and horizontal stabiliser pressure-distribution data to Mach 1.75.
May 19 1955	NACA flight 53, Crossfield. To gather lateral stability and structural loads data to Mach 1.6, but aborted when fire warning indicator came on.
June 8 1955	NACA flight 54, Crossfield. Lateral stability and aerodynamic loads data to Mach 1.67 at 60,000 ft. Subsequently, nozzle extensions removed from plane.
June 21 1955	NACA flight 55, Crossfield. Static and dynamic stability investigation to Mach 1.4. This marks end of pressure-distribution programme. Recording manometers removed from the aeroplane.
July 1 1955	NACA flight 56, Crossfield. Supersonic dynamic stability and structural loads investigation.
July 20 1955	NACA flight 57, Crossfield. Same as flight 56.
Aug. 3 1955	NACA flight 58, Crossfield. Same as flight 56.
Aug. 12 1955	NACA flight 59, Crossfield. Same as flight 56.

Aug. 24 1955	NACA flight 60, Crossfield. Same as flight 56.
Sept. 2 1955	NACA flight 61, Crossfield. Dynamic stability investigation. Beginning of vertical tail-loads research programme. One rocket chamber failed to ignite, so plane limited to Mach 1.25 at 40,000 ft.
Sept. 16 1955	NACA flight 62, John B McKay. Pilot familiarisation, but some data on stability and control and tail loads taken. McKay had to use emergency hydraulic system to lower landing gear on this flight.
Nov. 4 1955	NACA flight 63, Walker. Dynamic stability and structural loads investigation. Mach 1.34. Following this flight, nozzle extensions were again fitted to the LR-8 engine.
Nov. 10 1955	NACA flight 64, McKay. Structural heating survey.
Dec. 14 1955	NACA flight 65, McKay. Same as flight 64, Mach 1.2
Dec. 14 1955	NACA flight 66, McKay. Same as flight 64, Mach 1.25. Structural heating investigation programme cancelled after this flight.
Mar. 22 1956	NACA flight 67, McKay. Plane jettisoned in inflight emergency from P2B-1S (runaway prop on #4 engine). McKay jettisoned propellants and made safe landing on lakebed. P2B-1S required extensive repairs.
Aug. 24 1956	NACA flight 68, McKay. Vertical tail-loads investigation to Mach 1.1.
Sept. 25 1956	NACA flight 69, McKay. Same as flight 68. This marks end of vertical tail-loads research programme.
Oct. 9 1956	NACA flight 70, McKay. Static and dynamic stability investigation to approximately Mach 1.5.
Oct. 19 1956	NACA flight 71, McKay. Same as flight 70.
Nov. 1 1956	NACA flight 72, McKay. Same as flight 70.
Nov. 7 1956	NACA flight 73, McKay. Same as flight 70.
Dec. 14 1956	NACA flight 74, McKay. For dynamic stability data at Mach 1.4, and to obtain overall sound-pressure levels in aft fuselage at subsonic and supersonic speeds.
Dec. 20 1956	NACA flight 75, McKay. Same as flight 74. This was last NACA research flight on D-558-2 #2.

XIII

D-558-2 #3 (BuNo. 37975) FLIGHTS

A. DOUGLAS CONTRACTOR FLIGHTS

Fifteen Douglas flights completed before aircraft modified to air-launch configuration. Pilot Eugene F 'Gene' May.

Sept. 8 1950 Douglas flight 16; William B 'Bill' Bridgeman pilot. First airdrop. Flight aborted after launch due to airspeed system malfunction.

Sept. 20 1950 Douglas flight 17, Bridgeman. 2nd airdrop.

Sept. 29 1950 Douglas flight 18, Bridgeman. 3rd airdrop.

Oct. 6 1950 Douglas flight 19, Bridgeman. Airspeed calibration.

Nov. 17 1950 Douglas flight 20, Bridgeman. Airspeed calibration and air-launch demonstration.

Nov. 27 1950 Douglas flight 21, Bridgeman. Airspeed calibration and air-launch demonstration. turbojet engine malfunction, premature rocket shutdown.

Dec. 15 1950 Plane delivered to the NACA High-Speed Flight Research Station, designated NACA 145.

B. NACA FLIGHTS

Dec. 22 1950 NACA flight 1, A. Scott Crossfield. Pilot and instrument check, Jet engine only.

Dec. 27 1950 NACA flight 2, Crossfield. Same as flight 1.

Mar. 27 1951 NACA flight 3, Crossfield. Slat-loads investigation, jet only. Stalls, turns, rolls, to Mach 0.7.

April 20 1951 NACA flight 4, Crossfield. Dynamic longitudinal stability investigation with slats locked to Mach 0.75; elevator and stabiliser pulses.

May 17 1951 NACA flight 5, Crossfield. First NACA rocket-jet flight. Jet engine shut off due to flame instability. Mach 0.86 max.

July 17 1951 NACA flight 6, Crossfield. Jet only, rocket failed to fire due to valve failure. Mach 0.84 max.

July 20 1951 NACA flight 7, Walter P Jones. Pilot check, jet only. Mach 0.73.

Aug. 9 1951 NACA flight 8, Crossfield. Rolls and accelerated turns to Mach 1.14. Jet and rocket.

Aug. 14 1951 NACA flight 9, Brigadier General Albert Boyd USAF. Pilot check. Jet and rocket. Mach 1.05.

Aug. 22 1951 NACA flight l0, Jones. Jet and rocket, lateral and longitudinal stability investigation. Aileron rolls, elevator pulses to Mach 1.l0.

Sept. 18 1951 NACA flight 11, Jones. Jet only, rocket failure. Longitudinal stability investigation with an accelerated pitching manoeuvre in landing configuration. Pitch-up followed by spin and normal recovery.

Sept. 26 1951 NACA flight 12, Jones. Lateral control investigation. Jet and rocket flight to Mach 0.96. Rolls, sideslips, elevator pulses, accelerated turns.

Oct. 18 1951 NACA flight 13, Jones. Beginning of pitch-up investigation. Evaluation of outboard wing fences at Mach 0.7. Fences markedly aid recovery.

Nov. 9 1951 NACA flight 14, Jones, Same as flight 13. Mach 0.95. Fences subsequently removed.

June 19 1952 NACA flight 15, Crossfield. Jet only. Pitch-up investigation with slats locked open. Mach 0.7.

July 3 1952 NACA flight 16, Jones. Same as flight 15. Mach 0.96.

July 31 1952 NACA flight 17, Crossfield. Jet and rocket. Slat investigation, aborted in climb due to faulty cabin heating. Some low-speed data.

Aug. 8 1952 NACA flight 18, Crossfield. Jet and rocket. Same as flight 15. Mach 0.96. Inboard wing fences subsequently removed. Plane now in clean, no-fence configuration.

Aug. 14 1952 NACA flight 19, Crossfield. Slats still locked open. Flight to check effect of removing wing fences. Removal indicates inboard fences had little effect upon aeroplane behaviour. Following flight, slats moved and locked in 1/2-open position.

Oct. 8 1952 NACA flight 20, Crossfield. Jet and rocket. Evaluation of effect of slats 1/2 open upon pitch-up. Plane pitched to 36°. Mach 0.97. Slats subsequently restored to free-floating condition.

Oct. 22 1952 NACA flight 21, Crossfield. Jet and rocket. Plane in basic no-fence configuration. Longitudinal and lateral stability and control investigation. Pitch-ups encountered during turns. Chord extensions subsequently installed on outer wing panels.

Feb. 27 1953 NACA flight 22, Crossfield. Jet only. First flight with chord extensions. Mach 0.7. Wind-up turns and 1g stalls. Manoeuvres terminated when decay in longitudinal or lateral stability became apparent.

Apr. 8 1953 NACA flight 23, Crossfield. Jet only, rocket failed to fire due to frozen valve. Wind-up turns, aileron rolls, sideslips, 1g stalls.

Apr. 10 1953 NACA flight 24, Crossfield. Jet and rocket. Mach 1.03. Same as flight 23. Pitch-up not alleviated by chord extensions, so extensions removed after flight and slats reinstalled on wings.

June 15 1953 NACA flight 25, Crossfield. Jet and rocket. Slats locked open. Accelerated longitudinal stability manoeuvres performed with control bungee installed. Decay in stability noticed at all speeds except at Mach 1. Stiff bungee subsequently installed.

June 25 1953 NACA flight 26, Crossfield. Jet only. Slats locked open, stiff bungee. Aeroplane appeared quite controllable at high angles of attack; stability decay less objectionable.

June 26 1953 NACA flight 27, Stanley P Butchart. Pilot checkout. Slats locked open and stiff bungee installed. Jet only.

July 24 1953 NACA flight 28, Crossfield. Jet and rocket. Plane in basic configuration. Transonic lateral and directional stability and control. Mach 1.05.

July 28 1953 NACA flight 29, Lieutenant Colonel Marion Carl USMC. Pilot check out in D-558-2 #3 before flying all rocket D-558-2 #2. Jet power only.

July 30 1953 NACA flight 30, Carl. Jet and rocket. Same as flight 29.

Sept. 9 1953 NACA flight 31, Crossfield. Longitudinal, lateral, and directional stability investigation, from Mach 0.4 to Mach 1.08.

Sept. 14 1953 NACA flight 32, Crossfield. Same as flight 31.

Sept. 22 1953 NACA flight 33, Crossfield. Same as flight 31. Due to malfunction, only two rocket chambers fired.

Dec. 10 1953 NACA flight 34, Crossfield. Transonic longitudinal stability investigation. Turns, stalls.

Dec. 22 1953 NACA flight 35, Crossfield. Same as flight 34. Jet only, rocket did not ignite. Plane subsequently modified for external-stores programme.

May 7 1954 NACA flight 36, Joseph Walker. Pilot checkout, plane in basic configuration. Jet only.

May 12 1954	NACA flight 37, Walker. Same as flight 36. Jet and rocket. Mach 0.97.
June 2 1954	NACA flight 38, Crossfield. First flight with external-stores pylons. Jet only. Evaluation of handling qualities to Mach 0.72.
June 16 1954	NACA flight 39, Crossfield. First flight with external stores (1,000-lb. bomb shapes). Jet only. No apparent adverse effects. Mach 0.72.
July 8 1954	NACA flight 40, Crossfield. Jet and rocket. Stores cause decrease in transonic performance and increase in buffet. Mach 1.0. Stores shapes later removed as being too small.
July 19 1954	NACA flight 41, Crossfield. Jet and rocket. Plane in clean configuration. Transonic directional and longitudinal stability and control. Mach 1.05. Sideslips, elevator and rudder pulses.
July 23 1954	NACA flight 42, Crossfield. Jet and rocket. Transonic lateral stability and control investigation. Rolls from Mach 0.5 to 1.05.
July 28 1954	NACA flight 43, Crossfield. Jet and rocket. Same as flight 42. Mach 1.1.
Aug. 9 1954	NACA flight 44, Crossfield. Jet and rocket. Dynamic stability investigation from Mach 0.5 to 1.05. Elevator, aileron, and rudder pulses.
Aug. 11 1954	NACA flight 45, Crossfield. Jet and rocket. Same as flight 44.
Aug. 18 1954	NACA flight 46, Crossfield. Jet and rocket. Same as flight 44.
Aug. 30 1954	NACA flight 47, Crossfield. Jet and rocket. Slats unlocked, flight for longitudinal stability and control and buffet characteristics of the aircraft in this configuration.
Oct. 8 1954	NACA flight 48, Crossfield. Resumption of stores investigation programme. Handling qualities with 150-gallon tanks. Jet only. Mach 0.74.
Oct. 21 1954	NACA flight 49, Crossfield. Jet and rocket. Same as flight 48. No adverse effects, but pilot noted drag rise and heavier buffet in longitudinal manoeuvres. As a result of strain-gauge-loads measurements, stores programme again temporarily suspended while Douglas checks strength factor of pylon and wing.
Dec. 23 1954	NACA flight 50, Lieutenant Colonel Frank K Everest, Jr, USAF. Pilot checkout, Jet-and-rocket flight in clean

configuration in preparation for the Bell X-2 programme.

Dec. 28 1954 NACA flight 51, John B McKay. Pilot check in clean configuration, jet only.

April 27 1955 NACA flight 52, McKay. Jet and rocket. Underwing pylons installed. Sideslips, rolls, elevator and rudder pulses. For handling qualities, wing and pylon loads, and buffet data. Mach 1.0.

May 23 1955 NACA flight 53, McKay. Jet and rocket. 150-gal. stores attached. Same manoeuvres as flight 52. Buffet levels higher with stores than with pylons only.

June 3 1955 NACA flight 54, McKay. Jet and rocket. Pylons only. Same manoeuvres as flight 52. Mach 1.0.

June l0 1955 NACA flight 55, McKay. Jet and rocket. Same as flight 54.

June 17 1955 NACA flight 56, McKay. Jet and rocket. 150-gal. stores attached. Same manoeuvres as flight 52.

June 24 1955 NACA flight 57, McKay. Jet and rocket. Same as flight 56.

June 28 1955 NACA flight 58, McKay. Jet and rocket. Same as flight 56.

Aug. 30 1955 NACA flight 59, McKay. Jet and rocket. Same as flight 56. Plane damaged on landing when tail cone touched lake first.

Nov. 2 1955 NACA flight 60, Butchart. Jet and rocket. Same as flight 56.

Nov. 8 1955 NACA flight 61, McKay. Jet and rocket. Same as flight 56.

Nov. 17 1955 NACA flight 62, McKay. Jet and rocket. Same as flight 56.

Dec. 8 1955 NACA flight 63, McKay. Jet and rocket. Same as flight 56. This concludes the stores-investigation programme. Plane returned to clean configuration.

Feb. 1 1956 NACA flight 64, McKay. Jet and rocket. To obtain wing-loads data for comparison with external stores data previously acquired. Lateral, directional, and longitudinal manoeuvres. Mach 1.0.

Feb. 3 1956 NACA flight 65, McKay. Jet and rocket. Rocket engine pump overspeed prevents acquisition of data at Mach 0.9. Flight for same purpose as flight 64, so one more flight scheduled to complete the research programme.

Aug. 28 1956 NACA flight 66, McKay. Jet and rocket. Same as flight 64, Mach 0.96. This completes research programme on this aircraft.

Bibliographical Note

The basic research for this study was conducted at the Historical Office, National Aeronautics and Space Administration Headquarters, Washington, DC The bulk of the progress reports, pilots' notes, test-reports, memorandums and letters come from retired NACA records. Where applicable, the author made use of NACA *Technical Reports* and *Technical Memorandums* prepared on the various research aircraft, which are on file in the NASA Headquarters Library, and also at the Library of the NASA Dryden Flight Research Center, Edwards AFB, California. The single greatest source of documentation used in this work was a group of retired records located at the Federal Records Center at Bell, California, which are now located at the Federal Records Center, Laguna Niguel, California. These records are part of the National Archives and Records Service's Record Group 255, NACA-NASA records, and consist of:

NASA/Flight Research Center (FRC) Box 310b, X-1 reports.
NASA/FRC Box 312a, X-1 reports (1955–59).
NASA/FRC Box 361, 1949–57 D-558 series airplanes.
NASA/FRC Box 362, D-558 series airplanes.
NASA/FRC Box 366, D-558 series airplanes.

Some additional X-1 material is interspersed in memorandums and reports found in NASA/FRC Box 321, X-2 reports (1949–56). The material in these containers is not limited to NACA; it also covers participation by contractors and military services. Further, in 1986 I donated copies of many of these X-1 reports to the archives of the Air Force Museum, Wright-

Patterson AFB, Ohio. At the same time, I donated copies of key D-558 programme documents to the archives of the San Diego Aerospace Museum, Balboa Park, San Diego, California. Chuck Yeager's X-1 pilot reports (including the 14 October 1947 flight) are on file at the History Office, Air Force Flight Test Center, Edwards AFB, California.

Supplementary data came from the Record Group 255 holdings in the National Archives, Washington, DC. Particularly helpful were the papers of Walter T Bonney and the extensive NACA photographic collection. Useful information also came from the aircraft and personnel files maintained at the National Air and Space Museum of the Smithsonian Institution. Valuable material on the workings of the Research Airplane Projects Panel and the NACA High-Speed Flight Station came from the files of the NASA Langley Research Center. The author obtained additional information on the D-558 programme from the Naval Air Systems Command and through individuals at the McDonnell-Douglas Corporation. The Office of Air Force History, Washington, DC (now the Air Force History Support Office, Bolling AFB, Washington, DC) and the Air Force Systems Command History Office, Andrews AFB, Maryland (now the Air Force Material Command History Office, Wright-Patterson AFB, Ohio) supplied the Edwards AFB semiannual histories from 1946–58. Some material concerning early Army Air Forces supersonic research came from the files of the Air Force Museum, Wright-Patterson AFB, Ohio.

The assistance of participants in the research aircraft programme was vital to this work. Their interviews and correspondence provided little-known background information and personal insights unavailable in written records. They rounded out the work and added new dimensions to the subject. Those who were interviewed or consulted are listed in the Acknowledgements.

Contemporary sources furnished data on individuals, national developments and aircraft. Perhaps more importantly, however, they serve to indicate the framework in which the emergence of the research-aircraft programme occurred. The best sources in this vein are the aviation trade journals; and newspapers and magazines furnish valuable commentary.

The research aircraft programme is still a fruitful subject for historical research, though much has been done since the original publication of this book. The following are works that have special value:

Beamont, Roland, *Testing Early Jets: Compressibility and the Supersonic Era*. Airlife, Shrewsbury, England (1990)

Becker, John V, *The High-Speed Frontier: Case Studies of Four NACA Programs*, 1920–1950. NASA, Washington (1980)

Bridgeman, William & Hazard, Jacqueline, *The Lonely Sky*. Henry Holt, New York (1955)

Burnet, Charles, *Three Centuries to Concorde*. Mechanical Engineering Press, London (1979)

Chilstrom, Ken & Leary, Penn, *Test Flying at Old Wright Field*. Westchester Publishers,Omaha, Nebraska (1991)

Crossfield, A Scott, *Always Another Dawn: The Story of a Rocket Test Pilot*. World Publishing Co., Cleveland (1960)

Emme, Eugene M, *Aeronautics and Astronautics: An American Chronology of Science and Technology in the Exploration of Space 1915–1960*. NASA, Washington (1961)

Everest, Frank K, *The Fastest Man Alive*. Dutton, New York (1959)

Gray, George W, *Frontiers of Flight: The Story of NACA Research*. Knopf, New York (1948)

Hansen, James R, *Engineer in Charge: A History of the Langley Aeronautical Laboratory, 1917–1958*. NASA, Washington (1987)

Hartman, Edwin P, *Adventures in Research: A History of Ames Research Center 1940–1965*. NASA, Washington (1970)

Heinemann, Edward H & Rausa, Rosario, *Ed Heinemann: Combat Aircraft Designer*. Naval Institute Press, Annapolis (1980)

Hoover, R A 'Bob' with Shaw, Mark, *Forever Flying*. Orion, New York (1996)

Horwitch, Mel, *Clipped Wings: The American SST Conflict*. MIT Press, Cambridge, MA (1982)

Johnston, A M 'Tex' & Barton, Charles, *Tex Johnston: Jet-Age Test Pilot*. Smithsonian Institute Press, Washington (1991)

Lopez, Donald S, *Fighter Pilots' Heaven*. Smithsonian Institution Press, Washington (1995)

Lundgren, William R, *Across the High Frontier: The Story of a Test Pilot–Major Charles E Yeager*. USAF. William Morrow, New York (1955)

Matthews, Henry, D.H. 108: *The Saga of the First British Supersonic Aircraft*. HSM Publications, Beruit (1996)

Matthews, Henry, *Samolyot 346: The Untold Story of the Most Secret Postwar Soviet X-Plane*. HSM Publications, Beruit (1996)

Middleton, Donald, *Tests of Character: Epic Flights by Legendary Test Pilots*. Airlife, Shrewsbury, England (1995)

Miller, Jay, *The X-Planes: X-1 to X-31*. Aerofax, Arlington, TX: (1988)

Owen, Kenneth, *Concorde: New Shape in the Sky*. Jane's, London (1982)

Rotundo, Louis, *Into the Unknown: The X-1 Story*. Smithsonian Institution Press, Washington (1994)

Saltzman, Edwin J & Ayers, Theodore G, *Selected Examples of NACA/NASA Supersonic Flight Research*, SP-513. Edwards AFB, California (1995)

Thompson, Milton O, *At the Edge of Space: The X-15 Flight Program.* Smithsonian Institution Press, Washington (1992)

Winter, Frank H, '"*Black Betsy*" The 6000C4 Rocket Engine, 1945–1989', a paper presented at the 23rd Symposium on the History of Astronautics, 40th International Astronautical Congress of the International Astronautical Federation, Malaga, Spain, October 1989

Yeager, Chuck & Janos, Leo, *Yeager: An Autobiography*. Bantam, New York (1985)

Young, James O, ed., *The Men of Mach One: a Supersonic Symposium.* Edwards AFB, California (1990).

Notes

NOTES TO CHAPTER 1

1. Caldwell, F W & Fales, E N, *Wind Tunnel Studies in Aerodynamic Phenomena at High Speeds,* NACA Technical Report No. 83 (1920), p. 77; Parrish, Wayne W, ed., *Who's Who in World Aviation, 1955.* Washington, DC (1955), pp. 49, 99–100.
2. NACA Technical Report No. 83.
3. Briggs, L J, Hull, G F & Dryden, H L, *Aerodynamic Characteristics of Airfoils at High Speeds*, NACA Technical Report No. 207 (1924), p. 465; see also Dryden, Hugh L, 'Supersonic Travel Within the Last Two Hundred Years', *The Scientific Monthly*, LXXVIII, No. 5, (May 1954), pp. 289–95.
4. Briggs & Dryden, *Pressure Distribution Over Airfoils at High Speeds,* NACA Technical Report No. 255 (1926), pp. 581–82; Briggs & Dryden, *Aerodynamic Characteristics of Twenty-Four* Airfoils *at High Speeds,* NACA Technical Report No. 319 (1929), p. 346; Briggs & Dryden, *Aerodynamic Characteristics of Circular-Arc Airfoils at High Speeds,* NACA Technical Report No. 365 (1931).
5. 'How Fast Can We Fly?', *The Sunday Star*, 11 September 1932.
6. Stack, John, 'Effects of Compressibility on High Speed Flight', *The Journal of the Aeronautical Sciences,* Vol. I, January 1934, pp. 40–43; Hilton quote from Hansen, James R, *Engineer in Charge: A History of the Langley Aeronautical Laboratory, 1917–1958*, NASA SP-4305, a volume in *The NASA History Series*. Washington, DC (1987), p. 253. See also Bonney, Walter T, 'High-Speed Research Airplanes', *Scientific American*, Vol. 189, No. 4, (October 1953), pp. 36–41.
7. Dryden, 'Supersonic Travel Within the Last Two Hundred Years', p. 293. For a survey of the sweptwing story, see Hallion, Richard P, 'Lippisch, Gluhareff, and Jones: The Emergence of the Delta Planform and the Origins of the Sweptwing in the United States', *Aerospace Historian*, XXVI, No. 1, (March 1979), pp. 1–10. See also Theodore von Karman with Lee Edson, *The Wind and Beyond: Theodore von Karman, Pioneer in Aviation and Pathfinder in Space*. Boston (1967), pp. 219–25; Theodore von Karman, *Aerodynamics: Selected Topics in Light of their Historical Development*. New York (1963), pp. 133–34, and Jones' memoir 'Recollections from

an Earlier Period in American Aeronautics', *American Review of Fluid Mechanics* (1977), pp. 1–11. The influence appeared notably in the design of the North American F-86 and Boeing B-47.

8. Letter, Ezra Kotcher to author, 23 January 1972.
9. Air Corps Materiel Division Engineering Section Memorandum Report 50-461-351 (18 August 1939), pp. 5, 6, 11, 14; Content transmitted in letter, Ezra Kotcher to author, 23 January 1972; see also Schlaifer, Robert and Heron, S D *Development of Aircraft Engines and Fuels*. Boston (1950), pp. 378–79, 457–58. Despite its age, the Schlaifer and Heron study is still the most complete and useful history of aero engine development, particularly for the early years of jet engine research. See also Karman and Edson, *The Wind and Beyond,* p. 225; Robert L. Perry, 'The Ancestors of the X-l', p. 10 (Rough draft, June 1965, in NASA Historical Archives files, NASA HQ, Washington, DC).
10. The choking phenomenon was not alleviated until the introduction of the 'slotted throat' tunnel after the Second World War. For an excellent case study on the evolution of the slotted throat tunnel, see Becker, John V, *The High-Speed Frontier: Case Studies of Four NACA Programs, 1920–1950*, NASA SP-445. Washington DC (1980).
11. Interview of John Stack by the author, 19 May 1971; Dryden, 'Supersonic Travel Within the Last Two Hundred Years,' p. 294; Gray, George W, *Frontiers of Flight: The Story of NACA Research*. New York (1948), p. 213.
12. Jodlbauer story from Beamont, Roland, *Testing Early Jets: Compressibility and the Supersonic Era*. Shrewsbury, England (1990), pp. 5–6; see also p. 33. Caiden, Martin, *Fork Tailed Devil: The P-38*. New York (1971), p. 213.
13. Ethell, Jeff, *Lockheed P-38*, New York (1983), has a good discussion of the Virden accident and the P-38's problems. See also Weber, Le Roy, *The Lockheed P-38J-M Lightning*, Profile Publication No. 106. Surrey, England (1966), pp. 4–5. *Nosey* was serial 41-2048; the upswept boom Lightning was 42-1986; LeVier, Tony & Guenther, John, *Pilot*. New York, 1954), p. 133.
14. Hartman, Edwin P, *Adventures in Research*. Washington (1970), *passim*; LeVier, *Pilot*, pp. 142–148.
15. Schlaifer & Heron, *Development of Aircraft Engines and Fuels,* pp. 457–58, 461–62. The subsequent story is well-told in Carpenter, David M, *Flame Powered: The Bell XP-59A Airacomet and the General Electric I-A Engine* (a 50th anniversary commemorative publication sponsored by the Jet Pioneers of America, 1992); see also Neal, Ronald D, 'The Bell XP-59A Airacomet: The United States' First Jet Aircraft', *Journal of the American Aviation Historical Society,* Vol. XI, No. 3, (Fall 1966), pp. 155–78.
16. Minutes of Meeting of Special Committee on Jet Propulsion, 22 April 1941, p. 5, NASA Historical Archives; Stack interview, 19 May 1971; Interview of Harold Turner by the author, 11 November 1971.
17. Statement of John Stack at AIAA History Session, San Francisco, CA, 28 July 1965, NASA Historical Archives.; see also interview of Clinton E Brown by the author, 12 May 1971 (from which the 'barrelled' quote is drawn); Ellis, Macon C, Jr & Brown, Clinton E, *NACA Investigation of a Jet-Propulsion System Applicable to Flight,* NACA Technical Report No. 802 (1943), pp. 498–501. There is a good discussion of the Campini project in Hansen, *Engineer in Charge*, pp. 238–245; see also Schlaifer & Heron, *Development of Aircraft Engines and Fuels,* pp. 450–51, 468.
18. Interview of Milton B Ames, Jr., by the author, July 1971; Farren, W S 'Research for Aeronautics – Its Planning and Application', *Journal of the Aeronautical Sciences,*

Vol. XI, No. 2 (April 1944), pp. 95–109; 'Beginnings of the X-l', n.d., in NASA Historical Archives. For the origins of the Miles M.52 and its subsequent history, see 'The Miles Supersonic', *The Aeroplane*, LXXI, No. 1842, (13 September 1946), pp. 295–96; Charles Burnet, *Three Centuries to Concorde*. London (1979), pp. 43–52; and Wood, Derek, *Project Cancelled: British Aircraft that Never Flew*. Indianapolis (1975), pp. 28–32.

19. Letter, Robert A Wolf to author, 4 April 1972.
20. Quoted in 'Beginnings of the X-I', p. 3, as is Lewis' response.
21. Surprisingly, the DFS 346 effort is still not as well-known as it should be. See Smith, J R & Kay, Antony L, *German Aircraft of the Second World War*. Baltimore, MD (1989 ed.), pp. 636–638; and Matthews, Henry, *Samolyot 346: The Untold Story of the Most Secret Postwar Soviet X-Plane*. Beirut, Lebanon (1996). The Nazi quote is from the latter, which is a most impressive work that breaks important new ground in understanding postwar Soviet research activities.
22. This and the subsequent discussion is drawn from 'Beginnings of the X-1'; data transmitted from Ezra Kotcher to author, 2–3 January 1972; 'MX-324-Step to Manned Rocket Flight', *Aeronautics and Astronautics*, II (October 1964); Statement of Ezra Kotcher at AIAA History Session, San Francisco, CA, 28 July 1965, NASA Historical Archives; Letter, Ezra Kotcher to author, 24 July 1971; also, letter Kotcher to William Lundgren, 4 November 1953. Northrop redesigned the XP-79 as the jet-powered XP-79B, with no better luck; it went out of control and crashed on its maiden flight on 12 September 1945, killing company test pilot Harry Crosby.
23. The 'Mach 0.999' discussion is drawn from Letters, Ezra Kotcher to author, 10 September 1971 and 23 January 1972. Also, letter, Kotcher to Lundgren, 4 November 1953, and Kotcher statement at AIAA History Session, 28 July 1965. Details of the GALCIT study project are drawn from von Karman statement in his *Where We Stand*, reprinted in Gorn, Michael H, ed., *Prophecy Fulfilled: 'Toward New Horizons' and Its Legacy*. Washington, DC (1994), p.19. There is some confusion in von Karman's mind about when this request occurred. In *Where We Stand* (written in 1945), he states it was early in 1944. In his later *Wind and Beyond* (1967), pp. 233–234, he states 1943. I have accepted the 1944 figure because it was recollected closer to the event, and also because it ties more closely and logically with other events and activities surrounding AAF interest in supersonic flight than an earlier 1943 date. Information on or related to the Ellis-Brown study can be found in Becker John V, *The High-Speed Frontier,* pp. 93–95, 98, and in Hansen, *Engineer in Charge*, pp. 294–296; and in the interview of Ellis, Macon C, Jr, by the author, 12 November 1971; Brown interview, 12 May 1971; Stack interview, 19 May 1971; Interview with Floyd L. Thompson by author , 31 May 1972; Kantrowitz, Arthur & Donaldson, Coleman duP, 'Preliminary Investigation of Supersonic Diffusers', NACA ACR L5D20 (May 1945); Ellis, Macon C, Jr, and Brown, Clinton E, 'Analysis of Supersonic Ram-jet Performance and Wind-Tunnel Tests of a Possible Supersonic Ram-jet Airplane Model', NACA ACR L5L12 (December 1945); Ellis & Brown, 'Proposal of Supersonic Ram-jet Missiles', NACA Memorandum Report, 30 January 1945; Stack statement at AIAA History Session, 28 July 1965.
24. Thompson, F L, 'Flight Research at Transonic and Supersonic Speeds With Free-Falling and Rocket Propelled Models', Paper presented at the Institute of Aeronautical Sciences – Royal Aeronautical Society meeting, May 1949, pp. 1–4, NASA Historical Archives; Letter, Robert R. Gilruth to author, 27 January 1972; interview of Robert R.

Gilruth by Michael D. Keller, 26 June 1967, NASA Historical Archives; Gray, *Frontiers of Flight,* p. 336; interview of John V. Becker by the author, 12 November 1971; Thompson interview, 31 May 1972.

25. Martindale text is from his flight report, 'Compressibility Research: Final Dive on Spitfire E.N. 409,' Ref. CTP/K/1/73, 28 April 1944. I wish to acknowledge with grateful appreciation the assistance of the late Air Commodore Allen H Wheeler, RAF (ret.), the former Officer Commanding Experimental Flying at the RAE, for making this report available to me, and for putting me in touch with Sir Morien Morgan, Downing College, Cambridge; Professor W A Mair, University Engineering Department, Cambridge, and Mr R P Probert, the Director of the Royal Aircraft Establishment, all of whom kindly furnished much useful material on the Spitfire trials, particularly Aeronautical Research Council, *Reports and Memoranda 2222* (edited by W A Mair), 'Research on High Speed Aerodynamics at the Royal Aircraft Establishment From 1942-1945'. HMSO, London (1950). The Martindale claim is still hotly disputed by many who argue that altimetre lag likely led to overestimation of Mach number, but, as Mair himself noted in a letter to me on 14 February 1975, 'In some of the early dives, in 1941 (and perhaps 1942), for which I was responsible, this effect was not properly understood and there were errors in measurement of the Mach numbers. By the time the dives with the Spitfire were made the lag effects were reasonable well understood and corrections were introduced, so that the recorded Mach numbers were believed to be correct'. There is more detail on these trials in the previously cited Burnet, *Three Centuries to Concorde*, pp. 29–40.
26. *Air Force Supersonic Research Airplane XS-1 Report No. 1*, 9 January 1948, p. 5, NASA Historical Archives; Ames interview, 26 July 1971, and letter, Walter S. Diehl to author, 6 January 1972. Louis Rotundo, *Into the Unknown: The XS-1 Story.* Washington, DC (1994), pp. 11–13, has a good summary of these two meetings; Letter, Ezra Kotcher to author, 7 November 1971; Shortal, Joseph A, *History of Wallops Station: Origins and Activities Through 1949* (comment edition, n.d.), pp. 11–12, NASA Historical Archives.
27. Letter, Walter S Diehl to author, 6 January 1972; Stack interview, 19 May 1971; Ames interview, 26 July 1971.
28. Thompson interview, 31 May 1972; Ames interview, 26 July 1971; Becker interview, 12 Nov. 1971; Air Materiel Command Correspondence Summary of Project MX-653 History, 14 January 1947, p. 1 (Transmitted to author by Ezra Kotcher); letters, Ezra Kotcher to author, 23 January 1979, and 22 February 1972; 'Beginnings of the X-l', p. 3; and Lundgren, William R, *Across the High Frontier: The Story of a Test Pilot – Major Charles E. Yeager, USAF*. New York (1955), p. 31.
29. Data from Kotcher Biographical File, National Air and Space Museum, Smithsonian Institution (hereafter cited as NASM). See also Kotcher biographical file, Aeronautical Systems Center Historical Office, Air Force Materiel Command, Wright-Patterson AFB, OH; letter, Kotcher to author, 7 November 1971, and Sturm, Thomas A, *The USAF Scientific Advisory Board: Its First Twenty Years, 1944-1964*. Washington, DC (1967), p. 4; Letter, Kotcher to Lundgren, 4 November 1953. The commando anecdote is from a conversation with Kotcher's brother Harry Kotcher, on 4 October 1996.
30. Hyatt, Abraham, *Proposed High Speed Research Airplane*, Memorandum Aer-E-225-AH, 22 September 1944. Transmitted to author from A M 0 Smith, 17 April 1972. See also Parrish, *Who's Who in World Aviation, 1955,* pp. 154–55; and letter, A M O Smith to author, 31 March 1972.

NOTES TO CHAPTER 2

1. Bell biographical file, NASM; *Aeronautical Engineering Review*, XVI, No. 1, (January 1957), pp. 21–22; Works Projects Administration, *Who's Who in Aviation, 1942–1943*. New York (1942), p. 475; Stack interview, 19 May 1971.
2. Letter, Robert A. Wolf to author, 4 April 1972; Kotcher letter to Lundgren, 4 November 1953.
3. Letter, Ezra Kotcher to author, 22 February 1972; Kotcher statement at AIAA History Session, 28 July 1965; Lundgren, *Across the High Frontier,* p. 33.
4. Kotcher statement at AIAA History Session, 28 July 1965; letter, Kotcher to Lundgren, 4 November 1953; Letter and recording, Benson Hamlin to author, 19 April 1972; Winfield Scott Downs, *Who's Who in Engineering, 1948*. (New York, 1948), pp. 820, 1851–52, 1887; *Who's Who in Aviation, 1942–1943,* p. 408; Parrish, ed., *Who's Who in World Aviation, 1955*, p. 271.
5. Emme, Eugene M, *Aeronautics and Astronautics: An American Chronology of Science and Technology in the Exploration of Space, 1915–1960*. Washington, DC (1961), p. 49; NACA LMAL, 'Estimate of Instrument Requirements for Experimental Aeroplane', 27 December 1944, in Langley Research Center files (Hereafter cited LARC files); Rotundo, *Into the Unknown*, pp. 35–36; Hanson, *Engineer in Charge*, pp. 272–73. In the XP-79, the pilot flew the plane from the prone position. The AAF Aeromedical Laboratory at Wright Field concluded that the prone position would enable the pilot of the XP-79 to best withstand the expected 12g accelerations he would experience. Apparently the AAF, NACA, and Bell independently arrived at the 18g figure. Lieutenant Colonel Carl Reichert of the Aircraft Laboratory derived the figure for the AAF. Stack suggested it to the NACA. At Bell, the same figure was arrived at 'Pseudo-scientifically', according to former Bell personnel. See Bonney, Walter T, 'The Research Aeroplane', *Pegasus*, XVIII, No. 6, (June 1952), pp. 1–16; Lundgren, *Across the High Frontier*, pp. 34–35; Stack interview, 19 May 1971; Letter and recording, Benson Hamlin to author, 19 April 1972; Letter, Robert M Stanley to author, 1 March 1972; Letter, Paul Emmons to author, 31 May 1972.
6. Lundgren, *Across the High Frontier,* p. 33. It is unclear whether MX-524 referred only to the proposed Bell research aircraft or to high-speed flight in general. Ezra Kotcher attended a conference in June 1945 under MX-524 to discuss a Lockheed P-80 test programme. Yet the development project designation for the P-80 was MX-409. Additionally, by this time (June 1945) the XS-1 was being developed under project designation MX-653. Letter, Ezra Kotcher to author, 7 November 1971; Letter and recording, Benson Hamlin to author, 19 April 1972.
7. Quoted in 'Beginnings of the X-l', p. 6.
8. Letter and recording, Benson Hamlin to author, 19 April 1972. R M Stanley and R J Sandstrom, 'Development of the XS-1 Aeroplane', in *Air Force Supersonic Research Aeroplane XS-1 Report No. 1,* 9 January 1948, pp. 7–8. An earlier version of the Stanley and Sandstrom article appeared in the Institute for the Aeronautical Sciences journal; see Stanley and Sandstrom, 'Development of the XS-1 Supersonic Research Airplane', *Aeronautical Engineering Review*, VI, No. 8, (August 1947), pp. 22–26, 72. It is not as complete (presumably for security reasons) as the later version. The two versions are distinguished here by appending the 9 January 1948 date when citing the Air Force report version.
9. Letter, Robert Gilruth to author, 27 January 1972; Letter, Robert Stanley to author, 1 March 1972; Thompson interview, 31 May 1972; Shortal, *History of Wallops*

Station: Origins and Activities Through 1949, p. 13. Shortal states that Langley completed specifications for the Bell aircraft in January 1945. Hanson has a good discussion on the thick vs. thin debate in his *Engineer in Charge*, pp. 275–279.

10. Letter and recording, Benson Hamlin to author, 19 April 1972; Letter, Kotcher to Lundgren, 4 November 1953; George F Bush, 'Early American Rockets', *Aerospace Historian*, XVI, No. 4, (Winter 1969), pp. 28–32.
11. For a thorough history of RMI and the XLR-8/11 family, see Frank H Winter, '"*Black Betsy*": The 6000C4 Rocket Engine, 1945–1989', a paper presented at the 23rd Symposium on the History of Astronautics, 40th International Astronautical Congress of the International Astronautical Federation, Malaga, Spain, October 1989. I wish to thank Mr. Winter for making this available to me. See also: Perry, 'The Ancestors of the X-1', p. 24; statement of Edward N Seymour at AIAA History Session, San Francisco, California, 28 July 1965, NASA Historical Archives; Lloyd Mallan, *Men, Rockets and Space Rats* (New York, 1958), pp. 23–25 (The LR-8 engine, however, was not being developed for the D-558-2 at the time it was selected for the XS-1; as is discussed subsequently, the D-558-2 was not conceived until August 1945, and not developed until 1946); Emme, *Aeronautics and Astronautics, p.* 43; John Shesta, 'RMI's Rocket Engine Which Powers Supersonic XS-1', *Aviation*, XLVI, No. 1, (January 1947), pp. 44–46; Letter and recording, Benson Hamlin to author, 19 April 1972. Also Lundgren, *Across the High* Frontier, pp. 69, 159–60; Stanley and Sandstrom, 'Development of the XS-1 Airplane', 9 January 1948, p. 10.
12. Air Materiel Command, Correspondence Summary of Project MX-653 History, 14 January 1947, p. 1. Transmitted to author by Ezra Kotcher; Date from *Air Force Supersonic Research Aeroplane XS-1 Report No. 1,* p. 5; XS-1 cost figures are from Rotundo, *Into the Unknown*, p. 26. In February 1945, Jones discussed the sweep theory with Jean Roche of the AAF ATSC liaison office at Langley, when there was still time to influence the XS-1's design.
13. Letter, Robert T Jones to Ernest O Pearson, Jr, 2 February 1960; Navy Department Record of Invention, l0 April 1946; Jones, Robert T, 'The Shaping of Wings to Minimize the Formation of Shock Waves', 27 February 1945; Robert T Jones, Memo for Chief of Research (Lewis), 5 March 1945; all in NASA Historical Archives. Letter, Ezra Kotcher to author, 7 November 1971; Kotcher statement at AIAA History Session, 28 July 1965; Letter and recording, Benson Hamlin to author, 19 April 1972; Shortal, *History of Wallops Station: Origins and Activities Through 1949,* p. 74A, has the Thompson quote; Hallion, 'Lippisch, Gluhareff, and Jones: The Emergence of the Delta Planform and the Origins of the Sweptwing in the United States', pp. 1–10. For Jones' reports, see his 'Properties of Low-Aspect Ratio Pointed Wings at Speeds Below and Above the Speed of Sound', NACA Report No. 835, 11 May 1945; and his 'Wing Planforms for High-Speed Flight', NACA Technical Note No. 1033, published in March 1946, but issued at Langley laboratory on 23 June 1945. See also Jones, R T 'Recollections from an Earlier Period in American Aeronautics', pp. 1–11; and von Karman, *Aerodynamics*, pp. 50–57, and 132–134. The von Karman Aberdeen story is from his *Where We Stand*, reprinted in Gorn, *Prophecy Fulfilled*, pp. 23, 25.
14. Stack statement at AIAA History Session, 28 July 1965; Letter and recording, Benson Hamlin to author, 19 April 1972; Stanley and Sandstrom, 'Development of the XS-1 Airplane', 9 January 1948, p. 7.
15. Kotcher statement at AIAA History Session, 28 July 1965; Bell Aircraft Corporation specification for XS-1 Aeroplane, prepared by Benson Hamlin, 1 March 1945, Report

44-947-001, p. 1, transmitted to author by Ezra Kotcher, 21 August 1971; van Lonkhuyzen, J, 'Problems Faced in Designing Famed X-1', *Aviation Week,* LIV, No. 1, (1 January 1951), pp. 22–24; Letter and recording, Benson Hamlin to author, 19 April 1972; Stanley and Sandstrom, 'Development of the XS-l Airplane', 9 January 1948, p. 14; Letter, Paul Emmons to author, 31 May 1972.

16. Letter and recording, Benson Hamlin to author, 19 April 1972; Kotcher statement at AIAA History Session, 28 July 1965; Kotcher letter to Lundgren, 4 November 1953; letter, Robert M. Stanley to author, 1 March 1972. While in Germany, Woods headed an air technical intelligence team that investigated the Messerschmitt Oberammergau research facility. He discovered the incomplete Messerschmitt P 1101 developed by Dr Woldemar Voigt, and subsequently, shipped it to the United States. The P.1101 provided the inspiration for Woods' later variable-geometry research, serving as a static test article for the Bell X-5 variable-sweep aircraft. See Perry, Robert, 'Variable Sweep: A Case History of Multiple Re-Innovation', RAND study P-3459, October 1966, pp. 3–6; Interview with John V Becker, 12 November 1971; memo, J V Becker to Langley Assistant Director for Administration, 16 May 1968, transmitted to the author November 1971.
17. Becker interview, 12 November 1971. Letter, Robert R Gilruth to author, 27 January 1972. Also, interviews with Axel T Mattson, 10 November 1971 and 6 June 1972. The first sting support systems used external balances, which were not sensitive enough for the small models. This spurred development of internal balances. Mattson ran tests using the sting support system and a special plaster throat in the 8-foot high-speed tunnel; with this combination, he demonstrated that he could reliably test to above Mach 0.9, then jump to Mach 1.2. The previously cited Becker, *High-Speed Frontier*, and Hanson, *Engineer in Charge*, have valuable accounts of wind tunnel development in this time period, and Baals, Donald D & Corliss, William R, *Wind Tunnels of NASA*, SP-440, Washington, DC (1981) is an excellent technical reference.
18. Stanley and Sandstrom, 'Development of the XS-1 Airplane', 9 January 1948, pp. 10, Figure 6, 11; letter, Robert M Stanley to author, 1 March 1972; Air Materiel Command, Correspondence Summary of Project MX-653 History, 14 January 1947, p. 2, transmitted to author by Ezra Kotcher.
19. Stanley and Sandstrom, 'Development of the XS-1 Aeroplane', 9 January 1948, pp. 4, 9, 11–14. Also interview of Richard Frost by the author, 25 May 1972.
20. Theodore von Karman, *Where We Stand: A Report Prepared for the AAF Scientific Advisory Group,* (Wright Field, Dayton, OH: HQ Air Materiel Command, May 1946). This report was published in 1946, but issued in August 1945. It is reprinted in Michael H Gorn, ed., *Prophecy Fulfilled: 'Toward New Horizons' and Its Legacy* (Washington, DC: Air Force History and Museums Program, 1994).
21. Ltr, Arnold to TvK, 7 November 1944, copy in the files of the Air Force Scientific Advisory Board, HQ USAF, Washington, DC.
22. This and subsequent quotes from *Where We Stand* are drawn from the reprinted version found in the previously cited Gorn, *Prophecy Fulfilled*, pp. 19–20, 23 and 26, as this version is more readily accessible to readers than the original archival copies.
23. *Ibid.,* p. 26.
24. Date from information supplied to NASM, found in X-1 Series #2 File, NASM. Discussion of the XS-1's technical details is drawn from: Stanley and Sandstrom, 'Development of the XS-1 Aeroplane' 9 January 1948, pp. 3, 4, 9, 11–15; Robert McLarren, 'XS-1: Design and Development', *Aviation Week,* XLIX, No. 4, (26 July

1948), 22–27; letter, Robert M Stanley to author, 1 March 1972; letter, Paul Emmons to author, 31 May 1972; letter and recording, Benson Hamlin to author, 19 April 1972; letter, Robert M Stanley to author, 1 March 1972; letter, Paul Emmons to author, 31 May 1972; Stanley and Sandstrom, 'Development of the XS-1 Airplane', pp. 7, 14–15.

25. Letter and recording, Benson Hamlin to author, 19 April 1972; letter, Paul Emmons to author, 31 May 1972. Letter, Robert M Stanley to author, 1 March 1972. Also *Air Force Supersonic Research Airplane XS-1 Report No. 1,* p. 5; completion date from information supplied to NASM, found in X-1 Series #2 File, NASM.
26. J van Lonkhuyzen, 'Problems Faced in Designing Farned X-1', p. 22.

NOTES FOR CHAPTER 3

1. Letter, Walter S Diehl to author, 6 January 1972. Also BuAer Conf. letter Aer-E-23-WSD, Aer-E-24-WHM Serial No. C35304 dated 19 December 1944, cited in letter, Director, Engineering Division to Chief, BuAer, Aer-E-11-EWC, Serial No. C-12002, dated 26 April 1945. Found in D-558 file, Historian, Naval Air Systems Command (hereafter cited as NASC). Ames interview, 26 July 1971.
2. Letter and recording, L Eugene Root to author, 18 March 1972; Parrish, *Who's Who in World Aviation, 1955*, pp. 82, 85, 139, 264. L J Devlin, 'General Summary of Project', Model D-558 Navy-Industry Conference, Douglas El Segundo plant, 10 November 1947, pp. 1–2. Frank Malina, 'Origins and First Decade of the Jet Propulsion Laboratory', in Eugene M Emme, ed., *The History of Rocket Technology: Essays on Research, Development, and Utility*, (Detroit, 1964), pp. 48–50; and Karman and Edson, *The Wind and Beyond,* pp. 239, 242, 261.
3. This discussion is based on Edward H Heinemann, 'The Development of the Navy-Douglas Model D-558 Research Project', Douglas El Segundo plant, 17 November 1947, pp. 1–2. Letter and recording, L Eugene Root to author, 18 March 1972. Letter, Edward H Heinemann to author, 10 February 1972. Letter, L J Devlin to author, 12 March 1972. Letter, A M O Smith to author, 31 March 1972. Devlin, 'General Summary of Project', p. 3. Douglas letter B300-JWR-104, 13 April 1945, Rogers to Chief, BuAer. In NASC D-558 file.
4. Douglas general arrangement three-view blueprint prepared by R G Smith, 16 January 1945. Transmitted to author by A M O Smith. Letter, Smith to author, 31 March 1972, and trip notes. There is a good discussion of Navy–NACA reservations about the first proposed aeroplane in Hansen, *Engineer-in-Charge*, pp. 290–91.
5. NACA Conference letter to Chief, BuAer 9 March 1945, cited in letter, Director, Engineering Division to Chief, BuAer, Aer-E-11-EWC, Serial No. C-12002, dated 26 April 1945. In NASC D-558 file.
6. Douglas letter B300-JWR-104, 13 April 1945, Rogers to Chief, BuAer. Letter, Director, Engineering Division to Chief, BuAer, Aer-E-11-EWC, serial No. C-12002, dated 26 April 1945, (and notation on bottom of letter). Both in NASC D-558 file.
7. Ames interview, 26 July 1971. Van Wyen, A O, D-558 *Programme Notes,* NASC D-558 file. GALCIT wind tunnel data sheet transmitted to author by A M O Smith. Letter, A M O Smith to author, 31 March 1972. Letter and recording, L Eugene Root to author, 18 March 1972. Douglas El Segundo, 'Brief Specification for the Model 558 High Speed Test Airplane', 28 March 1945. Copy in D-558 file, Soule record collection, LARC.

8. Heinemann, 'Development of the Navy-Douglas Model D-558 Research Project', pp. 7–8. Also 'Brief Spec. for the Model 558 High Speed Test Airplane'. Donovan, Robert C, 'Escape Method Developed for Douglas D-558 Skystreak', *Aviation Week,* XLVII, No. l0, (8 September 1947), 34–37.
9. Ames interview, 26 July 1971. Devlin, 'General Summary of Project', p. 3. It should be noted that the D-558 design had an ultimate load factor of 18g, just like the XS-1; the limit load factor was 9g. Later, on the D-558-2, these figures were lowered to 12g ultimate and 7.33g limit.
10. Van Every, K E 'Aerodynamics of D-558', Model D-558 Navy-Industry Conference, Douglas El Segundo plant, 10 November 1947, pp. 2–6.
11. Douglas Aircraft Company, Summary *Report U.S. Navy Transonic Research Project Douglas Model D-558,* Report E. S. 15879, 31 August 1951. pp. 61–62, 67. Hereafter cited as *D-558 Summary Report.* Also letter, A M O Smith to author, 31 March 1972. GALCIT wind tunnel data sheet transmitted to author by A M O Smith. Also letter, Robert R Gilruth to author, 27 January 1972. Van Wyen, D-558 *Programme Notes,* p. 1. At a later date, 27 January 1947, the Navy revised the contract to provide for the deletion of three D-558 turbojet aircraft and the substitution of three D-558-2 sweptwing rocket- and jet-powered aeroplanes.
12. The discussion of D-558 technical details is drawn from Heinemann, 'Development of the Navy-Douglas Model D-558 Research Project', pp. 6–17. *D-558 Summary Report,* p. 24. Also Walter C Williams, Memo for Chief of Research, 'Visit to Douglas Aircraft Corporation, El Segundo Plant', 29 October 1946. R C Donovan, 'General Description Model D-558 Airplanes', Model D-558 Navy-Industry Conference, Douglas El Segundo plant, 10 November 1947, p. 6. Also, 'How New Douglas Skystreak Will Probe the Transonic', *Aviation,* XLVI, No. 5, (May 1947), pp. 54–56. Captain L D Coates, USN, to Chief, BuAer, 'Progress Report for Period Ending, 31 October 1946', n.d. Van Every, 'Aerodynamics of D-558', pp. 7–9. Also Williams Memo to Chief of Research, 29 October 1946.
13. Muroc Army Air Field Historical Report, 1 January 1947–30 June 1947, pp. 2, 23; Air Force Archives, Maxwell AFB.
14. Letter and recording, L Eugene Root to author, 18 March 1972. Letter, A M O Smith to author, 31 March 1972. Letter, Walter S Diehl to author, 6 January 1972. Devlin, 'General Summary of Project', pp. 3–4.
15. *D-558 Summary Report*, p. 22. Edward H Heinemann, 'The Navy Douglas D-558-2 Skyrocket Douglas El Segundo, n.d., p. 2. Devlin, 'General Summary of Project', pp. 4–5.
16. Van Every, 'Aerodynamics of D-558', p. 11. GALCIT wind tunnel data sheet transmitted to author by Smith. Devlin, 'General Summary of Project', p. 5.
17. Details on the D-558-2's design choices are drawn from Letter, A Smith to author, 31 March 1972. Letter and recording, L E Root to author, 18 March 1972. Van Every, 'Aerodynamics of D-558', pp. 12, 15, 16. Heinemann, 'The Navy-Douglas D-558-2 Skyrocket', p. 4. *D-558 Summary Report*, p. 18. Donovan, 'General Description Model D-558 Airplanes', p. 7.
18. Technical details for the following text from: Douglas El Segundo, 'Model D-558 Phase 11 Mockup Conference', 18 March 1946. In D-558 file, NASM. *D-558 Summary Report,* pp. 20, 22, 24–25. Donovan, 'General Description Model D-558 Airplanes', pp. 8–9. Also Heinemann, 'The Navy-Douglas Model D-558 Research Project', pp. 5, 12–13. Captain L D Coates to Chief, BuAer. Heinemann, 'The Development of the Navy-Douglas Model D-558 Research Project', pp. 10–12.

19. *D-558 Summary Report*, p. 22; Van Wyen, *D-558 Programme Notes,* pp. 1–2; Devlin, 'General Summary of Project', p. 5. Gardner, John J, Memo for Chief of Research, 'Visit to Douglas El Segundo Plant to discuss D-558 Phase 2 rocket system', 9 July 1947. Letter, J W Crowley (NACA Associate Director of Aero. Research) to Chief, BuAer, 18 November 1947. Donovan, 'General Description Model D-558 Airplanes', p. 9.
20. Letter, Walter C Williams to Robert Donovan, 25 August 1947. Also Heinemann, 'The Development of the Navy-Douglas Model D-558 Research Project', p. 10. Muroc Army Air Field Historical Report, 1 July 1947–31 December 1947, p. 26. In Air Force Archives, Maxwell AFB.

NOTES FOR CHAPTER 4

1. Letter, H J E Reid (Director, Langley Laboratory) to NACA HQ., 29 December 1945. The actual decision to use Pinecastle had been made about a month before the Reid letter.
2. Williams biographical data sheet in NASA Historical Archives. Also Stack interview, 19 May 1971. Truszynski biographical data sheet in NASA Historical Archives. Also interview of Gerald M. Truszynski by the author, 21 May 1971. Jack Woolams, 'How We Are Preparing to Reach Supersonic Speeds', *Aviation*, XLV, No. 9, (September 1946), 38–39. Ronald D Neal, 'The Bell XP-59A Airacomet: The United States' First Jet Aircraft', pp. 171–72.
3. Stanley and Sandstrom, 'The Development of the XS-1 Airplane', 9 January 1948, p. 15. Date of captive flight from information supplied to NASM. Found in X-1 Series #2 File, NASM. Also Lundgren, *Across the High Frontier,* pp. 64–65. Louis Rotundo, *Into the Unknown: The X-1 Story* (Washington, 1994), pp. 51–69. See also chronology of rocket research aircraft flights prepared by Robert W Mulac, Langley Research Center. Hereafter cited as Mulac, first XS-1 (or second) chronology. Copy in NASA Historical Archives.
4. Mulac, first XS-1 chronology.
5. Interview of Richard Frost by the author, 25 May 1972.
6. Letter, Chalmers H Goodlin to author, 19 August 1974.
7. Woolams, 'How We Are Preparing to Reach Supersonic Speeds'. p. 39.
8. *The New York Sun*, 3 September 1946. John H Newland, 'Stardust on his Boots', in Gene Gurney, ed., *Test Pilots* (New York, 1962), p. 124. A M 'Tex' Johnston with Charles Barton, *Tex Johnston: Jet-Age Test Pilot* (Washington: 1991), pp. 89–100, Goodlin letter.
9. Goodlin letter.
10. *Air Trails Pictorial*, XXX, No. 4, (January 1948) pp. 15, 127–28. Frost interview, 25 May 1972. Chalmers Goodlin, 'Test Pilot's Bail Out', in Gurney, ed., *Test Pilots,* pp. 71–74.
11. Walter C Williams, 'A Brief History of the High-Speed Flight Station'. Williams biographical data sheet. Both in NASA Historical Archives.
12. Air Force Flight Test Center, ARDC, *Experimental Research Aircraft,* pp. 2–4. .John Ball, Jr, *Edwards: Flight Test Center of the USAF*. (New York, 1962), pp. 12–15. Gladwin Hill, Dancing *Bear: An Inside Look at California Politics* (New York, 1968), p. 16. Wesley Price, 'They Fly Our X-Ships', *The Saturday Evening Post,* CCXXIII, No. 1, (July 1, 1950), pp. 26–27, 105. H H Arnold, *Global Mission,* (New York, 1949), p. 136.

13. Ball, *Edwards*, pp. 42–43. *Experimental Research Aircraft*, p. 2. Muroc AAF Air Field Historical Report, 1 July 1946–30 September 1946, pp. 1–2. In Air Force Archives, Maxwell AFB.
14. Two excellent and reliable sources of M.52 information are the previously cited Wood, *Project Cancelled*, pp. 29–38 (including Miles' statement), and Burnet, *Three Centuries to Concorde*, pp. 62–66.
15. See Lieutenant Colonel A A Izmaylov, ed., *Aviation and Astronautics in the USSR*, NASA TT F-600, a volume in the *NASA Technical Translation Series* (Washington, October 1969), p. 287. This is a translation of *'Aviatsiya I Kosmonavtika SSSR' Voyennoye Izdatel'stvo Ministerstva Oborony SSSR* (Moscow: 1968).
16. There is an indication that Perl was born Mutterperl. For general information, see Michael Dobbs, 'Julius Rosenberg Spied, Russian Says', *Washington Post*, 16 March 1997; and Harvey Klehr, *et. al.*, *The Secret World of American Communism*, a volume in the *Annals of Communism* series (New Haven, Yale University Press, 1995). The latter is an excellent documentary history of espionage ties between the Communist Party of the United States of American (CPUSA) and the NKVD and Comintern. For further information on the Perl connection to the Rosenbergs, see Ronald Radosh and Joyce Milton, *The Rosenberg File: A Search for the Truth* (New York, 1983), though at the time of their book, his guilt could not be conclusively stated as the critical message intercepts–the 'smoking gun', in post-Watergate parlance–possessed by the National Security Agency had not yet been declassified and released to the public.
17. For example, see declassified *Venona* messages 732 (20 May 1944), 1314 (14 September 1944), and 1797 (20 December 1944), from the files of the National Security Agency (NSA/S4J). I wish to thank Dr Diane Putney of Air Force History Support Office, Bolling AFB, and personnel of the National Security Agency for making the now-declassified Perl *Venona* intercepts available to me.
18. The Lewis quotes on Perl are from Virginia P Dawson, *Engines and Innovation: Lewis Laboratory and American Propulsion Technology*, SP-4306, a volume in *The NASA History Series* (Washington, DC, 1991), pp. 92, 102, 151–152. I also wish to thank Lee Saegesser, the NASA archivist, for his assistance in locating relevant materials on Perl.
19. Dawson, p. 92.
20. Izmaylov, p. 287.
21. This discussion on the DFS 346 is based on the previously cited Smith and Kay, *German Aircraft of the Second World War*, pp. 636–38, and Matthews, *Samolyot 346, passim*. The I-270 information is from Bill Gunston, *Aircraft of the Soviet Union* (London: 1983), pp. 172–173.
22. Muroc AAF Air Field Historical Report, 1 October 1946–31 December 1946, p. 11. In Air Force Archives, Maxwell AFB. Stanley and Sandstrom, 'Development of the XS-1 Aeroplane', 9 January 1948, pp. 15–16. Progress report, Project MX-653, Richard Frost to Commanding General, Air Materiel Command, 21 October 1946. (Hereafter cited as Frost, *MX-653 report*, and date). Also Lundgren, *Across the High Frontier*, p. 162.
23. Muroc AAF Historical Report, 1 October 1946–31 December 1946, p. 12. Frost, *MX-653 report*, 21 October 1946. Lundgren, *Across the High Frontier*, p. 163. Also Stanley and Sandstrom, 'Development of the XS-1 Airplane', 9 January 1948, p. 16. Frost, *MX-653 report*, 13 December 1946.
24. Chalmers H. Goodlin, *Pilots Report, Flight 5*, 9 December 1946. Frost, *MX-653 report*, 13 December 1946.

25. A J Marchese, *MX-653 report* (rough draft), 14 January 1947. Frost, *MX-653 report* (rough draft), 28 January 1947. Frost, *MX-653* report, (rough draft), 10 February 1947. Marchese, *MX-653 report,* l0 March 1947.
26. Marchese, *MX-653 report,* 1 April 1947. Frost, *MX-653 report,* (rough draft), 11 April 1947. Frost, *MX-653 report,* (rough draft), 28 April 1947. Frost, *MX-653* report, (rough draft), 12 May 1947. Frost, *MX-653 report* (rough draft), 12 May 1947. Frost, MX-653 *report,* 2 June 1947. Lundgren, *Across the High Frontier,* p. 67. Muroc AAF Historical Report, 1 Jan. 1947–30 June 1947, pp. 20–21. Frost interview, 25 May 1972.
27. Data on the 346 programme is from the previously cited Matthews, *Samolyot 346, passim* It should be noted that the 346's launch aeroplane was clearly one of the three B-29's interned in the Soviet Union, and not a Soviet-made Tupolev Tu-4 copy; the Tu-4 experienced great teething difficulties, and was still in developmental flight test itself in the spring of 1947. See 'The Billion Dollar Bomber', Parts 1 and 2, *Air Enthusiast* (July–August 1971), pp. 107–08, 160–63.
28. Frost, *MX-653* report, 16 June 1947. Lundgren, *Across the High Frontier,* pp. 67–68. Muroc AAF Historical Report, 1 January 1947–30 June 1947, pp. 31–32.
29. Goodlin letter.
30. Air Materiel Command, Correspondence Summary of Project MX-653 History, 14 January 1947, p. 2. Transmitted to the author by Ezra Kotcher. Letter, J W Crowley (Acting Director of Aero Research) to Brigadier General A R Crawford, 19 February 1947. In LARC files. NACA High-Speed Flight Station, 'X-Press: l0th Anniversary Supersonic Flight', 14 October 1957, pp. 11–12. Rotundo, *Into the Unknown*, pp. 209–211. Johnston and Barton, *Tex Johnston: Jet-Age Test Pilot*, pp. 116–117. Hartley A Soule, Memo for Chief of Research (NACA), 'AAF proposal for accelerated tests of the XS-1 to a Mach number of 1.1', 21 July 1947. In LARC files. Also see Charles E Yeager, 'The Operation of the XS-1 Airplane', in *Air Force Supersonic Research Airplane XS-1 Report No. 1,* 9 January 1948, p. 17. Walter C Williams, 'Instrumentation, Airspeed Calibration, Tests, Results and Conclusions', in *Air Force Supersonic Research Airplane XS-1 Report No. 1*, 9 January 1948, pp. 21–22. Statement of Walter C Williams at the AIAA History Session, San Francisco, California, 28 July 1965, in NASA Historical Archives.

NOTES FOR CHAPTER 5

1. Chuck Yeager and Leo Janos, *Yeager: An Autobiography* (New York: 1985), *passim.* Lundgren, *Across the High Frontier*, pp. 38–48, 92–151. Parrish, *Who's Who in World Aviation, 1955*, p. 342.
2. Don Downie, 'Aerobatics "Sell" Aviation', *Air Progress*, XV, no. 5 (October-November, 1963), pp. 14–15, 89–93. R A 'Bob' Hoover with Mark Shaw, *Forever Flying* (New York: 1996), *passim.* Ridley biographical data sheet in *Edwards AFB Historical Report*, Vol. 2, *1 January 1954–30 June 1954*. In Air Force Archives, Maxwell AFB. See Yeager and Janos, *Yeager*, pp. 107–108.
3. Lundgren, *Across the High Frontier*, pp. 48–55. Frost, *MX-653 Report*, 16 June 1947; *MX-653 Report*, 15 July 1947; *MX-653 Report*, 1 August 1947. Also *Muroc AAF Air Field Historical Report, 1 July 1947–31 December 1947*, p. 23. In Air Force Archives, Maxwell AFB.

4. Mulac, first XS-1 flight chronology. Yeager and Janos, *Yeager*, pp. 111–13. Lundgren, *Across the High Frontier*, pp. 172–81.
5. Two letters, Herbert H Hoover to Melvin Gough, 22 August 1947 and 17 September 1947. Both in LARC files. Also see Gray, *Frontiers of Flight*, pp. 162–63. Shortal, *History of Wallops Station: Origins and Activities Through 1949*, p. 104. Also Lilly biographical file, NASA Historical Archives.
6. Discussion of 1st powered flight and Boyd's reaction is from Charles E Yeager, Pilot's Report No. 1 Powered Flight. In LARC files. Frost interview, 25 May 1972. Yeager and Janos, *Yeager*, pp. 118–122. Lundgren, *Across the High Frontier*, pp. 194–211; Also letter, Herbert H Hoover to Melvin Gough, 17 September 1947. In LARC files.
7. NACA HSFS, 'X-Press: 10th Anniversary Supersonic Flight', p. 3.
8. Frost interview, 25 May 1972. Walter C Williams statement at AIAA History Session, 28 July 1965. Also NACA HSFS, 'X-Press: 10th Anniversary Supersonic Flight', p. 3
9. Walter C Williams, Memo for Chief of Research (NACA), 'XS-1 Progress Report for the period August 15–September 12 1947', 12 September 1947. In LARC files. NACA HSFS, 'X-Press: 10th Anniversary Supersonic Flight', p. 3. Letter, Herbert Hoover to Melvin Gough, 17 September 1947. Mulac, second XS-1 flight chronology. Also NASA-Edwards release, 14 January 1959, of NACA/NASA rocket aircraft flights, p. 4.
10. The best history of the Sabre is Ray Wagner's time-honoured *The North American Sabre*, a volume in the *MacDonald Aircraft Monographs* series (Garden City, NY, 1963).
11. Frost interview, 25 May 1972. Yeager and Janos, *Yeager*, p. 123.
12. Yeager, Pilot's Report No. 8 Powered Flight, in AFFTC historical archives. Yeager and Janos, *Yeager*, p. 123. Mulac, XS-1 #3 flight chronology. Also Lundgren, *Across the High Frontier*, pp. 222-24. NACA HSFS, 'X-Press: 10th Anniversary Supersonic Flight', p. 3. Exact performance on flight obtained from James A Martin, 'The Record Setting Research Airplanes', *Aerospace Engineering*, XXI, No. 12, (December 1962), pp. 49–54. This survey is one of the finest and most reliable summary documents on the early X-series technological accomplishments.
13. Yeager and Janos, *Yeager*, pp. 123–125. Frost interview, 25 May 1972.
14. Details on the first supersonic flight are drawn from Charles E Yeager, Pilot's Report No. 9 Powered Flight, in AFFTC historical archives. Yeager and Janos, Yeager, pp. 127–131. Lundgren, *Across the High Frontier*, pp. 226–43. 'Man in a Hurry', *Time*, LIII, No. 16 (18 April 1949), pp. 64–66, 69–71. See also account by Yeager in *Aviation Week*, XLVIII, No. 25 (21 June 1948), pp. 13–14.
15. Yeager, 'The Operation of the XS-1 Airplane', 9 January 1948, p. 19.
16. H M Drake and H R Goodman, Memo for Chief of Research (NACA), 9 December 1947. Progress report for second XS-1 aeroplane, 18 October to 5 December 1947. Hereafter cited as NACA XS-1 PR, followed by date. Also Hoover pilot reports, 21 October 1947 and 16 December 1947, in LaRC files.
17. *Muroc AAF Historical Report, 1 July 1947–31 December 1947. Air Force Supersonic Research Airplane XS-1 Report No. 1*, 9 January 1948, 'Foreword'. Robert McLarren, 'Bell XS-1 Makes Supersonic Flight', *Aviation Week* XLVII, No. 25 (22 December 1947), pp. 9–10. Charles L Hall, 'Future Programme', *Air Force Supersonic Research Airplane XS-1 Report No. 1,* 9 January 1948. Yeager and Janos, *Yeager*, pp. 150–153.
18. Mulac, NACA and Air Force X-1 flight chronology. Hoover pilot report 27 January

1948. Letter, Robert Hoover to author, 2 November 1971. Letter, Brigadier General Charles E Yeager to author, 12 October 1971. Hoover and Shaw, Forever Flying, pp. 123–125. Yeager, Pilot's report, flight 22, 26 March 1948. Martin, 'The Record-Setting Research Airplanes,' p. 51. Flight records for flight 35, 6 March 1948. Mulac, XS-1 #1 flight chronology. *Also Muroc Army Air Base Historical Report, 1 Jan. 1948–30 June 1948*, p. 21. In Air Force Archives, Maxwell AFB.

19. Mulac, NACA XS-1 flight chronology. NACA XS-1 PR's for 12 March 1948, 29 March 1948, 13 April 1948, 27 April 1948, 21 June 1948, 2 July 1948, 19 July 1948, 2 Aug. 1948, 13 August 1948, 1 September 1948, 13 September 1948, 27 September 1948, 12 October 1948, 26 October 1948, 5 November 1948, 22 November 1948, and 6 January 1949. See also 'Flying for the Future', *Flying Safety* (September 1951), pp. 16–17. Interview of A Scott Crossfield by the author, 2 February 1971. Interview of Robert A Champine by the author, 11 November 1971.
20. *New York Times*, 16 June 1948. 'Captain Yeager's Story', *Aviation Week*, p. 13. Statement by Dr Hugh L Dryden, 15 June 1948, NASA historical archives. 'Supersonic' (editorial) *Washington Post*, 15 June 1948. The story had apparently first appeared in *The New York Herald Tribune*. See Robert McLarren, 'Symington Confirms XS-1 Story', *Aviation Week*, XLVIII, No. 25 (21 June 1948), pp. 13–14. For foreign perspective, see Beaumont, *Testing Early Jets*, pp. 30–36; Gunston, *Aircraft of the Soviet Union*, pp. 161–62; and Henry Matthews, *D.H. 108: The Saga of the First British Supersonic Aircraft* (Beruit: 1996), pp. 56–60.
21. National Aeronautic Association news release, 17 December 1948. 'Highest Aviation Award Made for Supersonic Flight'. NAA release, 17 December 1948.
22. Lundgren, *Across the High Frontier*, pp. 252–61. Yeager and Janos, *Yeager*, pp. 154–155. Frost interview, 25 May 1972.
23. Flight records for flight 40, 11 March 1949. *Muroc AFB Historical Report, 1 July 1948-31 Dec. 1948*, p. 8. In Air Force Archives, Maxwell AFB. Flight records for flight 42, 21 March 1949. Mulac X-1 #1 flight chronology. Everest biographical details from Frank K Everest, *The Fastest Man Alive* (New York, 1959), *passim*.
24. Balloon data from Emme, *Aeronautics and Astronautics*, p. 162. Everest, *The Fastest Man Alive*, pp. 78–84. Also X-1 #1 PR, 9 May 1949. Frost interview, 25 May 1972.
25. X-1 #1 PRs, 21 June 1949, 15 July 1949. Also Everest, *The Fastest Man Alive*, pp. 85–92. Martin, 'Record Setting Research Airplanes', p. 52, Table 3.
26. Everest, *The Fastest Man Alive*, pp. 96–97. *Edwards AFB Historical Report, 1 July 1949-31 December 1949*, pp. 60–2. In Air Force Archives, Maxwell AFB.
27. Hartley A Soule, Memo for Chief of Research (NACA Langley), 'Research airplane projects–Visit to Wright Patterson Air Force Base on November 17, 1949', 29 November 1949.
28. Mulac, X-1 #1 flight chronology. Flight records for flights 51 and 52. *Edwards AFB Historical Report, 1 July 1949–31 December 1949*, p. 1. *Edwards AFB Historical Report, 1 Jan. 1950–30 June 1950*, pp. 102, 106. *Edwards AFB Historical Report, 1 July 1950–31 Dec. 1950*, pp. 124–25. In Air Force Archives, Maxwell AFB. Remarks by General Hoyt S Vandenberg, 26 August 1950, NASM files.
29. Champine interview, 11 November 1971.
30. Griffith pilot report 26 May 1950. X-1 #2 PRs, 9 May 1949, 24 May 1949, 7 June 1949, 4 May 1950, 10 May 1950, 24 May 1950, 8 June 1950, 19 June 1950, 8 July 1950, 18 July 1950, 2 August 1950, 17 August 1950, 11 October 1950, 9 November 1950, 10 April 1951.

31. A Scott Crossfield, *Always Another Dawn: The Story of a Rocket Test Pilot* (Cleveland 1960), pp. 125–34.
32. Mulac, X-1 #2 flight chronology. Walker biographical file, NASA Historical Archives. X-1 PRs, 15 August 1951, 7 November 1951, 21 November 1951, 4 December 1951. Mulac, X-1 #2 flight chronology. Walker pilot report, 24 October 1951.

NOTES FOR CHAPTER 6

1. Williams memo, 29 October 1946. Later, perhaps because of the availability of NACA personnel and better locale, Douglas agreed to conduct the contractor tests at Muroc. Edwin P Hartman, Memo for the Director of Aeronautical Research (NACA), 'Douglas flight tests of 558 at Mojave', 14 February 1947. Gene May and Guy Halferty, 'My Biggest Thrill', *Flying*, LII, No. 6 (June 1953), pp. 12–14, 51–54.
2. Letter, Bureau of Aeronautics to Bureau of Aeronautics representative, Douglas Aircraft Company (Naval Speedletter Aer-DE-237) 14 April 1947, in NASC D-558 file. Also Mulac, D-558-1 flight chronology. Mulac, D-558-1 flight chronology. Also May and Halferty, 'My Biggest Thrill', p. 52.
3. Paul E Purser, Memo for Chief of Research (NACA), 'Douglas D-558 Phase 2 airplane discussion with Douglas representatives concerning tests of rocket-powered models', 1 April 1947. Heinemann, 'Development of the Navy-Douglas Model D-558 Research Project', p. 18. *D-558 Summary Report*, p. 38. Naval Communications Message from BuAero AC-2 to NACA, received by NACA 22 August 1947. (This message was sent following Caldwell's flight, but before Carl's.) The figure of 650.796 is taken from Emme, *Aeronautics and Astronautics*, p. 158. For Carl's explanation of flight, see William Askins, 'The Ultimate Fighter Pilot', *Air Progress*, Vol. XXVII, no. 3 (September 1970), pp. 46–49, pp. 76–77. Both Carl and Caldwell later flew the D-558-2 Skyrocket, Carl setting an altitude record in the all-rocket aircraft.
4. Captain Frederick M Trapnell, Memorandum to Bureau of Aeronautics (Piloted Aircraft Division), D-558, Evaluation flight on 3 September 1947. William H Barlow, Memo for Chief of Research (NACA), 'Progress report on D-558-1-#2 airplane test program to December 8 1947', 8 December 1947. Hereafter NACA D-558-1 #2 Progress Reports are cited as follows: D-558-1 #2 PR followed by date. With the initiation of the NACA D-558-1 #3 programme, the individual airplane number (i.e., #3) will be dropped since this was the only Skystreak that NACA operated following the May 1948 accident to the D-558-1 #2.
5. Letter, Chief BuAer to NACA Headquarters and BuAer Rep., Douglas Aircraft Corporation, 4 November 1947 (letter Aer-AC-25). Heinemann, 'Development of Navy-Douglas Model D-558 Research Project', p. 1. Mulac, D-558-1 flight chronology. Also, D-558-1 #2 PRs for 8 December 1947 and 18 January 1948.
6. Letter, Walter C Williams to Robert C Donovan, 25 August 1947. D-558-1 #2 PRs for 2 February 1948, 13 February 1948, 2 March 1948, 29 March 1948, 13 April 1948, 12 May 1948. See also D-558-1 flight chronology. William H. Barlow, Memo for Chief of Research (NACA), 'Research Program for D-558 Phase I no. 2 Airplane', 17 February 1948.
7. The following discussion of the accident and its aftermath is based on M N Gough, A Young and H A Goett, Aircraft Accident Investigation Report Douglas D-558-1

Airplane, BuNo 37971, Muroc Air Force Base, Muroc California, 3 May 1948, *passim*. Biographical data on Howard C Lilly from the Lilly Biographical File, NASA Historical Division. Author unknown, Memo for D-558 Accident Board, 'Comments on Operation of Research Airplanes', 11 May 1948, pp. 1–2. D-558-1 #3 PR, 18 January 1948, and 27 September 1948. *D-558 Summary Report* pp. 22, 28, 50–51.

8. An interesting example of the close weight-and-balance tolerances of research aircraft came to light during the painting of this D-558-1: 'When the aircraft was sprayed with the white undercoat, as mentioned in my preceding report, the control surfaces were painted. I had seen a schedule of work, issued by the office of the chief engineer for this division, which called for a white paint coat on the entire ship, except the control surfaces. I called this to the attention of the aircraft project engineer and investigation proved that the shop order was in conflict with the original order. Further investigation showed that the reason for not painting the surfaces involved a problem of weight and balance. With the original red color the surfaces were just within the allowable margins and the addition of the white undercoat threw them over the limits. The solution was to remove the white, rubbing down into the red just slightly and then fog a light mist coat of red on the units.' From letter, R B Cox (Douglas Aircraft Company) to Walter C Williams, 11 October 1948. This explains why the NACA Skystreak flew its entire research program with red ailerons, elevators and rudder.
9. D-558-1 PRs, 27 October 1948, 5 November 1948, 11 January 1949, 28 April 1949, 6 May 1949, 24 May 1949. Letter, Eugene F May to Walter C Williams, 4 January 1949. *D-558 Summary Report*, pp. 22, 50–51. NACA placed the first D-558-1 in dead storage. In 1955, NACA presented it to the California Polytechnic Institute for instructional purposes. In 1965, the institute presented it to the Naval Aviation Museum, Pensacola, Florida, where it is today, repainted in red.
10. See D-558-1 PRs: 24 May 1949 to 30 October 1950. Also, Mulac D-558-1 flight chronology.
11. 'DRB' (Donald R Bellman), Status of Research with the D-558-1 Airplane (NACA, 14 September 1950), pp. 2–4.
12. *Ibid.*, pp. 4-5. See D-558-1 PRs from 30 October 1950 to 7 September 1951 .
13. D-558-1 PRs, 19 February to 14 May 1951.
14. D-558-1 PRs, 13 September 1951 to 26 August 1952. See also Mulac, D-558 chronology.
15. D-558-1 PRs, 26 August 1952 to 11 May 1953; Mulac, D-558-1 flight chronology.
16. D-558-1 PRs, 11 May 1953 to 6 August 1953. Mulac, D-558-1 flight chronology. Today this aeroplane is at the Marine Corps Air-Ground Museum, Quantico, Virginia.
17. D-558 Summary Report, p. 15. NACA Research Memorandum RM L52A08, *Handling Qualities of High-Speed Airplanes*, by W C Williams and A S Crossfield, 28 January 1952; data pertaining to the D-558-1 is on pp. 1, 2, 6, 9, 10, 17. NACA Research Memorandum RM H54K24, *Results of Measurements Made During the Approach and Landing of Seven High-Speed Research Airplanes*, by Wendell H Stillwell, 4 February 1955, pp. 6–7.

NOTES FOR CHAPTER 7

1. *Muroc AAF Historical Report, 1 July–31 Dec. 1947*, p. 24. Also Wagner, *The North American Sabre*, pp. 16–18. Hartman, *Adventures in Research*, pp. 166–67.

2. Parrish, *Who's Who in World Aviation, 1955*, p. 201. William Bridgeman and Jacqueline Hazard, *The Lonely Sky* (New York, 1955), pp. 63–64.
3. *D-558 Summary Report*, pp. 17–22. D-558-2 #2 PRs, 11 January 1949, 26 April 1949, 6 May 1949, 20 May 1949, 3 June 1949. Mulac, D-558-2 #2 flight chronology.
4. Champine interview, 11 November 1971. D-558-2 #2 PR's, 15 August 1949, 14 November 1949. Mulac, D-558-2 #2 flight chronology.
5. Letter, Hugh L Dryden to Chief, BuAer, 1 September 1949. In NASC D-558 file. *D-558 Summary Report*, p. 23. Also Mulac, D-558-2 #3 flight chronology. Also May and Halferty, 'My Biggest Thrill', p. 53.
6. Dryden letter to Chief, BuAer, 1 September 1949. Van Wyen, D-558 Programme Notes, p. 2. See also Bridgeman and Hazard, *The Lonely Sky*, p. 194.
7. *D-558 Summary Report*, pp. 20, 21. D-558-2 #2 PRs, 29 August 1950; 8 November 1950; 24 November 1950.
8. James Dugan and Carroll Stewart, *Ploesti* (New York: 1962), p. 161. Also Bridgeman and Hazard, *The Lonely Sky*, pp. 62–65, 211–16. Biographical data from Bridgeman and Hazard, *The Lonely Sky*; 'Bill and the Little Beast', *Time*, Vol. LXI, no. 17 (April 27, 1953), pp. 68–74; Parrish, *Who's Who in World Aviation, 1955*, p. 40; and Douglas company release 'Bill Bridgeman Douglas Test Pilot', n.d.
9. D-558-2 #3 PRs, 11 September 1950, 4 October 1950, 11 October 1950, 30 October 1950, 8 November 1950, 28 November 1950. Also Bridgeman and Hazard, *The Lonely Sky*, pp. 222–31. See also D-558-2 #3 PRs, 5 December 1950, 20 December 1950. The account of the first D-558-2 drop flight given by Bridgeman and Hazard in *The Lonely Sky* is actually an account of the D-558-2 #3 flight of 27 November 1950.
10. Bridgeman and Hazard, *The Lonely Sky*, pp. 254–78. D-558-2 #2 PRs, 30 January 1951, 2 February 1951, 19 February 1951, 23 February 1951, 16 March 1951, 30 March 1951, 10 April 1951, 26 April 1951, 14 May 1951. Bridgeman cites 5 April 1951, instead of 26 January, but the earlier date is correct, and is borne out by NACA and Douglas records.
11. Bridgeman and Hazard, *The Lonely Sky*, pp. 277–92. D-558-2 #2 PRs, 28 May 1951, 26 June 1951, 10 July 1951. William B Bridgeman, 'Supersonic Flight From the Pilot's Seat', paper presented at meeting of the Washington, DC, section of the Institute for Aeronautical Sciences, 2 October 1951.
12. Bridgeman incorrectly places the date for the top-off modification as May–June 1951, but it came after his 23 June flight, and the P2B-lS did not return until July. Bridgeman IAS talk, 2 October 1951. D-558-2 #2 PRs, 10 July, 1951, 21 August 1951, 30 August 1950. Bridgeman and Hazard, *The Lonely Sky*, pp. 279, 293–296, 304–05. Martin, 'The Record-Setting Research Airplanes', pp. 51–52.
13. Letter, A J Marchese to Commanding General, AMC, 1 May 1947. See also *Air Force Supersonic Research Airplane XS-1, Report No. 1*, 9 January 1948, p. 45. De E Beeler, Memo for Chief of Research (NACA), 'XS-2 Conference at Bell Aircraft, May 14, 1948', 15 June 1948. Hartley A Soule, Memo for Chief of Research (NACA), 'Visit to Bell Aircraft Corporation on January 6 and 7 1949', 8 January 1949. Hartley A Soule, Memo for Chief of Research (NACA), 'Research airplane projects-Visit to Wright-Patterson Air Force Base on 17 November 1949', 29 November 1949. Soule memo, 29 November 1949. H Arthur Carner, Memo for Head, NACA High-Speed Flight Research Station, 'Visit to Bell Aircraft Corporation', 18 November 1949. Mulac, X-1-3 flight chronology.
14. Bell Aircraft Corporation, 'Comparative Specification Sheet for the X-1 #1, X-1-3,

and X-1D aircraft', 23 July 1951, in NASA Historical Archives. Frost interview, 25 May 1972. AFFTC, *Experimental Research Aircraft*, p. 28.

15. Flight transcript, X-1-3 glide flight, 20 July 1951. *Annual Report for the NACA High-Speed Flight Research Station for 1951*, in LaRC files. *Edwards AFB Historical Report, 1 July–31 December 1951*, pp. 16–17, 30, in Air Force Archives, Maxwell AFB. Mulac, X-1D flight chronology.
16. Everest, *The Fastest Man Alive*, pp. 129–30. Also X-1D Accident Report, 10 September 1951. Also transcript of X-1D flight, 22 August 1951, by NACA HSFRS.
17. Report of Engineering Officer (Captain David M Sharp, USAF). NACA HSFRS 1951 annual report. X-1-3 Accident Report, n.d. *Edwards AFB Historical Report, 1 July–31 December 1951*, pp. 30–31.
18. *Ibid.* NACA HSFRS 1951 annual report. Crossfield interview, 2 February 1971.

NOTES FOR CHAPTER 8

1. *D-558 Summary Report*, p. 43. Van Wyen, D-558 Programme Notes, p. 2. Also Mulac, D-558-2 # 1 flight chronology. *NACA HSFRS 1951 annual report.*
2. *D-558 Summary Report*, p. 23. D-558-2 #3 PRs, 20 December 1950–31 August 1951. Crossfield, *Always Another Dawn*, pp. 45–52. Mulac, D-558-2 #3 flight chronology.
3. D-558-2 #2 PRs from 16 October 1951–7 January 1953. Mulac, D-558-2#2 flight chronology. *NACA HSFRS 1952 annual report.*
4. *Edwards AFB Historical Report, 1 January–30 June 1953*, Vol. 5, pp. 22–23. In Air Force Archives, Maxwell AFB. *NACA HSFRS 1953 annual report.*
5. Tape transcription of air-ground communications. In Tab C, X-2 #2 Accident Report, Vol. 1. See also X-2 #2 Accident Report, Vol. 1, esp. 'Memorandum Report', 26 May 1953. Also *NACA HSFRS 1953 annual report. Edwards AFB Historical Report, 1 Jan–30 June 1953*, Vol. 5, p. 26. *Edwards AFB Historical Report, 1 July–31 December 1953*, Vol. 1, pp. 87–88. In Air Force Archives, Maxwell AFB. Henry Matthews, *Jean 'Skip' Ziegler: Legendary American Chief Test Pilot* (Beruit: HPM Publications, 1996), pp 10–14.
6. D-558-2 #2 PRs, 9 April 1953, 11 May 1953, 10 June 1953, 14 July 1953, 2 September 1953, 6 October 1953. Also *NACA HSFRS 1953 annual report*. Mulac, D-558-2 #2 flight chronology. Martin, 'The Record-Setting Research Airplanes', p. 52. Also Askins, 'The Ultimate Fighter Pilot', p. 77.
7. The subsequent discussion of Crossfield's preparations and the Mach 2 flight is drawn from: D-558-2 #2 PRs, 12 November 1953, 7 December 1953. Also Mulac, D-558-2 #2 flight chronology, and *NACA HSFRS 1953 annual report*. Crossfield, *Always Another Dawn*, pp. 169, 173–79. Martin, 'The Record-Setting Research Airplanes', p. 51. Mulac, D-558-2 #2 flight chronology. Emme, *Aeronautics and Astronautics*, p. 194.
8. This discussion is based upon: *Edwards AFB Historical Report, 1 July–31 Dec. 1953*, Vol. 1, pp. 87–88. Also letter, Brigadier General Charles E Yeager to author, 12 October 1971; Hubert M Drake and Wendell H Stillwell, 'Behavior of the Bell X-1A Research Airplane During Exploratory Flights at Mach Numbers Near 2.0 and at Extreme Altitudes', NACA RM H55G25, 7 July 1955; Lundgren, *Across the High Frontier*, pp. 278–84. *NACA HSFRS 1953 annual report*. Martin, 'The Record-Setting Research Airplanes', p. 5; and Yeager and Janos, *Yeager*, pp. 196–203.
9. *NACA HSFRS 1953 annual report*. Mulac, X-1A flight chronology. Also Drake and

Stillwell, 'Behavior of the Bell X-1A Research Airplane During Exploratory Flights at Mach Numbers Near 2.0. and at Extreme Altitudes'. Martin, 'The Record-Setting Research Airplanes', p. 52. *NACA HSFS 1954 annual report*. X-1A PR, 6 October 1954.

10. Data on NACA pitch-up programme on D-558-2 #3 is in Mulac, D-558-2 #3 flight chronology, and D-558-2 #3 PRs, 28 September 1951, 6 November 1951, 21 November 1951, November 1951–December 1952, 12 March 1953, 9 April 1953, 11 May 1953. The conclusions were presented by A Scott Crossfield, Hubert M Drake, Jack Fischel and Joseph A Walker in 'Additional Investigation of the Handling Qualities of Airplanes at High Speeds', paper presented at the NACA Conference on Aerodynamics of High Speed Aircraft at the Ames Aeronautical Laboratory, Moffet Field, California, 8–10 July, 1953.
11. *NACA HSFS 1954 annual report*. Also D-558-2 #3 PRs, May 1954–September 1956. Also Mulac, D-558-2 #3 flight chronology.
12. *NACA HSFS 1953* and *1954 annual reports*. D-558-2 #2 PRs November 1953–March 1957. Mulac, D-558-2 #2 flight chronology. ARDC letter to NACA, 24 September 1954, cited in letter, J W Crowley to Commander, ARDC, 22 October 1954. Also Everest flight report for D-558-2 flight, 9 May 1955.
13. Interview of Neil A Armstrong by the author, 26 January 1972. D-558-2 #2 PR, 5 April 1956. Mulac, D-558-2 #2 flight chronology. Crossfield, *Always Another Dawn*, pp. 201–202.
14. Subsequent discussion is drawn from: *NACA HSFS 1954 annual report*. Mulac, X-lB flight chronology. Charles V Eppley, *The Rocket Research Aircraft Program 1946–1962*, TDR No. 63 3, February 1963, AFFTC, Air Force Systems Command, p. 11. Everest, *The Fastest Man Alive*, pp. 135–40. Additionally, see X-1B PR, 7 September 1955. X-1A PRs, January–September 1955. Walker pilot report, 20 July 1955. Mulac, X-1A flight chronology. NACA HSFS, 'Report of Investigation into the Loss of the X-1A Research Airplane on 8 August, 1955', November 1955. NACA release 26 November 1956, 'NACA Honors Staff Members in X-1A Incident'.
15. X-1B (and X-2) discussion drawn from: X-1B PR's, October 1955–June 1958. Mulac, X-1B flight chronology. Richard D Banner, 'Measurements of Airplane Structural Temperatures at Supersonic Speeds', NACA RM H57D18b, 7 June 1957. Hubert M Drake, 'Flight Research at High Altitude', Proceedings of the Seventh AGARD General Assembly, Washington AGARD Conference, 18–26 November 1957. Armstrong interview, 26 January 1972. Mulac, F-100A flight chronology. X-1B PR's, November 1957–June 1958. James E Love, Memo for Chief, HSFS, 'Unsafe Condition of Liquid Oxygen Tank of the X-1B Airplane', 14 August 1958. Walter C Williams, Memo for Research Airplane Projects Leader, 'Retirement of the X-1B from flight status', 27 August 1958. Letter, De E Beeler to Directorate of Systems Management, ARDC, 20 February 1959.
16. X-1E discussion from: X-1 #2 PRs, 6 March 1952–February 1954. X-1E PRs, March–May 1954, and January 1956–April 1959. (There were no PRs submitted from May 1954 until January 1956). For wing data see Norman V Taillon, 'Flow Characteristics About Two Thin Wings of Low Aspect Ratio Determined From Surface Pressure Measurements Obtained in Flight at Mach Numbers From 0.73 to 1.90', NASA Memo 5-1-59H, May 1959. Also 'Thinner Wing to Raise X-1 Mach Limit', *Aviation Week*, LXII, No. 17 (25 April 1955), 42. Letter, Walter C Williams to Walter T Bonney, 4 January 1956. Mulac, X-1E flight chronology. De E Beeler, Memo

for Research Airplane Projects Leader, 'First Supersonic Flight of the X-1E Airplane', 20 June 1956. Joseph A Walker, Memo for Files, 'X-1E Landing Accident', 29 June 1956. Letter, Walter C Williams to NACA HQ, 9 July 1957. Banner, 'Flight Measurements of Airplane Structural Temperatures at Supersonic Speeds', 7 June 1957. Hubert M Drake and Donald R Bellman, Memo for Chief, High-Speed Flight Station, 'Recommendations for Performance Improvements for the X-1 Airplanes', 20 November 1957. Letter, Walter C Williams to Richard H Frost, 4 June 1959. The anecdote about proposals to rebuild the X-2 after Apt's fatal crash is from Richard E Day, *Coupling Dynamics in Aircraft: A Historical Perspective*, NASA Special Publication 532 (Dryden Flight Research Center, 1997).

Epilogue Note

1 Sir Roy Fedden, *Britain's Air Survival: An Appraisement and Strategy for Success* (London, 1957), pp. 19–20.

Index